AMERICA'S SPACE SENTINELS

AMERICA'S

SPACE SENTINELS

The History of the DSP and

SBIRS Satellite Systems

Second Edition, Expanded

Jeffrey T. Richelson

UNIVERSITY PRESS OF KANSAS

Published by the University Press of Kansas (Lawrence, Kansas 66045), which was organized by the Kansas Board of Regents and is operated and funded by Emporia State University, Fort Hays State University, Kansas State University, Pittsburg State University, the University of Kansas, and Wichita State University

Library of Congress Cataloging-in-Publication Data

Richelson, Jeffrey.
  America's space sentinels : the history of the DSP and SBIRS satellite systems / Jeffrey T. Richelson.—2nd edition expanded.
    p.   cm. —(Modern war studies)
  Includes bibliographical references and index.
    ISBN 978-0-7006-1879-8 (cloth : alk. paper)—ISBN 978-0-7006-1880-4 (pbk. : alk. paper) 1. Space surveillance—United States—History. I. Title. II. Title: America's space sentinels, the history of the Defense Support Program and Space-Based Infrared Satellite systems. III. Title: History of the DSP and SBIRS satellite systems.
  UG1523.R535 2012
  355.3'432—dc23
                                                                    2012025940

British Library Cataloguing in Publication Data is available.

Printed in the United States of America

10 9 8 7 6 5 4 3 2 1

Contents

List of Figures and Tables   vii

Preface   ix

Acknowledgments   xiii

List of Abbreviations   xv

1. Bad News Travels Fast   1

2. MIDAS   11

3. Vindication   33

4. Nurrungar and Buckley   49

5. DSP's First Decade   63

6. Surviving World War III   85

7. Slow Walkers, Fast Walkers, and Joggers   95

8. Evolutionary Developments and Suicidal Lasers   123

9. Australia, Germany, and New Mexico   137

10. Desert Storm   157

11. False Starts   177

12. The Unwanted Option   189

13. High Now, Low Later   211

14. Future Missions   223

Epilogue   237

Appendices

   A. Space-Based Infrared Detection of Missiles   275

   B. Chronology   281

   C. Satellite Launch Listing   289

Notes   291

Bibliographic Essay   359

Index   363

Figures and Tables

FIGURES

MIDAS detection probabilities   15
Approximate Earth coverage of DSP satellites   70
View of Earth from F-7   80
Simplified Processing Station   90
Model of a MIDAS satellite   111
Early 1960s launch of a MIDAS satellite from Cape
     Canaveral Air Force Station, Florida   112
Sensor carried on the Program 461 satellites launched
     in 1966   113
DSP satellite in the cargo bay of the space shuttle
     Atlantis   114
Buckley Air National Guard Base, Colorado, 1986   115
Joint Defense Facility–Nurrungar   116
Soviet Tu-22 Backfire bomber in flight   117
Sensor Evolutionary Development sensor   118
Two Soviet-built Iraq-modified Scud-B missiles and their
     launchers   119
Patriot air defense system deployed in Saudi Arabia during
     Operation Desert Shield   120
SBIRS-Low system being proposed by Lockheed Martin
     and Boeing   121
SBIRS-Low system being proposed by TRW and
     Hughes   122
Survivable DSP-1 concept   132
MGT, MCT operations   152
Mono versus stereo coverage   172
BSTS operations concept   182
Launch of DSP-23   236
DSP vs. projected SBIRS capabilities   259
A Geosynchronous SBIRS Satellite   274

Ballistic missile flight trajectory   276
DSP onboard processing   279
DSP ground processing   280

TABLES

Satellite stations after launches F-4 through F-8   79
Stations and satellites after launch of F-10   100
DSP constellation, June 1, 1985   126
DSP constellation, August 10, 1988   128
DSP primary satellite positions, November 1, 1990   162
Scud Launches During Desert Storm   164
DSP-II Versus FEWS cost data   200
ALARM requirements comparison   214

Preface

On Friday, November 18, 1995, a group of present and former industry and Air Force officials gathered for a black-tie event at the Los Angeles Airport Marriott Hotel. The event was a twenty-fifth anniversary gala in celebration of the Defense Support Program (DSP).

The DSP satellites that had been launched into orbit for the previous twenty-five years could detect the infrared emissions of missile plumes from their stations 22,300 miles above the earth. They would have provided the first warning of a possible Soviet missile attack (as well as a means of confirming the accuracy of subsequent radar reports that missiles were headed toward the United States). By eliminating U.S. reliance on a single means of warning and extending the warning time that U.S. leaders would have in the event of a Soviet attack, they represented one of several technological achievements that had helped stabilize the precarious U.S.–Soviet nuclear standoff during the 1970s and 1980s.

The evening began at 6:00 P.M. with cocktails, followed by dinner. The attendees had been scheduled to hear a 9:00 P.M. address by General Thomas S. Moorman Jr., the Air Force vice chief of staff. Moorman had served in a variety of intelligence and reconnaissance assignments since joining the Air Force in 1962, including a stint as the staff director of the supersecret National Reconnaissance Office. More recently, in 1990, he had become commander of the Air Force Space Command, whose responsibilities included operating the DSP satellites.

Moorman was to be there to praise the DSP program. However, many of those who had been involved in that program were not happy about either his prospective presence or the event's focus. Some felt the anniversary event slighted the contributions of those involved in the Missile Defense Alarm System (MIDAS) and 461 programs, DSP's forerunners. Others were displeased at Moorman's presence, for they felt he had been among key Air Force officials who were denigrating the capabilities and accomplishments of DSP, in order to push for a new, costly infrared satellite program.

Because of events in Washington, Moorman never made it to the dinner. But the schisms that revolved around the event, including Moorman's planned address, represent part of the history of MIDAS and DSP—one of intense debates, disagreements over technical feasibility and proposed programs, and, at times, hard feelings.

America's Space Sentinels: DSP Satellites and National Security tells that story. But it also tells several other, far more important, stories. One of these concerns the creation and evolution of the MIDAS/DSP programs, which has involved design of the original satellites and sensors, the establishment of the original ground stations and subsequent fixed and mobile terminals, improvements in the sensors carried on the satellites, and upgrades of the ability to process and disseminate the data collected.

The consequences of these improvements and upgrades have included more accurate estimates of missile launches and impact areas, the reduction in coverage gaps (some of which resulted from changes in foreign missile capabilities), as well as the ability to employ DSP in support of theater conflicts such as Operation Desert Storm. In the near future the DSP system will be replaced by the Space-Based Infrared System (SBIRS), whose satellites will have different infrared sensors and a greatly reduced dependence on overseas ground stations.

As will become clear, much of the evolution of the U.S. satellite early warning system has occurred in response to changes in the international political and military environment, as well as changes in U.S. strategic policy. Thus, the development of improved sensors and mobile ground terminals was the direct result of a shift in U.S. strategic nuclear policy initiated by the Carter administration and vigorously pursued by the Reagan administration. More recent changes, such as the Air Force Attack and Launch Early Reporting to Theater (ALERT) and Army-Navy Joint Tactical Ground Station (JTaGS) programs, are the product of a world in which intermediate-range ballistic missiles (IRBMs) fired by Iraq, Iran, or North Korea, with their short flight times and theater targets, are considered a more pressing threat than intercontinental ballistic missiles (ICBMs) sitting in Russian or Chinese silos.

The history of the DSP program also serves as a case study of a system that performs well beyond expectations. It is not simply that DSP satellites exhibited far greater lifetimes than had been expected as well as greater accuracy in identifying the location of missile launches, even prior to sensor upgrades. DSP satellites also proved to be valuable in a variety of missions beyond their primary mission of detecting Soviet or Chinese intercontinental or submarine-launched ballistic missiles and their secondary mission of monitoring nuclear detonations in the atmosphere.

DSP satellites, even the earliest generation, would prove useful in detecting and monitoring the launch of IRBMs (such as Scuds), detecting aircraft flying on afterburner, monitoring the movements of other spacecraft, and providing data on events such as explosions at weapons depots, airplane crashes, the detonation of meteorites in the atmosphere, and raging forest fires. Some DSP capabilities are only now beginning to be fully exploited.

When a system proves to be so versatile and able to satisfy the data requirements of a much wider range of consumers than anticipated, complications can arise; this, in fact, did occur with DSP. More customers can produce fear on the part of the system "owner" (in DSP's case, the Air Force) that the system will be diverted to other missions and compromise the original, primary mission. As a result, the owner may prove reluctant to see the system's capabilities fully exploited. Thus, the Air Force was uninterested when one of the DSP contractors suggested employing DSP data to detect the movements of Soviet Backfire bombers that might threaten the U.S. fleet. Years later, the Air Force was also concerned that Army and Navy involvement in the DSP program might dilute DSP's primary mission of warning of Soviet missile launches.

The DSP's history also illustrates how such a program, because of the need to locate ground stations on foreign territory and the value of the data produced to other governments, can play a role not only in U.S. foreign relations but even in the domestic politics of a number of governments. A ground station may become a target for protesters and a national political issue, which the host government may seek to manage through a policy of secrecy. Issues concerning the host government's role in ground station operations may become a point of contention between the host and guest. At the same time, the strategic relationship with the host government may grow stronger—both because the need for a ground station allows the host government to successfully negotiate for a greater role in the program and because it becomes dependent on the data. Eventually, the guest government may seek to avoid the vagaries of depending on a foreign host.

A program such as DSP may also play a role in relations with nations that have no formal role in the program. The data produced by DSP may be offered by the United States as an inducement to other nations to pursue programs such as theater missile defense, or demanded by allies (such as Israel) who see it as an important means of reducing their vulnerability to missile attack. Thus, the history of the DSP can broaden a reader's understanding of subjects such as the impact of policy decisions and a changing international environment on weapons and space systems development, the impact of competing bureaucratic interests and priorities on the chances of fully exploiting a warning/intelligence system such as DSP, and the significant role a relatively unknown program such as DSP can have in relations with a number of nations.

But aside from its implications for such issues, the history of DSP is important because DSP was an important space program during the Cold War and remains one in the post–Cold War era. As noted, it added stability to the U.S.–Soviet strategic relationship by improving U.S. warning capability. In addition, it has enhanced U.S. intelligence capability with respect to foreign missile and nuclear programs, played a significant role in Operation Desert Storm, and is, to an increasing extent, being enlisted in the causes of disaster relief and prevention as well as scientific exploration.

★   ★   ★

Since 1999, the history and evolution of the DSP and SBIRS programs have added to the lessons that could be drawn from the programs' earlier years and have also provided additional lessons.

DSP's employment for a multitude of functions beyond the detection of strategic missile launches and atmospheric nuclear detonations became an accepted and valued part of its capability. As a result, those capabilities—including detection of theater missile launches, the detection of military and nonmilitary infrared events on the surface of the earth (such as aircraft flying on afterburner, forest fires, and detonations), and the monitoring of orbiting space systems—have become standard elements of the SBIRS mission.

The last few years of the DSP program, whose launch history concluded with DSP Flight 23 in 2007, also demonstrated the fragility of space systems, and how failure in orbit can quickly turn an extraordinarily sophisticated piece of hardware into little more than a piece of space junk. Such failures, coming at the end of a program, and after a decision to close down the production line, put additional pressure on the successor program to deliver its first spacecraft on time.

Unfortunately, SBIRS provided one of the primary examples of some of the problems plaguing several of America's national security space programs at the end of the twentieth century and the beginning of the twenty-first. Those problems included the expansion of requirements to satisfy potential customers seeking SBIRS data and an undisciplined requirements process that caused the continual addition of requirements long after the spacecraft entered the development phase.

Such problems helped kill another key national security space program, the electro-optical component of the National Reconnaissance Office's Future Imagery Architecture. SBIRS was almost another casualty, but it survived; as of early 2012, it has produced two highly elliptically orbiting payloads and one geosynchronous satellite. But it will still be several years before SBIRS has completed the takeover from DSP and before it will be possible to judge the extent of the program's success.

Acknowledgments

This book grew out of a project I headed for the National Security Archive on the Military Uses of Space. In the course of that project I began, via the Freedom of Information Act (FOIA), to accumulate a significant number of documents concerning the infrared launch detection satellites that the United States began work on in the late 1950s.

But those documents would represent only the tip of the iceberg of what I would acquire over the next several years, including yearly histories of all the government organizations involved in the research and development and operation of the spacecraft as well as a large number of memos from the files of the Aerospace Defense Command, the United States and Air Force Space Commands, the Air Force Space and Missile Systems Center (formerly known as the Space Systems Division, Space Division, and the Space and Missile Systems Organization [SAMSO]).

In addition, a number of valuable documents were released pursuant to FOIA requests to the Office of the Secretary of Defense, the Air Force, the Navy, the Central Intelligence Agency, the Defense Intelligence Agency, and other organizations. Therefore, I owe an enormous debt to the Freedom of Information offices at these organizations as well as to those at a number of other organizations.

My knowledge and understanding of MIDAS and DSP, as well as the search for a successor, are also the product of a number of interviews and conversations held with individuals presently or formerly involved with the programs. In particular I would like to thank Marvin Boatright, Harold Brown, Patrick Carroll, Jim Chambers, Dave Cohn, Sidney Drell, Daniel Fink, Alexander Flax, William G. King Jr., Ellis Lapin, Ed Mickalaus, Wolfgang Panofsky, Bob Richards, Jack Ruina, Bernard Schriever, and Ed Taylor. Others who cannot be acknowledged know who they are and that they have my thanks.

A number of colleagues also provided advice and information. Foremost are William Arkin, who generously shared relevant material that he had accumulated in researching his book on the Persian Gulf air war, and Dwayne Day of the George Washington University Space Policy Institute, who provided copies of a number of documents he found in the course of his research. Others who made significant contributions are Professor Desmond Ball of the Australian National University, Bill Burr of the National Security Archive, Robert Windrem of NBC Nightly News, Steven Zaloga, and John Gresham.

In addition, despite the extent to which this book relies on declassified documents and interviews, it also relies on the reporting that has appeared in a number of key aviation and space publications—particularly *Aviation Week & Space Technology*, *Aerospace Daily*, and *Space News*. I greatly appreciate the work these publications have done in reporting the DSP story.

Finally, the National Security Archive provided support in a variety of ways, and the Center for Defense Information's library proved to be a valuable resource.

Abbreviations

3GIRS	Third-Generation Infrared Surveillance Satellite
ABM	Antiballistic missile
ACDA	Arms Control and Disarmament Agency
ADC	Air Defense Command
ADCOM	Aerospace Defense Command
AEC	Atomic Energy Commission
AFBMD	Air Force Ballistic Missile Division
AFL-CIO	American Federation of Labor–Congress of Industrial Organizations
AFSATCOM	Air Force Satellite Communications System
AFSC	Air Force Systems Command
AFSCF	Air Force Satellite Control Facility
AFSPACECOM	Air Force Space Command
AFSPC	Air Force Space Command
AFTAC	Air Force Technical Applications Center
AIRSS	Alternative Infrared Satellite System
ALARM	Alert, Locate, and Report Missiles
ALERT	Attack and Launch Early Reporting to Theater
AMOS	Air Force Maui Optical Station
ANGB	Air National Guard Base
AR	Advanced RADEC
ARAM	Advanced RADEC Analysis Monitor
ARDC	Air Research and Development Command
ARPA	Advanced Research Projects Agency
ASAT	Antisatellite
ASIO	Australian Security Intelligence Organization
ATF	Activation Task Force
ATRR	Advanced Technology Risk Reduction
AWS	Advanced Warning System
BMD	Ballistic Missile Defense
BMDO	Ballistic Missile Defense Organization
BMEWS	Ballistic Missile Early Warning System
BSTS	Boost Surveillance and Tracking System
CENTCOM	U.S. Central Command
CEP	Circular error probable
CGS	CONUS Ground Station

CHIRP	Commercially Hosted Infrared Payload
CIA	Central Intelligence Agency
CINCAD	Commander in Chief, Aerospace Defense Command
CINCEUR	Commander in Chief, European Command
CINCPAC	Commander in Chief, Pacific Command
CINCSPACE	Commander in Chief, U.S. Space Command
CMO	Central MASINT Office (DIA)
COMINT	Communications Intelligence
CONAD	Continental Air Defense Command
CONUS	Continental United States
CSS	Communications Subsystem
CSTC	Consolidated Satellite Test Center
CSV	Crew support vehicle
CTPE	Central Tactical Processing Element
DAB	Defense Acquistion Board
DARPA	Defense Advanced Research Projects Agency
DCI	Director of Central Intelligence
DDC	Data Distribution Center
DDR&E	Director of Defense Research and Engineering
DEFSMAC	Defense Special Missile and Astronautics Center
DEW	Distant Early Warning line
DIA	Defense Intelligence Agency
DICBM	Depressed trajectory ICBM
DNI	Director of Naval Intelligence
DOD	Department of Defense
DPSS	Data Processing Subsystem
DRB	Defense Resources Board
DRM	Discoverer Radiometric Mission
DSARC	Defense Systems Acquisition Review Council
DSCS	Defense Satellite Communications System
DSMG	Designated Systems Management Group
DSP	Defense Support Program
DSP-A	DSP Augmentation
DTS	Detection Test Series
ECS	Engagement Control Station
EGS	European Ground Station
ELINT	Electronic Intelligence
EMP	Electromagnetic pulse
EPAC	Eastern Pacific
FBW	Fly-by-wire
FEWS	Follow-On Early Warning System
FOBS	Fractional Orbital Bombardment System
FOIA	Freedom of Information Act

FT	Fuel tanker
FTD	Foreign Technology Division (AFSC)
GAO	General Accounting Office
GAO	Government Accountability Office
GCN	Ground Communications Network
GPS	Global Positioning System
GRID	Group for the Investigation of Jet Propulsion
GRU	Chief Intelligence Directorate, Soviet General Staff
HRMSI	High Resolution Multi-Spectral Instrument
IAC	Intelligence Advisory Committee
IBM	International Business Machines
ICBM	Intercontinental ballistic missile
IDA	Institute for Defense Analyses
IOT&E	Initial operational test and evaluation
IRBM	Intermediate-range ballistic missile
IRAS	Infrared Augmentation Satellite
IUS	Inertial upper stage
JCS	Joint Chiefs of Staff
JDEC	Joint Data Exchange Center
JDF-N	Joint Defense Facility–Nurrungar
JDSCS	Joint Defence Space Communications Station
JROC	Joint Requirements Oversight Council
JSTARS	Joint Surveillance Target Attack Radar System
JTaGS	Joint Tactical Ground Station
KGB	Committee for State Security (USSR)
KY	Kapustin Yar
LADS	Low-Altitude Demonstration Satellite
LASER	Light Amplification by the Simulated Emission of Radiation
LCS	Laser Crosslink System
LCV	Logistics crew vehicle
LOCE	Limited Operations–Intelligence Center, Europe
LPS	Large Processing Station
LRS	Limited Reserve Satellite
MASINT	Measurement and Signature Intelligence
MCS	Mission Control Station
MCT	Mobile Communications Terminal
MDA	Missile Defense Agency
MDM	Mission Data Message
MEL	Mobile erector launcher
MENS	Mission Element Need Statement
MGS	Mobile Ground System
MGSU	Mobile Ground Support Unit
MGT	Mobile Ground Terminal

MIDAS	Missile Defense Alarm System
MILSTAR	Military Strategic and Tactical Relay
MOS	Multi-Orbit Spacecraft
MPA	Magnetospheric Plasma Analyzer
MPT	Multipurpose tanker
MRBM	Medium-range ballistic missile
MSP	Mosaic Sensor Program
MSTI	Miniature Sensor Technology Integration
MSX	Midcourse Space Experiment
MTBF	Mean time between failure
MUFON	Mutual UFO Network
NASA	National Aeronautics and Space Administration
NATO	North Atlantic Treaty Organization
NAVSPASUR	Naval Space Surveillance Center
NCA	National Command Authorities
NCMC	NORAD Cheyenne Mountain Complex
NDS	NUDET Detection System
NEACP	National Emergency Airborne Command Post
NIE	National Intelligence Estimate
NII	Scientific Research Institute (USSR)
NLT	No later than
NMD	National Missile Defense
NORAD	North American Aerospace Defense Command
NRO	National Reconnaissance Office
NSA	National Security Agency
NSC	National Security Council
NSD	National Security Directive
NSDD	National Security Decision Directive
NSDM	National Security Decision Memorandum
NSSD	National Security Study Directive
NTBJPO	National Test Bed Joint Program Office
NTPR	Nuclear Targeting Policy Review
NUDET	Nuclear Detonation
ONIR	Overhead Non-Imaging Infrared
OSI	Air Force Office of Special Investigations
PASS	Phased Array Subsystem
PATRIOT	Phased Array Tracking to Intercept of Target
PAVE PAWS	Perimeter Acquisition Vehicle Phased Array Warning System
PD	Presidential Directive
PIM	Performance Improvement Spacecraft
PMD	Program Management Directive
PRM	Presidential Review Memorandum
PSAC	President's Scientific Advisory Committee

PTSS	Precision Tracking Space System
RAMOS	Russian-American Observation Satellite
RADEC	Radiation detection
RGS	Relay Ground Station
RM	Radiometric mission
ROC	Required Operational Capability
RRG	Requirements Review Group
RTS	Research Test Series
SABRS	Space and Atmospheric Burst Reporting System
SAC	Strategic Air Command
SAIC	Science Applications International Corporation
SALT	Strategic Arms Limitation Talks
SAMSO	Space and Missile Systems Organization
SAVE	SABRS Validation Experiment
SBEWS	Space-Based Early Warning System
SBIR	Space-Based Infrared
SBIRS	Space-Based Infrared System
SBS	Space-Based Sensor
SCG	Security Classification Guide
SCS	Space Communications Squadron
SDI	Strategic Defense Initiative
SDIO	Strategic Defense Initiative Organization
SDS	Satellite Data System
SED	Sensor Evolutionary Development
SGT	Survivable Ground Terminal
SIOP	Single Integrated Operational Plan
SLBM	Submarine-launched ballistic missile
SLRA	Soviet Long-Range Aviation
SLV	Security/logistics vehicle
SMTS	Space and Missile Tracking System
SNA	Soviet Naval Aviation
SNIE	Special National Intelligence Estimate
SOC	Satellite Operations Center
SOPA	Synchronous Orbit Particle Analyzer
SOR	Specific Operational Requirement
SPACC	Space Command Center
SPS	Simplified Processing Station
SPS/R	SPS Replacement
SRAM	Short-range attack missile
SRAM	Sandia Radiation Analysis Monitor
SRBM	Short-range ballistic missile
SSBN	Nuclear ballistic missile submarine
SSD	Space Systems Division, Air Force Systems Command

STSS	Space Tracking and Surveillance System
SWRS	SLOW WALKER Reporting System
SWS	Space Warning Squadron
TACDAR	National Systems Tactical Detection and Reporting
TaGS	Tactical Ground System
TBMDS	Theater Ballistic Missile Defense System
TCE	Three-Color Experiment
TEL	Transporter-erector-launcher
TERS	Tactical Event Reporting System
THAAD	Theater High Altitude Area Defense
TIBS	Theater Intelligence Broadcast System
TRAP	Tactical Related Applications
TRD	Tactical Requirements Doctrine
TRE	Tactical Receive Equipment
TRUMP	Target Radiation Measurement Program
TRW	Thomson-Ramo-Woolridge
TSD	Tactical Surveillance Demonstration
TSDE	Tactical Surveillance Demonstration Enhanced
TSS	Transportation Subsystem
TVM	Target-via-missile
UHF	Ultrahigh frequency
USSPACECOM	U.S. Space Command
VHF	Very-high frequency
VUE	Visible Ultraviolet Experiment
VSS	Visible Light Surveillance System

AMERICA'S SPACE SENTINELS

ONE

Bad News Travels Fast

From September 8, 1944, to March 27, 1945, over a thousand German V-2 (Vengeance Weapon Number Two) missiles landed on London, killing and injuring thousands. The V-2, which had been known as the A-4 until renamed by Joseph Goebbels's Propaganda Ministry, was only one of a number of German missiles that had reached various stages of design, testing, or production before the Nazi Reich crumbled in May 1945.[1]

Nazi Germany's ultimate failure in the war did not prevent either the United States or the Soviet Union, allies on their way to becoming adversaries, from appreciating the accomplishments of German scientists in the aviation and missile fields—or from seeking to acquire their future services as well as the results of their wartime endeavors.[2] The U.S. Army had to move quickly, for many of the key sites in the German missile program were located in what would be the Soviet zone of occupation. The U.S. Army Ordnance's Special Mission V-2 had managed to determine the location of the archive for Peenemunde, the German missile research and development center, located on the Baltic. On May 27 U.S. Army trucks hauled away fourteen tons of documents. The United States also acquired parts for 100 V-2s, as well as Wernher von Braun and other scientific talent.[3]

German missile facilities located in the Soviet occupation zone included, in addition to Peenemunde, the Zentralwerke V-2 assembly facility and the Mittelwerke Gmbtl production plant in Thuringia. In its march through Poland, the Red Army had captured a facility at Lehesten, where the Nazis had conducted missile flight tests and static test firings of V-2 rocket engines.[4] With the potential value of such facilities in mind, the Soviets sent scientific intelligence teams to investigate. A 1944 visit to Poland by Soviet missile experts was followed, beginning in 1945, by a more extensive effort. The Soviets sought to collect as much V-2 hardware and as many launch facilities, blueprints, engineers, and technicians as possible.[5]

In September 1945 Sergei Korolev arrived in Germany to join the effort. Korolev was a rocket scientist who had fallen victim to Stalin's reign of terror, having been sentenced to ten years' imprisonment in September 1938 for his alleged subversion of military projects. After about a year of forced labor in the bleak arctic gold mines at Kolyma, he had been transferred to a *sharaga,* a penal institution for engineers and scientists who worked on military projects while serving their sentences.[6]

Like many scientists in the Soviet Union and the United States who would pioneer the development of ballistic missiles and satellites, Korolev had become engrossed in the romance of space and rocketry as a youngster. He had been inspired by the work of Konstantin E. Tsiolkovskiy, who wrote of spaceships and interplanetary travel, and became one of the leaders of the rocket enthusiasts who formed the Moscow-based Group for the Investigation of Jet Propulsion (GRID).[7]

Korolev and his fellow Russians were disappointed by what they found at Peenemunde. The Germans had stripped the center of much of its equipment when they evacuated the facility in early 1945, while the forces defending it had blown up many of the buildings. In addition, the invading Soviet forces had already made off with some of what was salvageable.[8]

What Korolev and the Soviets could not get from the remains of the German facilities they hoped to get from the scientists who had worked there. In the spring of 1946 they selected Helmut Groettrup, who had served as the liaison between the guidance development unit of the Wehrmacht's rocket program and the staff of program head Walter Dornberger, as the director of the German contingent. By September 1946 the 5,000 personnel under Groettrup and Korolev's guidance had produced a V-2 variant, designated the R-1 by the Soviets. Its range was about 167 miles, only marginally greater than that of the V-2.[9]

In the early morning of October 22, 1946, only hours after a symposium to consider German suggestions for further missile development, the Soviets abruptly deported all of the important missile specialists to the Soviet Union, along with their families and some household goods. Thousands of other specialists from the eastern zone were also relocated. In addition, the Soviets brought along the Nazi missile plants, some V-2 rocket engines, and blueprints and engineering studies for other advanced long-range weapons.[10]

After arriving in Moscow on October 28, a small number of scientists, including Groettrup, were settled about thirteen miles north-northeast of central Moscow, near Kaliningrad, where Scientific Research Institute-88 (NII-88) had been established. About 175 of the German scientists were shipped off to the institute's Branch 1, on Gorodomyla Island in Lake Seliger, approximately 150 miles to the northwest of Moscow.[11]

The Germans would be employed to help the Soviet Union move far beyond the capabilities embodied in the V-2s. On March 14, 1947, Soviet Premier Georgi Malenkov told a meeting of aircraft and rocket designers that "we cannot rely on such a primitive weapon; our strategic needs are predetermined by the fact that our potential enemy is to be found thousands of miles away."[12]

The next day, at a Politburo–Council of Ministers meeting, Soviet dictator Josef Stalin observed: "Under Hitler, German scientists have developed many interesting ideas. . . . a rocket [with intercontinental range] could change the fate of the war. It could be an effective strait jacket for that noisy shopkeeper, Harry Truman. We must go ahead with it, comrades. The problem of the creation of transatlantic rockets is of extreme importance to us."[13]

A few months later, work at NII-88 came to a halt when several German experts were ordered, without warning, to board a train. A week later the train arrived near Kapustin Yar, a small village southeast of Moscow, on the banks of the Volga River, which became the first Soviet ballistic missile test range. Awaiting their arrival were several thousand engineers from the Red Army as well as Soviet military officers. The Germans were to assist in test launches of the first R-1s. The first successful R-1 launch, in October 1947, landed a substantial distance from the intended target. By that time the Soviets were also at work developing the R-2, which had a projected range of 365 miles.[14]

In July 1949 Korolev was summoned to a meeting in Stalin's office to discuss the future of the missile program. Stalin's focus was still on a missile with a far greater range than that of the R-1 or R-2; he told the rocket designer, "We want long, durable peace. But Churchill, well he's warmonger Number One. And Truman, he fears the Soviet land as the devil's own stench. They threaten us with atomic war. But we're not Japan. That's why . . . things must be speeded up!"[15]

As part of that process a new missile program was initiated in 1950, one that involved three distinct designs. One was the R-3, a seventy-five-ton missile with a range of 1,860 miles. Although it could not reach the United States, it was conceived of as the first Soviet strategic missile, with the ability to reach American bases in England, Japan, and elsewhere.[16]

Stalin's death in March 1953 was followed by a general reevaluation of strategic weapons programs. At a summer 1953 meeting, attended by Vyacheslav A. Malyshev, head of the nuclear weapons program, and Dmitriy Ustinov, director of the strategic missile and bomber program, Korolev suggested that the R-3 program be canceled because of the missile's limited range. What was really required, he argued, was a missile with a range of 4,200 to 5,000 miles, which thus would be capable of delivering a nuclear warhead to U.S. territory. By this time, studies concerning the feasibility of such missiles had been under way for several years.[17]

Despite initial opposition to Korolev's suggestion, a study of his proposal led to approval, in May 1954, of a program to begin development of the *Semyorka*, or R-7 intercontinental ballistic missile (ICBM), which would be designated the SS-6 SAPWOOD by the U.S. intelligence community. Along with the new missile program the Soviets decided to construct a new and more intricate test center, choosing a site they code-named Tashkent-50, near the Tyuratam railroad station in the Kazakhstan desert.[18]

The first static test firings of the R-7 engines began in February 1956. After several postponements the first launch was finally scheduled for March 1957. Liftoff would finally take place on the third attempt, on May 15, 1957. But at T+50 seconds, the missile exploded over the test range.[19]

On August 21, 1957, several weeks after a U.S. Atlas missile test had ended in failure, another in a long line of failures, an R-7 lifted off from its Tyuratam launchpad and made a successful journey to the Pacific Ocean. All systems worked properly, resulting in the first flight of an ICBM. The missile's dummy warhead splashed down in the Pacific, after a journey of about 4,000 miles. On August 26 TASS, the official Soviet news agency, announced the August 21 test and observed that "a super-long-range multi-stage intercontinental ballistic rocket" had been successfully tested, demonstrating that "it is now possible to send missiles to any part of the world." A second successful test followed on September 7.[20]

Even before the first successful tests, Soviet ICBMs were operational politically. Beginning in 1955, First Secretary Nikita Khrushchev denigrated the capabilities of bombers, in which the United States had a significant advantage, and praised the utility of missiles. In a December 29, 1955, speech, Premier Nikolai Bulganin boasted of Soviet successes in missile development. Less than two months later, foreign ministry official Anastas Mikoyan asserted that, using planes or missiles, the Soviet Union could deliver a nuclear weapon to any point on earth. During his April 1956 state visit to Britain, Khrushchev claimed that the Soviet Union would soon have "a guided missile with a hydrogen warhead that can fall anywhere in the world." Later that year, in response to the British-French-Israeli invasion of Egypt, Khrushchev "rattled his rockets."[21]

The Soviet military was also trumpeting the value of missiles and discussing the strategy for their use. In his speech on Soviet military accomplishments to the Twentieth Party Congress in 1956, Defense Minister Georgi K. Zhukov referred to long-range and "mighty" missiles. Between July 1955 and December 1956, the number of Soviet technical personnel assigned to research and development programs increased by almost 25 percent. A 1957 Soviet article cited three advantages that would accrue to the owner of

ICBMs: the missiles could be used with mobile launchers; they could operate under all weather conditions and cut through air defenses; and they would be capable of launching surprise attacks from concealed positions.[22]

While Soviet scientists had been developing and testing their missiles, and Soviet leaders talked of rocket warfare, the U.S. intelligence community had been investigating, wondering about, and attempting to project the course of the Soviet missile program. In the early 1950s very little concrete information was available. Colonel Georgi Tokaty-Tokaev, deputy chairman of the Soviet state commission on missile production, had defected in 1948 and was debriefed shortly afterward. There were also the German scientists who had been repatriated. In 1954 a U.S. listening post in Turkey began intercepting signals from Kapustin Yar. But that was about all.[23]

The October 1954 National Intelligence Estimate (NIE) entitled *Soviet Capabilities and Probable Programs in the Guided Missile Field* reflected the sparse sources of information. Its authors noted that "we have no firm current intelligence on what particular guided missiles the USSR is presently developing or may now have in operational use."[24] However, based on evidence concerning Soviet interest in missile development, including the exploitation of German experience in the area, Soviet capabilities in related fields, and several other factors, U.S. intelligence concluded:

> We believe that the USSR, looking forward to a period, possibly in the next few years, when long-range bombers may no longer be a feasible means of attacking heavily defended US targets, will make a concerted effort to produce an [ICBM]. In this event it probably could have ready for series production in about 1963 (or at the earliest possible date in 1960) an [ICBM] with a high yield nuclear warhead.[25]

Over the next few years, additional intelligence was obtained from new aerial and ground collection operations. A British Canberra photo reconnaissance aircraft had overflown Kapustin Yar by 1955. Further coverage was obtained after U-2 overflight missions began in 1956. An intercept station at Peshawar, Pakistan, began operations in 1957, with Tyuratam as one of its targets. That same year an overflight of Tyuratam produced a picture of an ICBM on its launchpad.[26]

On March 12, 1957, the Intelligence Advisory Committee (IAC) approved a new NIE on Soviet missiles, relying on these sources of information. The estimate predicted that the Soviets might have an ICBM ready for operational use by 1960 or 1961.[27]

The tests of August and September provided additional data for the U.S. intelligence community to study. As a result, the forecast for Soviet missile development, contained in NIE 11-4-57, published on November 12, 1957, advanced to 1959 ("or possibly even earlier, depending upon Soviet requirements for accuracy and reliability") the date at which the Soviets could attain a marginal missile capability with up to ten ICBMs in the field. The estimate assumed the missiles would have a maximum range of 5,500 nautical miles and a 50 percent chance of landing within 5 nautical miles of the target.[28]

But virtually before the ink on the new NIE had dried, the Soviet Union, on October 4, 1957, placed the earth's first artificial satellite, *Sputnik* ("Traveling Companion") into orbit. The satellite weighed only 184 pounds, but it was substantially heavier than the 3.5-pound *Vanguard,* America's first satellite-to-be, and had a traumatic effect on the American public. Even sophisticated scientists were moved to expressions of panic. John Rinehart of the Smithsonian Astrophysical Observatory announced, "No matter what we do now, the Russians will beat us to the moon. . . . I would not be surprised if the Russians reached the moon within a week." Edward Teller, a father of the H-bomb, declared on national television that the United States had lost "a battle more important and greater than Pearl Harbor."[29]

Some prominent senators, including Henry M. Jackson of Washington and Stuart Symington of Missouri, were also alarmed. To Jackson, *Sputnik* was "a devastating blow to the prestige of the United States." Symington urged President Dwight Eisenhower to call a special session of Congress. In a telegram to Richard Russell, chairman of the Senate Armed Services Committee, Symington called *Sputnik* "proof of growing Communist superiority in the all important missile field." Russell agreed, telling his Georgia constituents that the "Russians have the ultimate weapon—a long range missile capable of delivering atomic and hydrogen explosives across continents and oceans."[30]

Compounding the fear, *Sputnik II,* weighing 1,121 pounds, was orbited on November 3; it carried research instrumentation and a living dog. The Soviets also orbited the entire second stage of the booster, bringing the total package in orbit to 4,000 pounds. The launchings seemed to reinforce the warning of the 1957 report of the Security Resources Panel (a subcommittee of the Scientific Advisory Committee of the Office of Defense Mobilization) that the Soviet Union had "probably already surpassed" the United States in ICBM development and that the Strategic Air Command (SAC) bomber force was threatened by the prospect of an early Soviet ICBM capability.[31]

A Special National Intelligence Estimate (SNIE) was commissioned in the wake of *Sputnik,* completed in December 1957, and approved by the IAC on December 17. The SNIE concluded that the "USSR is concentrating on the development of an ICBM which, when operational, will probably be capable of carrying a high-yield nuclear warhead to a maximum range of about 5,500 nautical miles." The estimate also predicted that the Soviets would probably have an initial operational capability of up to ten prototype ICBMs between mid-1958 and mid-1959. Within a year after achieving that initial capability, it could have 100 operational ICBMs, and 500 about one or two years after that.[32]

In the early 1950s, before the missile threat really arrived, the United States began a series of projects to allow detection of attacking Soviet bombers. In 1952, work on the Pinetree Line of radars in Canada commenced. In February 1954 President Eisenhower approved the Distant Early Warning (DEW) Line project. Construction of the radars began in the spring of 1955 and ended early in 1957. These ground-based radar systems would provide the SAC with one or two hours' tactical warning of any approaching Soviet bombers.[33] But one or two hours of warning that Soviet missiles had been launched would not be available. Only thirty minutes after the missiles left their launchpads, their warheads would detonate on U.S. soil. Maximum warning required placing detection systems where they could provide the earliest possible warning.

On January 14, 1958, in reaction to the *Sputnik* launch, Secretary of Defense Neil H. McElroy approved construction of the Ballistic Missile Early Warning System (BMEWS). BMEWS would consist of large ground-based radars at Clear, Alaska; Thule, Greenland; and Fylingdales Moor in Yorkshire, along with a complicated system of rearward communication lines to bring the information to commanders in the United States.[34]

But even before the *Sputnik* launch, concern over obtaining adequate warning of a Soviet missile attack had led to the investigation of the possibility of detecting missile launches even earlier than radars could manage—via heat generated by the missile and the infrared emissions of missile plumes.[35] In 1948 J. A. Curcio and J. A. Sanderson of the Naval Research Laboratory produced a report that discussed the use of lead sulfide detectors to distinguish the infrared signal from the rocket motor plume. In 1955 two RAND Corporation scientists, Sidney Passman, an expert on infrared technology, and William Kellogg, an expert on high-altitude earth observation, commented:

It appears to be a basic characteristic of an ICBM that aerodynamic heating causes it to get hot during takeoff and even hotter during re-entry into the atmosphere. Since hot metal is a good emitter of infrared radiation, it would be expected that the missile could be detected by infrared detectors during its flight through the atmosphere. The emission of infrared radiation by the rocket flame during the boost stage further increases this expectation, with the added possibility that there may be enough radiation to permit infrared detection during that part of the powered flight which occurs above the atmosphere.[36]

The RAND experts explored techniques for detecting ICBMs during and after their boost phase, including a fleet of "picket airplanes" that would patrol the Soviet periphery in search of missile launches.[37] There was, however, according to Kellogg and Passman, a serious problem with any plan to use aircraft for the infrared early warning mission. Because of the curvature of the earth, much of the boost phase would be unobservable. There would then be a serious risk of detection failure once the powered phase of the missile's flight had ended. Kellogg and Passman noted that

> during the early stages of the [ICBM] takeoff there is more than enough infrared emission, but the earth gets in the way. . . . After burnout there is not nearly enough infrared signal to give detection at any useful range. . . .
>
> The figures . . . lead one to speculate on the increased warning time and perhaps more accurate trajectory prediction that might be possible by getting around this geometrical limitation with a very-high-altitude search station—perhaps with a satellite-borne infrared search set.[38]

In June 1956 the Air Force selected Lockheed's Missile Systems Division to build a military photographic reconnaissance satellite. Lockheed also proposed a number of additional systems, such as electronic and weather reconnaissance satellites, largely following suggestions made by RAND experts since the late 1940s. Joseph Knopow, a young Lockheed engineer, proposed using a satellite equipped with an infrared radiometer and telescope to detect both the hot exhaust gases emitted by long-range jet bombers and large rockets as they climbed through the atmosphere.[39]

Such thinking was the motivation for Lockheed's proposal for a constellation of accurately positioned polar orbiting satellites, which would sweep over the vast Sino-Soviet land mass and instantly report any detection of missile launches to one of three strategically located ground stations. If the idea proved workable, satellites equipped with infrared "eyes" would signal the departure of large rockets as soon as they left their launchpads.[40]

Two individuals who played an important role in persuading military officials of the feasibility and value of such a system were Colonel William G. King Jr. and Lieutenant Colonel Quentin Riepe of Detachment 1 of the Air Research and Development Command (ARDC). They frequently traveled from their Wright Field, Ohio, headquarters to espouse the potential of satellites for reconnaissance and surveillance to often skeptical Air Force officers. In addition, the work of Kellogg and Passman "had captured the attention of various science advisory committees" and created support for an infrared warning satellite.[41]

As a result, before the end of 1957, Lockheed's proposal became Subsystem G of Weapons System 117L (WS-117L), the overall Defense Department space-based reconnaissance and surveillance program. WS-117L consisted of three programs: the SENTRY radio-return reconnaissance satellite project, the Discoverer experimental satellite (which served as a cover for the Central Intelligence Agency's (CIA's) CORONA film-return reconnaissance satellite project), and Subsystem G.

Although Lockheed would be the prime contractor for the spacecraft, it would not design the critical payload. Among the companies Lockheed turned to was Aerojet-General of Azusa, California, which had been formed in 1942 by California Institute of Technology scientist Theodore von Karman and four associates. Aerojet had been involved in the development of rockets and in studying the possible infrared detection of ballistic missiles for a number of years (and had provided some of the data that had been used by Curcio and Sanderson). By the end of 1956, a number of scientists and engineers who would be heavily involved in the infrared detection program in the years ahead had joined the staff.[42]

Subsequent to the October *Sputnik* launch, Aerojet received a follow-on contract. But Aerojet would not be alone in trying to design a suitable spaceborne detection system. Its competition came from Baird-Atomic of Cambridge, Massachusetts. Of course, at that time it had yet to be demonstrated that a reliable payload could be designed by anyone. It would be several more years before the issue would be settled.

TWO

MIDAS

On November 5, 1958, the Advanced Research Projects Agency (ARPA), which had been assigned responsibility for Air Force space progams earlier that year, decreed, in a classified order, that Subsystem G would become an independent project. That information was made public during a December 3, 1958, press conference. ARPA director Roy Johnson told reporters that after he became director in March, he "broke out the Discoverer series and this summer broke out the infrared early warning series so that we could better work with the separate projects." He also announced that "internally we have given it a name. . . . we are calling it MIDAS [for Missile Defense Alarm System]. I suppose there is no reason why I shouldn't tell you that." [1]

Throughout 1959 the Air Force and ARPA would produce a number of plans specifying the characteristics of a MIDAS constellation. A January plan recommended an operational constellation of twenty spacecraft operating at 1,000 miles. Later that year a revised plan envisioned a constellation of twelve spacecraft at 2,000-mile altitudes. [2]

Achieving an operational constellation was a major objective of Lieutenant General Bernard A. Schriever, head of the ARDC, who regarded the program as of the greatest national urgency. Schriever, one of the senior officers influenced by Riepe, recalls that "there was no question in my mind [concerning the importance of] reconnaissance and early warning." In early 1960 testimony before the Senate Committee on Aeronautical and Space Sciences, Schriever specified two benefits of a MIDAS system: it would provide confirmation of warning from radars, which was particularly valuable since "there are always things that you see on radar that you might think are something else," and it would also provide almost double the warning time of BMEWS. [3]

But despite Schriever's enthusiasm two related factors served as roadblocks. One was limited funding; the other was skepticism about the short-term ability to overcome technical hurdles. A February 1959 development plan drawn up by the Air Force, calling for a ten-launch research and development program designed to produce an initial operational capability in July

1961, was submitted by ARPA to the director of defense research and engineering (DDR&E)—with mixed results.[4]

In a letter to General Thomas S. Power, commander in chief of the SAC, Air Force Chief of Staff General Thomas D. White explained that

> the Office of the Secretary of Defense has approved in principle the development program. . . . These R&D launches will be accomplished on the fastest possible schedule that is feasible. . . .
>
> The operational program, both as to scope and timing, does not enjoy the same status. . . .
>
> There has been and continues to be strong opposition to the Operational Plans. Approval of the MIDAS plan is being delayed because of the doubt . . . by DDR&E in our ability to achieve necessary system reliability. . . . There are no operational funds in the FY 1961 budget.[5]

Among the specific concerns of the research and engineering organization was that it would take some time to develop reliable key components, such as the satellite power source and sensitive infrared detectors that would have to survive the rigors of space for at least a year in an operational system.[6] The research and development program sought to address these and other worries. Objectives included the collection and evaluation of the infrared signals detected by the sensor; maximizing geographic readout coverage, particularly in high latitudes; and determining how additional ground stations—particularly in Turkey and Japan—could improve overall system performance and geographic coverage. Of particular concern was the ability to assure uninterrupted data transmission, by 1962, from Turkey and Japan to the continental United States.[7]

The ten launches that were to make up the space segment of the research and development program would take place in two phases. The first phase would consist of two launches from the Atlantic Missile Range beginning in January 1960, and two polar orbit launches from the Pacific Coast beginning around July 1960. The second would involve at least six launchings into polar orbit.[8]

The following month the status of early warning programs, including MIDAS, was considered by the President's Scientific Advisory Committee's Early Warning Panel, chaired by Jerome Weisner. The panel observed, "It is imperative that some degree of early warning operational capability be achieved at the earliest possible date." In addition, the science advisers argued that "to achieve unequivocal warning of an actual attack, and at the same time avoid the danger of accidents and provide a degree of protection against counter-measures, it is most desirable that the warning system actually consist of a spectrum of systems, complementing one another in technique, space and time."[9]

The panel had in mind an infrared system as a complement to the BMEWS radar system—but not MIDAS. Rather, it recommended "immediate procurement of a complementary airborne infrared detection system which can first achieve limited capability in early 1960." Such a system had been proposed by Lockheed in a thick report entitled *ICBM Detection System.* It involved devoting eighty-four U-2 aircraft and 1,333 total personnel to the mission. The U-2, flying 3,000-nautical-mile missions at 65,000 to 70,000 feet, would provide considerably better coverage than lower-flying aircraft because of its broader view of the earth. Eighteen U-2s would be deployed to Misawa, Japan, allowing for three to be in the air at any time. Thirty-three U-2s would operate from Nome, Alaska, with six planes constantly on patrol. Another thirty-three U-2s would be based at Tromso, Norway, with six on patrol at all times.[10]

With respect to MIDAS, the panel was skeptical; it concluded that there was "insufficient evidence concerning the effective implementation of such a system to justify its construction at this time." It noted, "calculations indicate that such infra-red detection systems can detect burning rocket motors at great ranges. There still remain, however, many unanswered questions." The questions concerned whether cloud cover and atmospheric absorption of infrared radiation would permit signals to be detected from space, whether it would be possible to distinguish between the infrared emissions of missiles and those of high-altitude jet aircraft engines, electronic reliability, and the design of orbital control equipment.[11]

Aside from questions over the technical future of MIDAS, there were also questions concerning which organization would control the program. In a letter to Air Force Chief of Staff Thomas White, SAC commander General Thomas Power made his bid, stating that he was particularly interested in having the MIDAS program in SAC, along with SENTRY and a simultaneous readout with the North American Air Defense Command (NORAD) of BMEWS data. He went on to tell White:

It seems to me that these three systems, integrated at a central point [SAC], give the best promise of realizing warning during the missile era. Because of the reaction time involved, I think we would deflect the purpose of these systems to do otherwise. Further, the inter-relationship of the tracking, scheduling and control of identical space vehicles, except for payload, convinces me that the MIDAS program must be employed operationally in conjunction with the SENTRY system.[12]

Control over MIDAS was transferred later that year—but not to SAC. As part of a general transfer of satellite project management from ARPA back to the Air Force, the MIDAS program once again came under management of Air Force headquarters on November 17, 1959.[13]

By this time the Air Force had come to envision MIDAS as a network of eight satellites at 2,000-nautical-mile altitudes, with four satellites each in two planes at right angles to each other. Such a configuration was believed to provide maximum coverage of the Soviet Union with a minimum number of satellites.[14] The Air Force expected that, within its line of sight, a MIDAS satellite would be able to detect the approximate launch location, the number of missiles launched, and the approximate direction of the missiles' flight. Sighting information would be relayed not to stations in Turkey or Japan but to three readout stations strategically located in the United Kingdom, Greenland, and Alaska (at Donnelly Flats and Fort Greeley)—collocated with the BMEWS stations to the extent possible.[15]

The probability that the MIDAS system would detect an ICBM launch would vary from .85 to 1.00, depending on the launch area in the USSR. The probability would reach 1.00 for launch points over 60 degrees, where some satellite would be certain to detect a launch and be within sight of a ground station, would decline minimally for longitudes between 50 and 60 degrees, and would fall significantly at longitudes between 40 and 50 degrees.[16] Figure 2.1 shows MIDAS detection and reporting and the probabilities of detection.

Of course, these probabilities were based on the optimistic assumption that the system would operate as intended—which could be determined only when infrared payloads were tested in orbit. The first attempt to do just that took place on February 26, 1960, from the Atlantic Missile Range. But the spacecraft, which carried an Aerojet Series I payload, never reached satellite status. Shortly after liftoff, radio contact with the booster and its payload was lost. It would be several hours before the Air Force discovered that the second stage did not separate but instead burned up along with the spacecraft as they reentered the atmosphere and landed in the Atlantic Ocean—2,500 miles downrange.[17]

The Air Force had more luck three months later, on May 24, 1960, when MIDAS 2 (Series I, Flight 2) was launched on an Atlas–Agena A from the Atlantic Missile Range, into a 33-degree inclined near-circular orbit of approximately 300 miles. The launching was considered of sufficient impor-

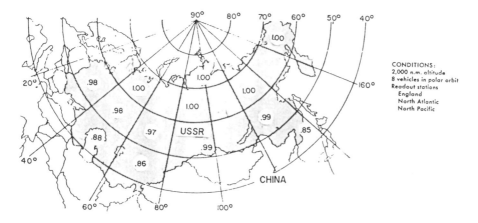

2.1. MIDAS detection probabilities

tance for the *New York Times* to award it front-page coverage, with the head-line "SATELLITE TO SPOT MISSILE FIRINGS PUT INTO ORBIT IN AIR FORCE TEST." The article reported that, to test the satellite's sensors, plans had been made to ignite large sodium flares at Edwards and Vandenberg Air Force Bases as the spacecraft passed overhead, thus providing an infrared source of known intensity against which the sensors' performance could be checked—much as the camera systems of reconnaissance satellites would be calibrated by having them take pictures of targets of known dimensions. In addition, a Titan ICBM was to be launched from Cape Canaveral in the hope that it would be detected by the satellite.[18]

The spacecraft functioned with precision and, according to the Air Force, entered the "most perfect circular orbit achieved by the United States." However, transmission of infrared data to a ground readout station lasted for only a short time before the satellite communication link failed. After two days, somewhere between the thirteenth and sixteenth orbits, the spacecraft fell permanently silent. As a result, it detected neither the flares nor the Titan.[19]

The mission did establish that infrared radiation could be detected at sufficient distances by the sensors, for while the satellite tumbled it picked up a variety of infrared returns, including stellar returns. A key question that remained unresolved, however, was whether it would be possible to differentiate between returns from a rocket's emissions and the earth's natural infrared background and radiation.[20]

★   ★   ★

The MIDAS test launches took place against a background of what appeared to be a growing Soviet ICBM threat to the United States. Telemetry intercepts in the first half of 1958 indicated that the Soviets had conducted four full-range ICBM tests. In January and July 1960 the Soviets test-fired ICBMs with somewhat lighter nose cones on 8,000-mile journeys that ended in the Pacific, 1,000 miles southwest of Hawaii. Soviet Premier Nikita Khrushchev cited the launchings as examples of Soviet capabilities and threatened the United States with "rocket war" if it intervened militarily in Cuba. In an August 1, 1960, NIE, the chief of Air Force intelligence stated that the Soviets could have 200 ICBMs by mid-1961, 450 a year later, and 700 by mid-1963. Not surprisingly, NORAD and the Air Defense Command (ADC) urged General White to accelerate the MIDAS program as much as possible.[21]

But on August 11 White replied that acceleration was premature because of the many unanswered technical problems that "must necessarily pace the program." He also observed that the Secretary of Defense had not yet approved the original Air Force development plan, nor had the Joint Chiefs of Staff (JCS) acted on the preliminary MIDAS operational plan. In addition, he noted the definite possibility that questions and reservations about system reliability in the Defense Department's research and engineering organization might forestall early approval of either.[22]

Dissatisfied with this situation and convinced that the threat facing the nation required faster and more dramatic action, Lieutenant General J. H. Atkinson, commander of the ADC, and General Laurence S. Kuter, NORAD commander, recommended on August 16 that additional MIDAS launchers and backup be provided and that the development program be compressed to demonstrate feasibility and reliability as soon as possible. White replied on August 27 and assured the two commanders that he would support any reasonable request to the Air Force Ballistic Missile Committee and Defense Research and Engineering (DDR&E).[23]

Meanwhile, to expedite experiments to obtain essential technical data on the earth's infrared background radiation, the Air Force asked the DDR&E to authorize two radiometric flights (RM-1 and RM-2) aboard Discoverer vehicles. The spacecraft would carry background radiometers, which would provide infrared background measurements for a wide variety of conditions that might exist between the arctic and tropical regions. The resulting data would supplement that already collected on background radiation by U-2s and help determine the magnitude of background radiance at 2.7- and

4.3-micron wavelengths (the significance of which is discussed in chapter 5 and Appendix A), as well as establish the spatial and spectral background characteristics that had to be known for current and future MIDAS requirements. On August 19 Herbert York, director of defense research and engineering, gave his approval.[24]

Before those flights could take place, a group from the President's Scientific Advisory Committee (PSAC), headed by W. K. H. Panofsky of Stanford University, undertook a review of the program from September 6 to 9. It would be the first of many outside reviews of the program over the next several years. The group included future Secretary of Defense Harold Brown, then director of the Lawrence Livermore National Laboratory, Jerome Weisner of MIT, and physicist Sidney Drell, who had become interested in the question of the "red-out"—whether infrared detectors would continue to function properly after a high-altitude nuclear explosion.[25]

Pankofsky's group had been charged with studying the problem for fear of Air Force bias and had produced a very qualified recommendation to continue work on the program. The committee concluded that the MIDAS concept was basically sound and that the program should proceed, but that more infrared measurements were needed, more than ten research and development flights would be necessary, and a greatly increased effort on simplification and reliability would be required.[26]

The Panofsky report represented one more salvo in the battle within the Defense Department and the Air Force over the future of MIDAS. The upper hand was held by those who favored focusing on system development, who continued to rein in those pressing for early deployment of an operational system. Acting on Air Staff instructions, on October 24, 1960, the Air Force Ballistic Missile Division (AFBMD) issued a revised development plan that called for an accelerated flight test program of eighteen flights to ensure confidence and system reliability.[27]

On November 4 the Air Force Ballistic Missile and Space Committee was briefed and generally approved the revised development plan. However, it directed that all references to operational funding or capabilities be removed and a new plan be submitted concentrating almost wholly on research and development. As a result, on January 3, 1961, the AFBMD issued a development plan calling for six additional research and development launchings, boosting the total to twenty-four. After reviewing the plan, the Air Force Ballistic Missile and Space Committee requested additional changes, including deferring construction of a MIDAS tracking and control center. The "final"

MIDAS development plan, which raised the total of developmental launches to twenty-seven, was completed by the AFBMD on March 31, 1961.[28]

But Defense Secretary McNamara was giving no assurances that 27 or even 270 test launches would produce results that could lead to an operational system. That April he told the Senate Armed Services Committee, "There are a number of highly technical, highly complex problems . . . associated with this system. The problems have not been solved and we are not prepared to state when, if ever, it will be operational." MIDAS was, McNamara reported, subject to repeated false alarms, mistaking the infrared radiation of sunlight reflecting off clouds for missile plumes, and it had poor reliability.[29]

While the generals and Pentagon scientists and planners battled over hypothetical MIDAS constellations and dates when those hypothetical constellations could or would become operational, basic research and development continued. That research was largely unaffected by the battles in Washington, for both sides agreed that the feasibility of a space-based infrared warning system was worth serious investigation.

On December 20, 1960, a Discoverer satellite was launched on the first MIDAS radiometric mission (RM-1), to obtain background information, which it did successfully for four and a half days and seventy-four orbits, until the satellite lost transmitting power. Specially modified U-2 aircraft, designated U-2D and nicknamed "Smoky Joe," were also used between January and June 1961 to obtain background infrared data in preparation for the launching of MIDAS 3. On February 18, 1961, Discoverer XXI orbited the RM-2 package to obtain additional background infrared data, which it did for six stable orbits.[30]

The July 12, 1961, launch of MIDAS 3 marked the beginning of MIDAS test launches designated Series II, for which the sensor payload was provided by Baird-Atomic of Cambridge, Massachusetts. The launches were intended to demonstrate the capabilities of the satellites when in their intended operational orbit. An Atlas–Agena B lifted off from Point Arguello and placed its cargo into a near-circular 1,900 nautical-mile-orbit. As in the MIDAS 2 case, the payload operated for just a short time. Data were transmitted for only five orbits, and, despite the advanced guidance devices aboard the satellite, it began tumbling after the first orbit.[31]

Less than a week after that disappointing result, on July 18, the United States and the United Kingdom exchanged notes announcing their intent to cooperate in the construction of a MIDAS ground station at Royal Air Force Station, Kirkbride, near Cumberland. The U.S. Air Force had first raised the issue in 1959, and a site survey had resulted in the provisional choice of Kirkbride by the end of the year. By-mid 1961 a formal U.S. proposal had been made and was approved by the Cabinet Defence Committee on June 28.[32]

That approval was given despite certain skepticism that the system would provide much benefit to Britain, even though the United States offered to provide any warnings produced by MIDAS to the United Kingdom. In a memo to the Defence Committee, Air Secretary Julian Amery noted that the "Americans claim that improved apparatus capable of detecting the Medium Range Ballistic Missiles, which are expected to constitute the main threat to Britain, should be operational by mid-1965." However, the prime minister's science adviser, Sir Solly Zuckerman, was "agnostic about the scientific claims on the system's likely performance," although he thought cooperation worthwhile because of the access it would provide to information in a key sector of military research. And Air Ministry experts, while confident in the ability of MIDAS to detect ICBMs, envisioned "great problems" in the detection of intermediate-range ballistic missiles (IRBMs) and submarine-launched ballistic missiles (SLBMs). However, the air secretary recommended approval on the grounds that it was in Britain's interest for her allies to have the best possible warning of attack and that warning from MIDAS's promised ability to detect medium-range missiles would allow the British bomber force to take to the air prior to any preemptive strike on their bases.[33]

On July 29, eleven days after what would prove to be a premature exchange of notes, Harold Brown, now DDR&E, reviewed the MIDAS situation at great length for McNamara. The presentation was very much in tune with McNamara's spring statement to Congress. Brown, who believed that while MIDAS was an important program it was "a program in trouble," told his boss that formidable technical and operational problems still remained in the areas of infrared detection and reliability, although they could be solved over a long time span. Brown believed that an effective system might be obtained by 1965–66, although the Air Force continued to believe it could achieve a limited operational capability in 1964.[34]

Brown stated that, at best, MIDAS would provide an additional five to twenty minutes of warning of a liquid-propelled ICBM attack and would be marginal against Soviet Minuteman or submarine-launched Polaris solid-fuel types of missiles. He estimated a cost of about $500 million for construction of an operational system, and annual operating expenses between $100 and $200 million. The primary question, the Pentagon's chief scientist told his boss, appeared to be: Was the extra five to twenty minutes of warning worth the expense and effort?[35]

He noted that the additional warning to alert aircraft was worth something, but the question was how much. He also noted that there was greater certainty of providing warning with MIDAS plus BMEWS than with the latter alone. However, this raised the question of when and how the United States would respond to an enemy attack. If the United States would not retaliate even on receiving warning from the BMEWS or MIDAS systems—a possible consequence of the U.S. program to harden its missile silos and deploy Polaris submarines—the value of warning would be the value of getting additional bombers off the ground. And if the number of additional aircraft alerted by MIDAS was small, earlier warning would be of little value.[36]

One Air Force counterargument—strongly supported by NORAD—was that early warning was essential to ensure a credible deterrent and the survival of the counterforce and defense forces. The Air Force claimed that with ten minutes of warning, 14 percent of the SAC force could become airborne; with fourteen minutes, 66 percent.[37]

Brown informed McNamara that he planned to form a task force to examine in detail MIDAS technical capabilities and the usefulness of early warning. He conjectured that the results of the study would not lead to termination, but he suggested that "a substantial reduction of the R&D program might appear desirable." To head the MIDAS study group, Brown chose Jack P. Ruina of the ARPA. Two Air Force representatives were selected to serve under Ruina along with other governmental and outside members.[38]

The Ruina group, formally the DDR&E Ad Hoc Group on MIDAS, was one more MIDAS study group that, many in the Air Force felt, unnecessarily delayed its development.* On September 5, 1961, the Space Systems Division (SSD) observed that despite past scientific reviews—which it said had found no technical problems to preclude successful development—

---

*This perspective was not shared by those who served on the scientific review panels. One, Sidney Drell, who had participated in the PSAC committee, has written that the committee "was an excellent example of the importance of a strong science advisory mechanism that provides objective, independent and informed advice. . . . Our contribution was to ensure that before the program went to space with big budgets, hardware, and all the gizmos, one understood the basic signals and background against which one was designing this system." (Letter to the author, February 24, 1995.)

there continued to be serious doubts "in the minds of certain people regarding the technical feasibility and operational capability of MIDAS." The division proposed that the Air Force establish an in-house group to prepare a report that could dispel expected criticism. However, at the suggestion of Air Force Under Secretary Joseph Charyk, this was postponed pending completion of the Ruina study.[39]

A few weeks later General Schriever reported to Air Force Chief of Staff General Curtis LeMay on actions that the Air Force Systems Command (AFSC), SAC, ADC, and Office of Civilian Defense Mobilization (responsible for the protection of the general population in the event of a Soviet attack) had under way to defend MIDAS. These included a reassessment of the military and national requirements for MIDAS and verifying the system's technical and operational feasibility. Schriever reported SSD's view that MIDAS was technically feasible and that steps were being taken to simplify the system for improved reliability. He concluded, however, that "complete satisfaction can only be achieved by a conclusive demonstration of system feasibility through an orbital flight test that detects and reports the launch of a ballistic missile and has a reasonable orbital life."[40]

The Ruina group began its evaluation of MIDAS in late September 1961. By this time a new national intelligence estimate, based on data obtained from recent satellite reconnaissance missions as well as from Soviet Colonel Oleg Penkovskiy, sharply reduced the near-term projections of Soviet ICBM strength. However, when Air Force officials addressed the group in October they argued, consistent with the Air Force's dissent to the estimate, that there had been no lessening of the Soviet threat or the need for detection of enemy missile launchings, and they urged an accelerated effort to achieve early operational capability. Thus, in an early October memo, Robert M. Lee, commander of the ADC, wrote, "The additional warning of attack provided by MIDAS is of vital importance to the nation and MIDAS development should be given sufficient emphasis to insure a January 1964 initial operational capability date."[41]

But even as the Air Force took this position, McNamara deleted all but developmental funds for fiscal year 1963 and withheld approval of an operational system. The Air Staff decided not to appeal the decision until the Ruina recommendations were received.[42]

In the meantime, on October 21, 1961, MIDAS 4 (Series II, Flight 2) was launched from the Pacific Missile Range into an approximately 2,200-nautical-mile near-circular polar orbit. Along for the ride was a seventy-five-pound payload consisting of thousands of copper wires, marking the beginning of Project WEST FORD, designed to test the efficiency of a metallic space belt for relaying communications. Although the MIDAS payload, at 3,500 pounds, dwarfed the piggybacked payload, it was the the latter that got the lion's share of attention in the next day's *New York Times*, which noted, "Almost lost in the radio-astronomical excitement of the project was the Air Force's pride in launching MIDAS [4]."[43]

But that pride would be short-lived. The satellite proved to be extremely unstable, operating for fifty-four orbits and nearly seven days before the main power source failed. Although it initially was reported that the satellite had detected a U.S. Titan missile, launched from Cape Canaveral on October 24 to coincide with the pass of MIDAS 4, the report proved to be false. Indeed, according to one individual familiar with the MIDAS program, the returns produced by Baird-Atomic's optical sensor, which had a design defect, were the product of "esoteric ideas that didn't work" and looked like a television set without an antenna—"nothing but snow."[44]

On November 18, 1961, the Ruina group completed the evaluation in the form of a draft report, which became the basis for the final report of November 30, *Evaluation of the MIDAS R&D Program.* A major conclusion was that MIDAS probably could detect the infrared signals of liquid-propelled missiles, but that gaps in knowledge of target and background radiation made this less than certain. The group thought that MIDAS would be unable to detect what they believed would be the less intense infrared signal from solid-fueled missiles of the Minuteman and Polaris class. Finally, the group asserted that the Air Force's preoccupation with an early operational capability had contributed to the neglect of the research and development required to develop an effective operational system.[45]

In the interim version of the report, dated November 1, 1961, the Ad Hoc Group had noted a "series of technical problems with the MIDAS concept of sufficient severity to make it impossible to predict with confidence that an operationally significant version of MIDAS can be obtained within the next decade."[46] The main problems the group predicted were false targets (caused primarily by sunlight reflected from high clouds), missed targets (the result primarily of possible enemy use of fuels with low radiance), and equipment reliability, the latter being a problem for all space systems of the time. In addition to such problems, there were several operational problems of concern: assurance of continuity of coverage (which might be limited because of the size of the constellation or satellite failure), vulnerability to enemy countermeasures, and the integration of MIDAS infrared data with BMEWS radar and other systems data.[47]

The Ruina group agreed, nevertheless, that there were good reasons for continuing the program, suggesting that an operational system could meet significant military and political needs and that "a simplified MIDAS" might have a good chance of achieving an acceptable level of reliability. The group therefore recommended drastic reorientation toward a less ambitious MIDAS and a larger research and measurements effort. It also recommended that until there was full confidence in the system's capabilities, schedules, and cost estimates, no thought be given to an operational capability.[48]

On December 8, 1961, Brown forwarded the Ruina report to Air Force Secretary Eugene Zuckert, noting that he agreed with its conclusions and recommendations. He directed the Air Force to implement the recommendations and asked for a revised development plan by February 1, 1962. Meanwhile, he would hold in deferred status $45 million of the fiscal year 1962 allocation.[49]

Not surprisingly, the conclusions and recommendations of the Ruina report, with its harsh criticism of the existing MIDAS effort, upset top Air Force officials. On December 22 Chief of Staff LeMay directed the AFSC to prepare a response to the "serious allegations." He also directed the Air Staff to prepare some convincing arguments that would support the urgent requirement for a MIDAS warning system. Finally, in case Secretary Zuckert might decide not to challenge the report or the Air Force would be overruled, LeMay ordered the preparation of a development plan based on the Ruina recommendations.[50]

On December 29 Air Force headquarters issued detailed guidance to the AFSC and asked for several alternate plans—one reflecting the Ruina recommendations and two others containing specific initial operational capability dates and oriented toward an operational "go-ahead." The SSD promptly formed a special advisory group headed by Cal Tech's Clark Millikan to analyze the Ruina study.[51]

On January 11, 1962, Vice Chief of Staff General F. H. Smith asked Secretary Zuckert to defer action on DDR&E's December 8 directive, pending completion of the analysis and development plans. After a five-page review of current conditions, General Smith concluded:

The need for warning of Soviet ICBM surprise attack exists today—and will grow more compelling as this Soviet ICBM threat steadily increases. The present BMEWS warning system, initially adequate to the threat, can now be overflown, underflown, skirted, jammed or removed.

The proposed MIDAS system can offset these inherent limitations and provide added credibility, reliability, more warning time, plus an intelligence readout. Additionally, MIDAS can strengthen the free world posture of deterrence, bolster U.S. resolve, and provide the U.S. with tangible, effective arms control measure.[52]

Zuckert agreed to withhold action pending receipt of AFSC's evaluation.[53]

On February 28 the Systems Command forwarded its report, based on the review of the Millikan group, and questioned the validity of the Ruina report, to the Air Staff. The report asserted that the Ruina group had misunderstood the scope of MIDAS research already under way and was unaware of the amount and content of the actual test data available; also, the report's cloud background clutter analysis—a key factor in the Ruina group's doubts about the feasibility of the infrared payload—was said to be in error. In addition, they charged that the report's reliability estimates had failed to take into account advances being made in system reliability.[54]

According to the AFSC, the simplified system under development could be operational before 1966. It submitted three development plans for consideration. One, requiring substantially increased funding, was intended to produce an initial operating capability in 1964. A second, somewhat less costly, plan envisioned operations beginning in 1965. The third plan, which the AFSC considered partly responsive to the Ruina report and was strongly supported by Schriever, emphasized research and development and a larger number of test flights. It would cost approximately $330 million during fiscal years 1962 and 1963 and would allow operations to begin in 1966.[55]

On March 12, 1962, Assistant Secretary of the Air Force for Research and Development Brockway McMillan and Air Staff officials met with Ruina and other Defense officials. At the conclusion of the discussion, McNamara's representatives asked the Air Force to submit a revised MIDAS plan and to conduct further informal discussions with the Defense research and engineering technical staff. These meetings were held during the next several weeks and led Brown to tentatively accept the third plan.[56]

On March 29 the SSD completed a revised MIDAS development plan that envisioned launching as many satellites as possible to demonstrate that the system was feasible and reliable. MIDAS was to achieve an initial operational capability between mid-1965 and mid-1966 at an estimated cost of $334 million for fiscal years 1962 and 1963 (versus the existing programmed amount of $290 million).[57]

During the spring of 1962, while the Air Staff reviewed the plan, data collection from air and space continued. Under the Target Radiation Measurement Program (TRUMP), the Air Force fired Nike-Javelin rockets simultaneously with ICBM and IRBM test launchings from Cape Canaveral to observe and measure the infrared plume radiation from above the earth's atmosphere during the ballistic missiles' powered and early coast phases. MIDAS 5, the third and last of the Baird-Atomic/Series II flights, was launched into a 1,500- to 1,800-nautical-mile polar orbit from the Pacific Missile Range on April 9. According to an Air Force history, although there was a power malfunction on the seventh orbit, great quantities of background information were obtained during the first six orbits. While the data obtained increased Air Force optimism about MIDAS detection capability, failure of the telemetry command function after the seventh orbit raised doubts about system reliability.[58]

The day after the launch, McMillan forwarded the March 29 development plan to DDR&E. Brown not only had his own staff review the March 29 plan but also requested the assistance of a special group of the Advanced ICBM Panel of the PSAC, headed by W. K. H. Panofsky. Panofsky, who had reviewed MIDAS in September 1960 and concluded that the basic concept was sound, did not reach the same conclusion this time. His panel's report, made without benefit of the data from MIDAS 5, noted that proposed flights were still conceived as evaluations of operational prototypes. It expressed doubts about the system's ability to detect any missiles other than those such as the SS-6, Atlas, and Titan, which employed liquid oxygen/kerosene fuel, the brightest-burning of all the fuels. Moreover, the report foresaw only limited success for MIDAS and declared that the value of early warning was decreasing. The panel recommended that Brown should make his decision in light of these findings. Jerome Wiesner, the president's scientific adviser, endorsed both the conclusions and the recommendations.[59]

On April 20 Brown met with McMillan to discuss his previous "general agreement" on the March 29 plan and the Panofsky panel's conclusions. He observed that recent infrared measurements, made during the first Titan II launch on March 16, indicated that MIDAS's performance against dimmer-burning, advanced liquid-fueled ICBMs might be marginal. As a consequence, he directed the Air Force to examine the most logical and expeditious way of introducing improved detection payloads that were effective against low-radiance missiles. Pending this action, he would withhold approval of the MIDAS flight test program and defer construction of the planned data readout center at Ottumwa, Iowa. Since the flight test program would not involve operational prototype satellites, there was no need for the center.[60]

The Panofsky report and Brown's guidance distressed the Air Force, but it had little choice but to comply. The Air Staff dispatched instructions to the AFSC on April 30 to prepare a revised development plan. Meanwhile, Harold Brown added to the general gloom when he commented, in connection with a Defense Department review of major program change proposals, that the apparent inability of MIDAS to detect low-radiance missiles raised doubts about whether full-scale development was justified, even along the lines proposed by the Ruina group.[61]

The Air Force nevertheless still considered MIDAS an essential "hard-core" item and continued to push for an operational system. On June 8, 1962, the AFSC published two new plans in conformity with Brown's instructions. Both emphasized the vital importance of MIDAS to national defense and the need to support it accordingly. Plan A reaffirmed objectives of the March 29 plan but provided for increased infrared measurement research and testing. Plan B, preferred by the Systems Command, satisfied Brown's mandate that the Air Force develop a low-radiance detection capability and called for a multisatellite flight series to support an accelerated research program. If a "go-ahead" were given July 1, the Systems Command estimated the Air Force could still achieve a MIDAS initial operational capability by late 1965.[62]

On June 12, however, Secretary McNamara officially told Zuckert probably the last thing the Air Force Secretary wanted to hear—that he was ordering another "full-scale study of the MIDAS program." He had given the Ad Hoc Group on Ballistic Missile Early Warning, under Dr. H. R. Skifter, the president of the Airborne Instruments Laboratory, its terms of reference on June 1. The group, which included several outside scientists as well as representatives of the Air Force and Defense Research and Engineering, was to review the value of early warning, the implications of a Soviet capability to launch ICBMs "the long way around" (that is, from the south rather than by a polar route), warning of SLBM attacks, the use of low trajectories to reduce warning time, and the growth potential of MIDAS. Brown, meanwhile, chastised the Air Force about reports he had received, indicating continued Air Force preoccupation with an early initial capability against Soviet Atlas-type missiles. He felt that the Soviets would not have many of these and said that answers to basic questions about low-radiance, high-noise background and reliability were still lacking. He reaffirmed his view that MIDAS "must remain an R&D program oriented toward developing . . . techniques."[63]

On June 25, 1962, Brown sent a memorandum to Air Force Assistant

Secretary McMillan, in which he noted, "As I have pointed out, the MIDAS system should not be oriented toward an operational system at this time. Nevertheless, informal reports indicate that there continues to be some Air Force sentiment urging that MIDAS be made operational as soon as possible, based on a capability of providing early warning on ATLAS-type ICBMs."[64]

Brown went on to specify his reasons for opposing early employment of MIDAS as an operational system: Soviet ICBM operational deployment might contain only a small quantity of ICBMs having high infrared radiance; the bulk of their operational ICBMs might employ storable propellants with a lower value of infrared radiance. Under such circumstances there would be little value, Brown argued, in deploying an operational MIDAS system that might have the capability to detect only a small portion of the Soviet ICBM force.[65]

He also mentioned that the results from the MIDAS program had not yet provided satisfactory answers to several fundamental questions concerning the basic performance characteristics of the design concept. He reaffirmed that MIDAS would have to remain a research and development program "oriented toward developing the techniques necessary to resolve the remaining basic issues and must not contain program elements which are time-oriented toward a specific operational date." Spelling it out, so there would be no leeway to deploy an operational system under the guise of research and development, Brown continued that "there should not be a series of six almost identical flights followed by six identical ones which, if they operated, would constitute a 'system.' " Such an approach, Brown wrote, "would make it almost impossible to solve the design and test problems which have so far resulted in the acquisition of very little in-flight data. By inhibiting the design of new payloads, it would also be likely to present us with a 'system' which generally did not work, and when it did, could see only the few missiles of high radiance."[66]

On June 29 McMillan forwarded to McNamara's office a Plan B supplement to the March 29 development plan, which he said would meet Defense research and engineering's requirement for early flight testing of payloads to detect low-radiance missiles.[67]

On the last day of July, Skifter presented his group's report, "Evaluation of Ballistic Missile Early Warning Systems," which had been completed earlier that month. The report's basic recommendation was that, given the technical uncertainty involved in MIDAS, the expectation that it could not reach operational status until 1966, and the political problems in placing the appropriate radars in Pakistan, Australia, or the Philippines, radar systems such as BMEWS should be relied on for early warning during the 1962–66 period.[68]

At the end of the month General Schriever told a House subcommittee that Pentagon had pushed MIDAS too rapidly into development, with the result that it would have to be reoriented. The overoptimism "on the part of the Air Force and Pentagon" occurred, he said, in two areas: the reliability that could be expected from components, and the technical features. He also told the subcommittee that "there is no question on the technical feasibility of the system, but it will take longer to develop than originally forecast."[69]

Not surprisingly, on August 6 Secretary of Defense McNamara directed the Air Force to drop all deployment plans; to reorganize MIDAS as a research and development effort directed toward attaining reliability, high detection probability, and low false-alarm rates; and to submit a new development plan.[70] McNamara listed four reasons for his decision: the expected late deployment of MIDAS; the expected high cost of about $1 billion to complete development and deployment, plus annual operating expenses; the existence of other techniques (such as over-the-horizon radar) to augment early warning capabilities; and the lessening worth of early warning in view of the strategic shift from manned aircraft to hardened missile silos.[71]

General Schriever, despite his public acknowledgment of the need for reorientation, was, in private, very unhappy with McNamara's decision. In an August 13 memo to Air Force Secretary Eugene M. Zuckert he wrote, "I am . . . deeply concerned about the impact that this course of action will have on the MIDAS Program [and] feel an even greater concern about the broad implications with regard to our defense philosophy." Schriever believed that the repeated studies were delaying MIDAS and that if a program was of sufficient importance, as he believed MIDAS was, it was necessary "to take a certain degree of risk."[72]

Schriever's concern with regard to defense philosophy stemmed from McNamara's statement about the declining value of early warning. Schriever objected that McNamara's approach "seriously limits the flexibility with which national strategy can be established and military forces utilized. It fails to recognize the deterrent value associated with unequivocal warning which is not attainable with radar type warning alone. It also gives no value to added warning associated with protection of the civilian populace."[73] Schriever also argued that unless MIDAS research and development were conducted with the objective that it would lead to an operational system at some *specific* time in the future, "a useful capability will always remain in the distant and indefinite future."[74]

Schriever went on to defend the way the MIDAS program had been conducted and the results obtained. He noted that, from its beginning, "the MIDAS Development Program has been based on first demonstrating the feasibility of detecting ballistic missiles by orbiting infrared sensors, then demonstrating that a practical and useful system could be evolved. The basic design parameters for the infrared sensors were chosen with extreme care and conservatism after several years of experiment and study."[75]

Turning to the ability of MIDAS sensors to detect the signatures of missiles employing different kinds of fuel, Schriever wrote that as a consequence of such care, "measurements have shown conclusively that infrared radiation from ATLAS and TITAN I (LOX-RP propellants) missiles exceeds by at least a factor of ten the basic design sensitivity of the MIDAS sensors. Radiation from TITAN II missiles, using 'storable' fuels exceeds MIDAS sensitivity by a factor of three or more."[76]

Schriever also stated that the fear that high-altitude sunlit clouds or particular cloud formations could result in "background clutter," which could obscure or simulate actual missile attacks, had proven unfounded. Such a fear, he wrote, had been based on theoretical extrapolations of limited data gathered by U-2 aircraft. However, actual data collected by MIDAS and Discoverer radiometric flights showed that "very little 'background clutter' exists under real conditions except, perhaps, for a very small area in the direct azimuth of the sun."[77]

Schriever went on to argue that "radar is not an alternative to MIDAS in either an operational or technical sense." MIDAS could guarantee detection of any launch of a liquid-fueled missile from any part of the world, whereas changes in missile launch sites could negate the value of land-based radars. Further, the addition of MIDAS to a radar system would "significantly [complicate] an enemy's means to achieve surprise." Therefore, he recommended that "Secretary McNamara be urged to reconsider the approach he has outlined."[78]

On August 17 McNamara sent a memo to Maxwell Taylor, chairman of the JCS. In addition to asking for comments on the Skifter report, McNamara notified Taylor that he had ordered upgrades of the BMEWS system that would allow operators to recognize when the system was being jammed and provide additional confidence in the detection of a missile attack. McNamara also noted that he requested the Air Force to complete studies and prepare proposals on three recommendations in the Skifter report: installation of a BMEWS gap filler station on Iceland or the eastern coast of Greenland; installation of a ballistic missile and satellite detection station or stations in the southwestern Pacific or Indian Ocean; and establishment of a detection system based in the continental United States to detect SLBMs using radars.[79]

Thus, during the summer of 1962, Air Force hopes for a space-based early warning system to detect enemy ICBM launches seemed unlikely to be realized anytime soon. But the Air Force refused to give up without a fight. On August 22 Air Force Chief of Staff LeMay, echoing many of Schriever's arguments, informed Secretary of the Air Force Zuckert that he disagreed with McNamara's decision because it would reduce MIDAS to a research and development effort with no useful military capability for the forseeable future, because he was not convinced that the value of early warning would decrease with time, and because he did not agree with McNamara's appraisal that MIDAS's technical feasibility was still doubtful.[80]

Schriever followed up with another letter, this time to Chief of Staff LeMay, on September 19. He repeated many of the points of his August 13 letter and responded specifically to McNamara's August 17 memo to the JCS as well as the Skifter report. A sense of what they were up against was obtained on September 22 when McNamara, Under Secretary of Defense Roswell Gilpatric, Harold Brown, Brockway McMillan, General Schriever, and other Defense Department and contractor representatives met to consider the MIDAS program. It became clear that the future of MIDAS would be determined by decisions on requirements, and McNamara ordered the preparation of a comprehensive study on early warning in qualitative and quantitative terms, systems to provide such warning, development plans, and cost estimates to achieve the required capability.[81]

In a postmortem meeting on September 25, Secretary Zuckert observed that extraordinary measures would be required to prepare an Air Force position on early warning, and he directed formation of a special Air Staff group to do the work. An October 3 memo from Zuckert put that suggestion in the form of a memo to Curtis LeMay and attached recommended terms of reference for the study. These included focusing on tactical warning requirements for the forthcoming decade, evaluating the capabilities of the MIDAS system in relation to the spectrum of technical requirements for tactical warning, preparing a development plan for MIDAS commensurate with the tactical warning objectives, and considering possible developments to be used instead of or in addition to MIDAS.[82]

Possibly providing some intellectual ammunition was a RAND study published in July entitled *Military Uses of Tactical Warning of an Attack on CONUS*. It discussed circumstances under which the BMEWS system would provide "zero warning," including firing of missiles from portions of the Soviet Union (e.g., the Kola Peninsula) outside the BMEWS radar fans or firing the missiles to the south—"the wrong way around." In addition, the study suggested the need for tactical warning of sea-launched ballistic missiles, which was not provided by BMEWS or other radars of the time but could be provided by a satellite system.[83]

On November 27 the *Air Force Tactical Warning Study* was submitted by Zuckert, along with a covering memo, to the Department of Defense.[84] While the study was intended to refute the Skifter report, it challenged the assessment not only of scientists and Defense Department civilians but also of the JCS. After examining the Skifter report they informed McNamara that they considered "the recommendations generally valid when considered in view of the specific problems addressed by the study group; that is what can be done to improve early warning within the next two years, and what can be done within the next three to four years." (However, they disputed the notion that early warning would not be critical beyond a few more years.)[85]

Not surprisingly, the next words heard from the Defense Department about MIDAS were discouraging. On December 5 Brown advised the Air Force that MIDAS funding for the 1963 fiscal year should be planned toward a $75 million ceiling, with $35 million for the 1964 fiscal year, rather than the expected $100 and $50 million, respectively. On December 7 Zuckert suggested that perhaps the Air Force's tactical warning study had not been reviewed by the Secretary of Defense.[86]

Six days later Brown replied that the Secretary of Defense had approved the reduction in MIDAS funding. Six days after that, Air Force Under Secretary Charyk asked for a specific response to the tactical warning study. Finally, on December 28 Roswell Gilpatric replied, stating that the Defense Department could not accept the recommendations of the tactical warning study.[87]

Gilpatric wrote that "despite the fact that [MIDAS] can probably be made effective against ICBM's, and eventually, SLBM's its high cost and the existence of reliable and less costly alternatives combine to make it seem unwise to go forward with a high priority program." He also pointed out that the reliability of satellite systems, particularly one as advanced as MIDAS, was uncertain at best, and it would probably be years before the design goals would be realized.[88]

With respect to the alternatives, consisting of radars at various locations, Gilpatric noted that while none of them "will do all that a MIDAS system might do, taken in combination a system can be envisioned which will approach MIDAS potentiality." Specifically, Gilpatric observed that over-the-horizon radars might be able to give the same warning time as MIDAS, radars could provide more accurate impact prediction than MIDAS, radars in the Pacific or Indian oceans could detect extended-range ballistic missiles, and other radars could be used to detect SLBMs.[89]

       Eleven days earlier, on December 17, the attempted launch of MIDAS 6 served as a bad omen of the forthcoming Gilpatric response. An Atlas-Agena rocket lifted off from Cape Canaveral carrying the first Series III payload, an advanced infrared sensor developed by Aerojet. At the time of the launch Atlas-Agena reliablity was only 50 percent. Thus, it was not terribly surprising that the launch vehicle lost flight control about eighty seconds after liftoff, after which the booster broke up and exploded.[90]

THREE

Vindication

Many in the Air Force and at Lockheed and Aerojet had hoped that 1963 would see the beginning of an operational MIDAS program. which by then was also known as Program 239A as a result of a 1962 Defense Department directive that mandated numerical designations be assigned to all space programs for security reasons. By March Program 239A would be become Program 461.[1]

But the changes in designation did not change the basic reality of MIDAS's situation. At the start of the year it appeared that it would be a long time, if ever, before MIDAS, by whatever name it was known, became operational. On January 8 Air Force headquarters acknowledged the implications of Roswell Gilpatric's December 28 memo by ordering a reduction in the scope and funding of MIDAS. The SSD was directed to drop all deployment plans, eliminate associated expenditures, and reorganize the program as a strictly research and development effort. The program's goal would be to develop an effective capability for infrared and visible-light detection of missile and space vehicles during the first and second stages of flight. Specific objectives included the development of a long-lived stabilized spacecraft, whose sensor had a high probability of detecting missiles and a low false-alarm rate. Full use was to be made of piggyback opportunities and cheap launch vehicles.[2]

In a memo to the Assistant Secretary of the Air Force, Harold Brown approved the March 8, 1963, "Space Program Development Plan—Program 461," but he expressed concern that only three radiometric satellite flights were planned for early 1965. If all three failed, "the program may be slowed down to an unacceptable degree." In any event, the Air Force was requested to furnish an analysis of the effect on the program of providing one additional flight, assuming no change in planned funding levels.[3] Brown also urged the Air Force to supplement the research and development program by studying how Program 461 fit into the total early warning environment and its possible use for additional missions beyond early warning.[4]

But Brown's memo did not mean that he was convinced that there was a future for MIDAS. By May 1963 the program was at its nadir, according to Daniel Fink, who began work in Harold Brown's office the next month. Alexander Flax, who became Assistant Secretary of the Air Force for Research and Development in May 1963, recalls that by that time Brown was "on guard" against Air Force attempts to push MIDAS. While Flax had known Brown at Lawrence Livermore National Laboratory and had always been on cordial terms with him, Brown, he recalled almost thirty years later, would become "cold and distant" when discussions of MIDAS took place.[5]

Among those who had reinforced Brown's skepticism was Eugene Fubini, a future DDR&E. Fubini strongly believed that space-based infrared sensors would never be able to discriminate between missiles and other infrared signals—including the earth's background radiation and the reflection of sunlight off clouds.[6]

Equally discouraging analyses came from Alain Enthoven and his Office of Systems Analysis. The systems analysts, who were skeptical that a reliable system could be developed, questioned whether such a system would actually be worth the cost even if it could be produced. There were long-standing concerns over MIDAS's ability to operate in the presence of high clouds and thunderstorms. In addition, the BMEWS and over-the-horizon systems appeared promising. The systems analysts considered the cost of MIDAS too high for a single-mission program, particularly since it appeared there would be little difference in warning time between MIDAS and the over-the-horizon system.[7]

Brown recalls that cancellation may have been close, and that while deploying a space-based warning system was important, it is "sometimes better to cancel [a program] and start over." His views were reflected in his testimony to the House Appropriations Committee on May 6, 1963, in which he stated that half of the $423 million spent on MIDAS through fiscal year 1963 had been wasted on "premature system-oriented hardware." The other half was spent "on gaining data which are quite necessary for any system of this kind that we develop in the future." Brown told the congressmen that the MIDAS program had been severely cut back because "the way the program was going, it would never produce a reliable, dependable system." He hypothesized that within a year or two the reorientation of the program might result in obtaining basic information that would allow development of some other system, but that "it would be a very different system from the one originally proposed." Congressman Daniel Flood observed that Brown's remarks sounded "like the kiss of death."[8]

Brown also addressed the question of the value of warning time, suggesting a reduced importance of the ten extra minutes of warning that would be available with a perfected MIDAS system: "As you go to an increasing fraction of your capability in missiles which can ride out an attack, you are less likely to be completely surprised and wiped out. . . . the extra 10 minutes was particularly useful for getting airplanes off the ground. That was very important. It is still important but it is not quite so important now when we have a substantial missile force as well."[9]

Brown also questioned the utility of MIDAS warning data for two other uses, remarking that "a number of people . . . feel [it] is quite important to civil defense and to preserving the command structure as well. The studies that I have seen that look in detail at what people would do with the extra time do not support that view." The extra warning time was much more valuable in getting bombers off the ground, he asserted, than in getting people into shelters or trying to preserve the command structure.[10]

Brown's view was challenged at the hearing by Congressman George Mahon, who asserted that ten minutes would be important to "the saving, probably, of tens of millions of lives in connection with shelters and the safety and security of our people. I do not think you can very well overestimate the importance of this extra 10 minutes if you can get it." Brown disputed Mahon's claim, responding that "the difference between what you do with 10 minutes warning, 20 minutes warning, 15 minutes or 25 minutes, is really not that great, I think."[11]

Over thirty years later, Brown would still dispute Mahon's thesis. While a big civil defense push early in the Kennedy administration called for the building of home shelters, Brown thinks it was "pretty dubious" that the extra fifteen minutes would have made a difference in getting "anyone to do anything." He also recalls that "every president I was associated with said he would go down with the ship."[12]

Clearly, MIDAS did not enjoy the same strong high-level support as another program that had emerged from WS-117L—the CORONA satellite reconnaissance program. In that case, no one doubted the urgency of developing a capability to photograph the Soviet interior, particularly Soviet production and deployment of missiles and other strategic weapons systems. And there was no alternative to a space-based system.

Lacking a similar strong commitment, it was not likely that MIDAS could survive the string of twelve initial failures that had plagued the CORONA program. There were too many doubters and too many who would just as soon see the funds shifted to the radar programs. But only three days after Brown's downbeat testimony, those who believed in MIDAS received vindication and new hope.

On May 9, 1963, MIDAS 7 (Series III, Flight 2) was successfully launched from Vandenberg Air Force Base on an Atlas–Agena B into a near-circular 87-degree inclined orbit with an apogee of 2,282 miles and a perigee of 2,234 miles. This time the sensor, designated the W-37, was produced by Aerojet-General and was based on what state-of-the art technology would allow. Objectives of the launch were to obtain infrared background data in the MIDAS environment and to test the sensor against cooperative missile targets. On May 10 and 17 the Navy provided the cooperation by timing launches of Polaris missiles from Cape Canaveral to coincide with the orbital passes of the MIDAS satellite over the launch site. A Minuteman launch on May 18 and a Titan II launch on May 24, both from Cape Canaveral, were similarly timed. The Minuteman malfunctioned and destroyed itself at T+45 seconds, while the Titan launch went as planned, with impact taking place approximately 6,000 miles downrange.[13]

The sensor, along with the entire spacecraft, operated successfully for forty-seven days. The launch and its aftermath were judged to be "the first completely successful demonstration in Program 461." Altogether, during its six weeks of operation, MIDAS 7 detected nine U.S. ICBM launches, including several solid-fueled Minuteman and Polaris missiles as well as the liquid-fueled Titan and Atlas missiles. The Atlas, Minuteman, and Titan detections were witnessed on a real-time display.[14]

The results of the test were exactly what MIDAS supporters would have wanted, for readout data "unequivocally established actual detection of each of the launches," while in the case of the Titan II launches eight "good strong returns were witnessed in real time on a scope display." The results were particularly encouraging because they demonstrated the ability of MIDAS to detect solid-fueled missiles—targets expected to be of much lower radiance and shorter burning time than liquid-fueled missiles. Confirmation of the Polaris detections required subsequent manual data analysis because of the short burn time, but it was believed that a solution to the real-time detection of Polaris launches did not present serious technological problems. Thus, Special Assistant to the President Timothy J. Rearden Jr. was informed that "results of the present infrared payload have demonstrated a much higher level of performance than was generally believed obtainable." Specifically, they indicated that previous estimates of the dimmest missile launch that could be detected against a severe background were conservative. The mission also indicated that the reliability problem could be solved—with mission termination occurring as predicted, when the auxiliary power from the satellite's solar array decreased below critical levels because of seasonal variation in the array's exposure to the sun.[15]

Marvin Boatright, who had become the program manager for MIDAS at Aerojet during its earlier, troubled days, recalls the launch and the detections that followed shortly after as "a moment in history" that demonstrated MIDAS "was a solid system, reaching substantially beyond original expectations." The launch and its aftermath also meant that the "detractors had been left behind."[16] The May launch, recalls Al Flax, also represented the culmination of efforts to correct problems or settle concerns associated with the spacecraft, such as reliability, the effect of radiation on solid-state electronics and solar cells, the degradation of thermal control surfaces, and the capability to provide the required stability in orbit.[17]

The next MIDAS flight, on June 12, failed because of a booster malfunction and was terminated approximately ninety-four seconds into the flight. But on July 18, 1963, MIDAS 9 (Series III, Flight 4) was placed into a circular orbit by an Atlas–Agena B with a perigee of 2,275 miles, an apogee of 2,311 miles, and an 88.4-degree inclination. The satellite performed satisfactorily for twelve days, through ninety-six orbits, but it began tumbling on its ninety-seventh orbit because of a power source malfunction. At 3:18 P.M. eastern daylight time on July 26, the new satellite detected and immediately reported the launch of an Atlas E from the Pacific Missile Range, the only detection of a cooperative launch during the operation. It also succeeded in acquiring data about Soviet missile activity.[18]

The successes of MIDAS 7 and MIDAS 9 confirmed Air Force and Aerojet faith in the program. The launches demonstrated that, despite the fear of skeptics, earth background phenomena would not interfere with the real-time detection of ballistic missile launchings. In addition, the successful repetition of the MIDAS experiments greatly increased confidence that space-borne systems could be used for military and peacetime surveillance. Air Force Secretary Zuckert urged the Secretary of Defense to reappraise MIDAS. In June, Harold Brown's office requested another review of the project in light of the new data, and Secretary McNamara asked the JCS to again evaluate the uses of tactical warning, particularly MIDAS.[19]

The Air Force responded eagerly. By late September, Air Force headquarters was considering a revised AFSC plan that listed four alternative approaches to achieving an operational system. The Air Defense Panel, meeting on September 23, rejected all four and recommended a fifth alternative.[20] In adopting the fifth alternative, which called for a developmental effort from which there would also be some immediate operational benefits, the Air Staff considered four major factors: cost-effectiveness, the need for continued research and development, the potential of development satellites to provide operational data, and the possibility of extending MIDAS to perform missions other than missile detection. With these in mind, the Air Staff identified a number of objectives to guide the MIDAS effort. First, it proposed that the AFSC seek to improve the satellite's ability to track and locate missile launchings and identify launch sites precisely. AFSC was to measure and evaluate earth background phenomena to confirm and increase existing knowledge.[21]

The board also felt the AFSC should emphasize advanced detector technology, with particular reference to increasing the system's ability to detect low-radiance, short-burning solid-fueled missiles in real time. It suggested cost-effectiveness could be enhanced by developing simplified and highly reliable spacecraft and payloads with long orbital lifetimes and operating at higher altitudes, thus reducing the number of satellites needed to provide the required coverage. Finally, it proposed that the AFSC study ways to employ the satellites for technical intelligence collection and other defense applications.[22]

On October 1, following a presentation to the Air Force Council, the chief of staff approved the fifth alternative and a MIDAS program comprising three series of satellite flight tests. Six DISCOVERER Radiometric Mission (DRM) launches were programmed in 1964–65, seven Detection Test Series (DTS) flights were to be spread over the years 1966–70, and eight Research Test Series (RTS) flights were programmed for 1967–69. Fiscal year 1969 costs were estimated at $100 million.[23]

Al Flax agreed with the Air Staff's recommendation, considering it preferable to the AFSC proposals. However, because of financial restraints, he recommended eliminating of one RTS flight and adjusting schedules to reduce 1965 funding to $75 million. Fiscal year 1964 funding would remain at $35 million. Secretary Zuckert accepted this adjusted proposal and on October 22, 1963, forwarded it to the Secretary of Defense, urging project approval. On October 29 the JCS—after being briefed on the Air Force plan—unanimously endorsed 461 continuation and advised Secretary McNamara of their views.[24]

Almost immediately after submission of the new proposal, the project was again reoriented on the basis of informal guidance from Harold Brown. On November 7 Brown told Flax that developments of the past six months (i.e., the successful flights and analyses of their results) had made it clear that no additional Discoverer radiometric flights were required. Consequently, pending a final decision on objectives and funding level of a reoriented MIDAS program, he asked that the effort devoted to the radiometric series of flights be diverted into a detection payload that could attain a capability against SLBMs and medium-range ballistic missiles (MRBMs).[25]

These concerns were echoed by an appendix to a memo McNamara received the next day from Maxwell Taylor, chairman of the JCS. That memo, which certainly reflected what the Joint Chiefs had told McNamara shortly after their October 29 decision to support MIDAS continuation, noted:

> As the USSR develops an increased SLBM capability, ballistic missile submarine operations off the US coasts can be expected to commence. [A satellite early warning capability] might tend to reduce their confidence in the over-all success of an SLBM strike. The capability to detect MRBMs could provide [theater] commanders . . . with warning of attack and permit launch and/or dispersal of theater quick reaction strike forces.[26]

Several weeks later, though, MIDAS suffered a sharp setback when, on November 26, McNamara's office reduced its 1965 funding level to $10 million.[27]

On January 14, 1964, Harold Brown released $9 million in 1964 funds to the program—about half of the 1964 money that had been deferred. He continued to withhold the remaining $8.5 million pending the Air Force's submission of a revised program plan to his office in February. The next day Brown took some pains to explain a new decision to reduce the 1965 program from $10 to $8.9 million, with $1.1 million being reallocated to satellite tracking and control. Brown knew this constituted a "drastic reduction" of the Air Force request, and he thought it worthwhile to explain his decision in some detail.[28]

In his memo Brown noted the "outstanding" Air Force adherence to previous Defense Department guidelines, particularly its effort to place MIDAS in the context of other actual and potential early warning systems. He added that it was the very existence of the alternate means that, when combined with the estimated costs of MIDAS deployment, resulted in the low funding support of MIDAS in recent years. However, the success of the 1963 flights and a joint technical review conducted by DDR&E and the Air Force had clearly identified the potential value of a space-based infrared detection capability. But Brown again suggested that in addition to its tactical early warning function, MIDAS ought to aim at several additional important goals, some of which he had previously mentioned. The additional objectives he proposed—

launch-point determination, detection in real time of nuclear detonations, nth country launches, submarine-launched and medium-range missiles, worldwide coverage, and improved reliability—would make the system far more than a simple "bell-ringer."[29]

Brown wrote that these multiple objectives represented a key departure from either the single-system development to detect a massive Soviet ICBM attack or a program designed solely to gather research information for that application. They constituted a prime reason for funding the 1965 program at such a low level and would permit orderly project transition. He asked the Air Force to revise its program on the basis of the broadened set of objectives he had suggested.[30]

While Brown may have been pleased, many elements of the Air Force were not—particulary those most directly involved with MIDAS. They viewed Brown's additional requirements as drag on the program and accepted such requirements only as a means of keeping MIDAS alive. Some of those requirements had actually originated with Air Force Assistant Secretary Flax (and Daniel Fink) with the express purpose of ensuring that the system would be judged cost-effective. Thus, a MIDAS nuclear detection capability would, in theory, allow termination of the VELA HOTEL nuclear detonation detection satellite program.[31]

On January 28, 1964, the Department of the Air Force issued Specific Operational Requirement (SOR) Number 209 for a Missile Defense Alarm System. The requirement noted that a "limited operational capabilty is required as soon as practicable to detect ballistic missiles launched from Soviet regions, for the purpose of investigating in detail and without incurring additional investment costs, the operational application of a satellite-borne missile defense alarm system." Four results were expected from the limited effort: development and deployment of operational prototype vehicles that would incorporate the latest advances in emerging technology from the development program; provision of limited additional early warning capabilities to complement other systems; derivation of data for determining the optimal fully operational system; and acquisition of technical intelligence on foreign missile and space activities. The requirement also specified that full operational capabilities were required by 1970.[32]

After analyzing the impact of reduced funding for 1965, the Air Force was apprehensive about the project's future. Among other effects, the lack of funds would preclude continuation of the previously approved RTS and the proposed DTS, which would result in an approximate four-year gap between launches. It would require terminating the majority of the existing contracts and a probable delay of between eighteen and twenty-four months in realizing the project's goals.[33]

On March 5 Assistant Secretary Flax proposed an alternate program that, he informed Brown, would maintain the 1965 funding level at a minimum figure, preserve current investment, take advantage of the technological progress to date, and advance the program's development in a better way. With respect to the last point, Flax emphasized that his alternative would retain the important reliability objectives, whereas any program based on Brown's proposed $10 million ceiling, with its lack of spacecraft orbital launches, would deal a severe blow to such goals. The Flax counterproposal would retain the RTS launches and their technical objectives but would accept some slippage. It would require $26.9 million, an increase of $16.9 million, but would avoid termination.[34]

On March 18 Brown agreed that the Air Force proposal would involve lower risk and, on technical merit alone, be preferable to his proposal. However, the Defense Department was still restricted by financial limitations, and the Air Force proposal involved a significant departure from approved funding levels. As a result, he informed the Air Force, he could make no decision until the final appropriations level was established and the Office of the Secretary of Defense knew what other projects would be affected if additional funds were allocated to MIDAS. Therefore, the question could not be resolved until the process of finalizing the Defense budget commenced.[35]

In the meantime, however, Brown wanted the Air Force option kept open and no action taken to preclude its ultimate selection. Thus, he asked the Air Force to prepare a development plan based on Alexander Flax's March 5 proposal. At the same time, since the Air Force plan had satisfied the Defense Department requirements for additional information, he released the remainder of the MIDAS funds to the Air Force.[36] At the fiscal year's end, Air Force Secretary Eugene Zuckert forwarded the new program documents to Brown's office for approval. The revised program was to consist of two overlapping phases, still labeled RTS and DTS.[37]

The first set of Program 461 satellites was designated the Research Test Series-1 (RTS-1). They were to be launched between January and May 1966 by Atlas-Agenas into 2,000-mile-altitude orbits. It was expected that during their six-month lifetimes the satellites would demonstrate an ability to detect SLBMs and MRBMs, show the capability of identifying launch points to within eight to ten miles, and exercise a limited radiometric capability.[38]

The second series, initially designated Detection Test Series-1 (DTS-1), would be launched between January 1967 and April 1968. The test satellites would perform essentially the same as the RTS-1 vehicles but would have a quasi-operational status. They would demonstrate the ability of the satellites, on average, to operate for a year before failure, define system operations requirements, and furnish data to a readout station at Fort Greeley, Alaska. They would provide approximately 20 to 25 percent coverage of areas of interest and a degree of technical intelligence and warning, in addition to giving the Air Force operational experience with such a system.[39]

The third series, initially designated RTS-2 (and later Program 266), would begin in January 1968 and end in July of that year. The spacecraft would be put into 6,000-mile orbits by improved Atlas-Agena boosters. They would demonstrate the ability to detect SLBMs and MRBMs, improved launch-point determination, onboard data processing, and nine-month lifetimes.[40]

RTS-2 would be followed by DTS-2 between March and November 1969, launched by whatever appropriate boosters were available, probably Titan IIIs. The DTS-2 satellites were expected to incorporate improvements demonstrated by the RTS-2 series. The final series would be RTS-3, to take place during the first seven months of 1970. The spacecraft was to be orbited at 6,500 miles, have a one-year lifetime, be able to locate launch points to within two to four miles, possess SLBM-MRBM detection capability, conduct a satellite-to-satellite communication link test, and demonstrate advanced attitude control.[41]

Although no formal Defense Research and Engineering response had been received by the end of June 1964, Air Force officials—on the basis of Brown's March 18 paper and informal discussions with his staff—were reasonably confident that the reoriented Program 461 would be approved and funded essentially as outlined in the Air Force proposal.[42]

On August 27, 1964, McNamara, in effect, approved the Air Force proposal, including the $26.9 million funding level. But he also directed the Air Force to reduce the number of DTS-1 launches from four to three, based on, among other things, the predicted high reliability of the series (at least 75 percent chance of launch). The defense secretary also ruled that the Fort Greeley receiving station—which the Air Force proposed to reactivate for tracking and data reduction—should be limited to data recording only and not include real-time data processing and displays. In addition, he refused to approve any funding beyond 1967 because he thought the Air Force had not sufficiently defined the satellite series planned after RTS-1 and DTS-1.[43]

★   ★   ★

It was not long, however, before the plans were undercut by new budget reductions. During fiscal year 1966 budget discussions in early November 1964, the DDR&E tentatively proposed funding reductions to $40, $80, $50, and $20 million (from $50, $100, $80, and $50 million) for fiscal years 1966 through 1969, respectively. Despite an Air Force plea for reconsideration on November 20, the reductions were confirmed with the publication of the Department of Defense Five-Year Force Structure of December 21, 1964. The primary impact of these decisions, as the Air Force had predicted in its appeal, was the elimination of the three DTS-1 launches. In the Air Force's view, this would introduce extremely high risks in achieving those objectives that depended on both RTS and DTS satellites. Of particular concern were the few vehicles available to test the radically changed design and the potential loss of already large investments. Moreover, the Air Force could not immediately establish the ability to keep the satellites in operation for a year, it would not acquire operational experience, and the operational fallout would be considerably less than that anticipated in August.[44]

Beginning on December 24, 1964, the Air Staff and the AFSC sought means of achieving the program goals within the reduced funding. In directing the AFSC to evaluate ways to carry on the program, the Air Force explicitly suggested that the three-launch RTS-1 schedule be protected. The SSD presented a revised proposal on February 23, 1965, and the Air Staff and DDR&E immediately approved it. The following day, the AFSC was authorized to proceed with contractor and in-house studies to establish an RTS-2 configuration and program within the financial restrictions. In mid-April eight firms responded with proposals that encompassed a wide variety of approaches. Two were oriented toward synchronous orbit, three toward medium altitude, and three were flexible. They also presented numerous concepts in attitude stabilization, power sources, data reduction, sensors, and communications. In June the Air Force selected Hughes Aircraft Corporation, Space Technology Laboratories, and Lockheed Missile and Space Company, along with TRW and Aerojet, to undertake the studies, scheduled to be completed by September 14, 1965.[45]

Following elimination of the planned DTS-1 launches, the AFSC evaluated the ground station requirements for the RTS-1 missions and determined that the Satellite Control Facility Tracking Station at Kodiak, Alaska, could not meet MIDAS requirements. The station provided support to a number of other space programs, and the RTS-1 phase of the program would require full use of the Kodiak station. However, the Fort Greeley readout station—intended to support the canceled DTS flights—would serve equally well for the RTS-1 operations. Air Force headquarters concurred on May 20 and authorized the AFSC to activate the station on a "record only" basis.[46]

Work also progressed on developing a nuclear detonation detection capability on MIDAS similar to that on ARPA's planned VELA HOTEL satellites. Defense Research and Engineering recommended that a coordinated research and technology program be established. With ARPA's agreement, the AFSC investigated the possibility and presented its findings along with an abbreviated technical development plan to Air Force headquarters during the week of December 14. The Air Staff concurred with the AFSC recommendation that a nuclear detonation detection package be flown on the second and third Program 461 satellites scheduled in 1966. On December 22, 1964, Air Force Assistant Secretary Flax forwarded the proposal and abbreviated plan to DDR&E, which approved it early in January 1965. A June 22, 1965, Air Force directive specified that the capability to detect nulcear detonations be included on future launch-detection satellites.[47]

By June 1966 the first test launch for Program 461 was imminent. And on June 9 an Atlas–Agena D rocket lifted off from Vandenberg Air Force Base carrying the infrared detection payload. The liftoff was successful, but the Atlas–Agena D vehicle failed to burn a second time. The result was a 90.5-degree inclined, elliptical (108 by 2,246 mile) orbit. The Agena misfire and some equipment failures prevented the mission's objective from being achieved. The satellite remained in orbit for 178 days, until December 3, 1966, when it reentered the atmosphere near Australia.[48]

A little over two months later, on August 19, 1966, an Atlas–Agena D left Vandenberg carrying a second Program 461 payload. Again the spacecraft was boosted into a 90-degree orbit, this time with virtually identical perigee and apogee of 2,285 and 2,298 miles—similar to the MIDAS orbits. Unlike the first Program 461 mission, this one was successful. A secondary payload, including an Army geodetic satellite designated SECOR, was also ejected safely.[49]

On October 5, 1966, the third of the Program 461 satellites was launched from Vandenberg on an Atlas–Agena D into a polar orbit with perigee and apogee of 2,287 and 2,299 miles, respectively. In an indication of things to come, the second and third RTS satellites exceeded their projected six-month lifetimes by considerable margins—the first operating for almost a year, and the second not ceasing operation until October 13, 1967. Together they detected forty-five launches, including a number of Soviet solid-fueled missile launches. Some of the Soviet missile tests detected by the satellite sensors may have been among the eleven August launches of liquid-fueled SS-11 ICBMs from Tyuratam to the Kamchatka impact area. In addition, there were four SS-11 launches from Tyuratam to the Pacific Ocean test area between August 27 and September 6. The Program 461 flight also may have detected an August 27 reconnaissance satellite launch from Tyuratam, a September 27 launch of an SS-9 from Tyuratam, or the missile firings on September 9 and 27 from Kapustin Yar. In November it was reported that data analysis indicated that the missiles burned liquid propellants. One or both of the satellites may have detected the launch of a solid-fueled three-stage ICBM on November 4, from the Plesetsk test center in the northwestern USSR to Kamchatka. The missile apparently was a SS-13 SAVAGE, the first Soviet solid-fueled ICBM.[50]

One or both of the satellites also detected the infrared signature of the SS-N-6 missiles, which would be operationally deployed in 1968 on Yankee-class submarines, during 1967 testing. Analysts looking at the infrared data discovered that the SS-N-6 signal was extraordinarily dim—indeed, much dimmer than the infrared signatures associated with intermediate-range missiles. What made the SS-N-6 so dim, and its dimness so surprising, was the Soviets' use of liquid amine propellant, for conventional wisdom held that submarine-launched missiles had to be solid-fueled. In contrast to the liquid oxygen/kerosene (LOX-RP) propellant used with the Atlas, Titan, and SS-6 missiles, the liquid amine propellant used in the SS-N-6, and all other Soviet liquid-fueled missiles, produced a very faint signature—even fainter than that produced by solid-fueled missiles. Beyond providing new information on the future Soviet SLBM, the detection also demonstrated the ability to detect not only solid-fueled missiles but also missiles with even dimmer signatures—something many had not believed possible.[51]

It is not clear whether the second and third 461 satellites carried a special nuclear detonation package, as had been intended. However, even without that package the satellites' infrared sensors were able to detect such blasts. And nine atmospheric nuclear tests did occur between the August 19, 1966, launch of the second 461 vehicle and the termination of the third satellite's mission in October 1967—three tests in China and six French tests in the Pacific.[52]

Marvin Boatright, the Aerojet 461 program manager, believed that, on the basis of the August and October flights, Program 461 could have "gone operational," using the design for the 461 spacecraft and sensor. Reflecting on the program, he has observed that "the system generated worldwide performance data; established reliable performance against lower target levels; exhibited service lifetimes that were almost a factor of two greater than requirements; [and] provided new applications that were unanticipated."[53]

But it remained the Air Force's intention, as detailed earlier, to conduct a number of additional test flights—although not as promptly as originally planned. On September 23, as a result of funding limitations, the RTS-2/Program 266 launches were pushed back to September and December 1969 and March 1970. And, on November 1, 1966, Air Force headquarters designated Program 949 as a follow-on to Program 461, with the intention of producing an enhanced system.[54]

Over four years would pass between the October 1966 Program 461 launch and the first operational launch. However, neither the RTS-2 launches, which had originally been intended to demonstrate SLBM and MRBM detection capability and later a missile strike/attack assessment capability, nor any other test launches would take place in that period.*

In July, between the first and second of the 1966 launches, Lockheed and a TRW-Aerojet team had submitted proposals to serve as contractors for a future launch-detection satellite. In December 1966 TRW and Aerojet were selected as associate prime contractors, being selected over Lockheed after six weeks of negotiations. TRW would be responsible for the spacecraft, while Aerojet would design and produce the payload. Together they would produce four Program 949 developmental satellites, relying on data from the MIDAS and Program 461 tests, the 1965 design study, and a comprehensive TRW-Aerojet design study. If the satellites proved successful, they would be converted to operational status.[55]

On August 14, 1967, Air Force headquarters requested the program office to study possible acceleration of the program to deal with the threat of a Soviet Fractional Orbital Bombardment System (FOBS), which

---

*It has often been reported that Program 949 test launches occurred on August 6, 1968, April 12, 1969, and August 31, 1970. In fact, the launches, in which Atlas–Agena D boosters placed spacecraft in orbits of approximately 19,000 by 25,000 miles with a 9.9-degree inclination, involved the first three U.S. COMINT satellites—designated by the BYEMAN code name CANYON and known by the unclassified designation Program 827. (See Christopher Anson Pike, "CANYON, RHYOLITE and AQUACADE: U.S. Signals Intelligence Satellites in the 1970s," *Spaceflight*, November 1995, pp. 381–83.)

had emerged as a concern in the previous year. NIE 11-8-66, *Soviet Capabilities for Strategic Attack,* published on October 20, 1966, noted that during the past year "the Soviets have been conducting from Tyuratam what we believe to be feasibility tests involving a new and quite different system, which we designate the SS-X-6" (which was, in fact, a variant of the R-36 or SS-9 ICBM).[56]

Based on the examination of data from tests during May and September 1966, the estimate concluded that "these tests could lead to the development of a fractional orbit bombardment system (FOBS) or a depressed trajectory ICBM (DICBM)." Such weapons "could serve to degrade the value of US missile detection systems and complicate the US problem of developing effective ABM defenses."[57]

A FOBS system would put an ICBM payload in orbit for a short time before heading toward its targets, while a depressed trajectory ICBM would fly at a much lower angle toward its target. The potential consequences were that "many DICBM trajectories could avoid the ballistic missile early warning system (BMEWS) radar fan [while] a FOBS could attack from many angles, and possibly pass unidirectional warning or defense systems undetected."[58]

Prompt study, according to an Air Force history, revealed that the program could be accelerated to provide warning of Soviet FOBS missiles as early as July 1969. This could be accomplished by developing and producing one research and development satellite, minus sensor elements not essential to the tactical warning mission yet capable of providing real-time data to a station located in the Eastern Hemisphere.[59]

Air Force concern was shared by the Secretary of Defense. An August 29, 1967, memo from McNamara to the chairman of the JCS stated that the "recent step-up in the Soviet FOBS test program creates concern" and asked the chairman for his "appraisal of the utility of the warning data from [over-the-horizon radars] and other sensors as they become available."[60]

However, even before the requests to study possible acceleration of the program were made, it was evident that the Soviets were having difficulties with the FOBS. A June 1967 European Command intelligence report stated that "the Soviets continued to experience apparent problems with their SS-X-6 program with the vehicle launched on May 17, probably ending in failure. The SS-X-6, believed to be an orbital bombardment vehicle, has now been fired a total of eight times, with six of the operations ending in probable failure."[61]

The FOBS program would actually involve twenty-four attempted launches and would be declared operational after the twentieth launch, with sporadic testing continuing until 1971. The system would eventually be dismantled in 1983 as a result of the SALT II agreement. But on November 8, 1967, as concern undoubtedly ebbed, the AFSC directed Program 949 to revert to the original plan of preparing three developmental satellites for Titan IIIC launches in June, September, and December 1970.[62]

FOUR

Nurrungar and Buckley

A key decision, made at the time the contract was awarded to TRW and Aerojet-General in late 1966, was to rely on a satellite in geostationary orbit to detect Soviet ICBM launches. That decision had come after prolonged discussion among Air Force officials. Research and development chief Al Flax, who was among those involved in debating the virtues and problems associated with candidate orbits, recalls having a folder full of notes on the pros and cons of different possibilities.[1]

One was to rely on satellites in Molniya orbit—an orbit named for the Soviet communications satellites that rose to 24,000 miles above the earth when they were over the Arctic and dipped to 200 miles as they zoomed over the lower part of the Southern Hemisphere. An alternative was to place a satellite in geostationary orbit—positioned 22,300 miles above a point on the equator, with a speed around the earth that would match the earth's rotation about its axis. As a result, a properly located satellite would have a constant view of the Eurasian landmass.

Flax and his colleagues reviewed the advantages of employing a Molniya orbit: they could use a Titan IIIB booster instead of the more costly Titan IIIC and could eliminate the costs and security risks associated with a foreign ground station. The satellite's orbit would ensure that it was in view of a U.S. ground station while it was over the Soviet Union. On the other hand, any geostationary satellite that was in position to monitor Soviet missile launches would not be within sight of any ground station located in the United States, regardless of where that ground station was built.[2]

But the deciding factor was money. Whereas a single geostationary satellite could continuously cover the Soviet Union and China, two elliptically orbiting satellites would be needed to maintain a constant watch on those countries. Given the cost of a single satellite, approximately $300 million, it would be much less expensive to build and operate a foreign ground station than to maintain a two-satellite constellation devoted to monitoring Eurasia.[3]

* * *

Given the decision to deploy a satellite in geostationary orbit, it became necessary to find a suitable location outside the United States for the ground station. Some territories controlled by the United States could fulfill the first function—most prominently Guam. Similarly, the island of Diego Garcia, controlled by the United Kingdom, could serve as the home for a ground station.[4]

But a tentative decision was made in fiscal year 1968 to locate the overseas ground station in Australia, which was simultaneously a close ally and a vast continent. In August 1968 a party headed by Colonel R. J. Duval, chief of the U.S. Air Force Air Defense Systems Office, arrived to examine a number of potential sites around Australia. Although the possibility of establishing the ground station in the United Kingdom was broached in September, in December the Air Force formally decided to locate the station in Australia—a decision followed by diplomatic negotiations and a variety of actions needed to establish such a facility on foreign soil.[5]

In January 1969 a contract was awarded to Aerojet for construction of the overseas ground station. Aerojet was to design the mission center and the Data Reduction Center; it would also specify requirements for security, antenna installation, hardware, software, and logistics, as well as produce the operations and maintenance manual.[6]

Two months later an Air Force official, Mike Yarymovych, arrived to gather additional information about a site about 300 miles northwest of Adelaide and 6 miles from Woomera village. Woomera had been built in 1947 as the support base for newly established British missile test ranges. From Woomera missiles could be fired and tracked northwest across Australia and then recovered on land. Yarymovych and his associates examined communications, water, roads, and life support requirements. He was also authorized to enter into an agreement with the Australian government on establishment of a site.[7]

However, final agreement was not reached until Prime Minister John Gorton visited Washington in April 1969. That month the Air Force informed the Aerospace Defense Command that "the Prime Minister of Australia will announce the 949 installation in Australia as a Joint United States/ Australian Defence Space Communications Facility." That the facility was the Overseas Ground Station (OGS) for the early warning satellite program would not be disclosed.[8]

The official announcement to Parliament of the establishment of a facility came on April 23, when the Australian prime minister informed the House of Representatives that the Government has accepted proposals which were made by the United States Government for the establishing in Australia of a joint United States–Australian defense space communications facility. The facility will be located at Woomera. . . . the installation should be ready to operate by late 1970. The joint staff of Australian and United States personnel required to operate the station will be between 250 and 300.[9]

An annual history of the Air Force's Space and Missile System Organization (SAMSO), written not long after the announcement, noted that "such matters had to be handled carefully for a large secretive installation in a foreign country was not without political hazards."[10] And Gorton's announcement that the United States would be establishing another major facility in Australia (in addition to the facilities at Pine Gap and Northwest Cape for intelligence and communications, respectively), along with the secrecy surrounding it, would result in the new facility being the subject of political controversy in Australia for the next three decades. Over the ensuing years, objections to the facilities would be based on a variety of concerns: that Australian sovereignty was being compromised, that the facilities would be nuclear targets in a U.S.–Soviet war, and that the bases facilitated objectionable foreign and defense policies such as nuclear war fighting. Questions would be asked in Parliament, demonstrations would be held outside the facilities, newspaper stories and the occasional book would explore the activities at the facilities and their implications for Australia.

Thus, Labor Party leader Gough Whitlam objected to the blanket of secrecy being thrown over the functions of Woomera, as well as those of the Joint Defence Space Research Facility at Pine Gap—a policy that left not only the Australian public but also its parliamentary representatives in the dark.* In addressing Parliament, Whitlam observed that while "it is right to co-operate with the United States in using the advantage of our size and site for communications and research in the space age . . . it is utterly wrong to withhold from Parliament and the public the general purposes of this and the whole dozen other facilities we are making available." Whitlam went on to argue that all the Russians had to do was read American magazines and the

---

*The Pine Gap facility was established to serve as the ground control station for a CIA-NRO satellite program initially code-named RHYOLITE (subsequently AQUACADE). The satellites, stationed over Borneo and the Horn of Africa, were designed primarily to intercept the telemetry from Soviet and Chinese missile tests, and secondarily to intercept VHF and UHF communications. Subsequently, their role as COMINT collectors was expanded.

Congressional Record to find out what was going on at the joint facilities. Other Laborites also objected to the secrecy, one referring to the prime minister's statement as among the "greatest acts of contempt that one could imagine." Some raised the specter that Australia would become a nuclear target as a result of hosting such installations.[11]

Among those on the government side was Minister for Defence Allen Fairhall, who observed that "these stations [Pine Gap and Nurrungar] . . . are to work in the field of space and therefore are to be concerned with satellites." With regard to the secrecy of the installations, the minister argued that "secrecy is the only thing which renders this proposal of any value at this time," and "it is the nature and the purpose of the station at Woomera . . . which constitute the essential military secret." On the other major subject of concern, he asked whether "anyone in his right senses really believes that any individual target in Australia would be singled out for nuclear attack in anything short of a global nuclear war."[12]

With regard to the issue of secrecy, a member of the prime minister's party suggested that the assumption that the "Russians already know what is going on" might not be valid. He went on to observe that "if the research that is being undertaken is at the ultimate limit of knowledge I take issue with [the assumption] and say that it is most unlikely that the Russians know what is going on."[13]

Among those speaking out at the time were two Labor backbenchers who would subsequently hold ministerial positions that required involvement in the bases issue. Kim Beazley, the representative from Freemantle, did "not think we are entitled to know the technology" but did want to know two things about Woomera: "Is it likely to involve us in war without our being consulted?" and "Is it likely to involve us in nuclear retaliation?" Bill Hayden, then the representative from Oxley, observed that "Australia is altogether too deeply involved in the chain of spy in the sky bases which the United States is operating."[14]

There were also objections in the Australian press. Two days after the prime minister's announcement, the *Sydney Morning Herald* expressed its concerns in an editorial entitled "Total Secrecy." The paper remarked: "Even the stoutest supporter of the alliance between Australia and the United States must be perturbed by the proliferation of secret and semi-secret American military bases in this country. . . . It is impossible to believe that the purpose of these bases could not be explained in a way which would not reveal military secrets to the Russians and Chinese."[15] The paper went on to observe that such secrecy would fuel rumors "which the government could not deny because it cannot tell the truth." In addition, it suggested that if "the main purpose of the installation at Pine Gap and the new station to be built at Woomera is to track reconnaissance satellites and enemy missiles, then it would be better to say so than to have people imagining nameless horrors of space warfare."[16]

While Australians debated the advisability of U.S. bases and the secrecy surrounding them, the Air Force organizations that would be involved in building and operating Woomera—the AFSC, the ADCOM, and the Logistics and Training Command—were busy establishing an Activation Task Force (ATF).[17] The task force's director-designate was Colonel Jerry F. Flicek, who had been on a intermittent tour of duty in Australia since May 1969. Flicek's initial title reflected the fact that a final agreement between the United States and Australia with regard to Woomera had yet to be concluded.[18]

The cause was not parliamentary or public protest in the wake of Gorton's announcement, although the delay was related both to Australia's democratic institutions and to its continued relationship to the British Crown. Before Minister for External Affairs Gordon Freeth could sign the agreement with the United States he needed approval of the Federal Executive Council. That approval had to be signed by Governor-General Sir Paul Hassluck, who decided that in light of the federal elections scheduled for November, which the Labor Party had a good chance of winning, no agreement should be concluded until after that time.[19]

Formal approval for the facility was not given until November 10, in the aftermath of Labor's defeat. That same day, Freeth and the U.S. ambassador, W. R. Rice, signed the agreement to establish an overseas ground station in the Woomera area.[20] The agreement contained eighteen articles, none of which directly specified what type of activities were to be conducted at Woomera or the targets of those activities. Most of the agreement was concerned with an assortment of issues connected with the construction and operation of the station, such as the tax-exempt status of property imported for use in constructing or operating the station. Only Article II, which specified that "information derived from the activities conducted at the station shall be available to the two Governments," indicated that surveillance or intelligence was involved.[21]

The November 10 agreement was followed by an implementing agreement on December 4, 1969. On December 20 Colonel Flicek became commander of the Systems Command's Space and Missile Systems Organization, Detachment Number 2, Woomera, Australia.[22]

★ ★ ★

The joint facility was to be known as the Joint Defence Space Communications Station (JDSCS). Consistent with the station's formal title and cover, the site for the new facility was named Nurrungar, which means "to hear" in Aborginal dialect. Presumably, to name it after the Aboriginal words for "fire watcher" or "heat seeker" would have compromised security. Nurrungar was eventually built about eleven miles from Woomera, in a small valley surrounded by low hills next to the ruins of an old staging point for coaches between Adelaide and Alice Springs. The site, like other sites in inland Australia, would offer unique protection against seaborne jamming and Soviet electronic intelligence ships, in comparison to other potential Pacific sites. It was also a "pretty dreary place," according to one member of the site survey team, who noted that if it were not for the location's immunity from Soviet vessels and their electronic equipment, the facility "presumably would have [been located in] some more civilized place." That view was shared by John McLucas, Under Secretary of the Air Force from 1969 to 1973, who characterized the site as being "at the end of the world." One trip to Woomera was all McLucas could stomach.[23]

Construction began in November 1969. The first radome, installed in 1970, housed the antenna that transmitted the command and control signals to the satellite and also received early warning and other telemetry data from the satellite. By the end of June 1970, contractors had the ground station buildings nearly ready for occupancy. Equipment began to be moved into the station's technical building on July 20, 1970. The data reduction center was equipped by August 24, but complete checkout and integration of all station components—the satellite relay station, tactical operations room, and Data Reduction Center—along with trained people to operate the equipment required several more months, and was one reason the Military Airlift Command flew a special weekly flight to Woomera from August to November 1970. External structures such as the sixty-foot antenna would be completed in time to monitor the first satellite launch in early November 1970.[24]

Through the fall of 1970, station personnel rapidly increased. Philco-Ford, responsible for operating the station's technical equipment, had all of its operational and management personnel on hand by the end of December 1970. Other arrivals included seven people from TRW Systems, an unknown number from Aerojet-General, two from the Air Force Technical Applications Center (AFTAC), three from the Atomic Energy Commission (AEC), and two from Sandia Laboratories. The presence of the latter five worried the Air Force, which was concerned about possible Australian political reaction to the presence of employees from U.S. organizations that dealt with nuclear weapons. In any event, by June 30, 1971, the overseas ground station popu-

lation included 92 U.S. military, 208 contractor, and 137 Australian employees. A total of 263 dependents brought the station population to 700.[25]

As the planned launch date neared, two problems arose between
the U.S. and Australian governments. With the influx of personnel, the problem of where to put everybody became acute, threatening continued U.S. operations from the station. As early as July 27, 1970, a team of SAMSO and
other Air Force personnel went overseas to begin housing negotiations with
the Australian government. That government, specifically its Department of
Supply, appeared reluctant to meet all U.S. housing requirements.[26]

Another Air Force housing assessment group visited Woomera in February 1971 and recommended spending $5.88 million on solving the problem.
But the Air Force and its leadership, displaying little enthusiasm for funding
expensive housing overseas, searched for alternatives. In an April 29, 1971,
memo, Under Secretary McLucas wrote that "we ought to have proved to
ourselves that we can meet the cost effectiveness criteria of going for the full
manning overseas versus bringing the data back and processing it here in
the U.S." However, by late June 1971 the realities of time, both for developing a viable satellite relay system and SAMSO's schedule for turning over
an operational station to the ADCOM in one year, led Air Force headquarters to agree "that the . . . housing program should be separated from any
communications relay decision and that action should be taken to provide the
necessary housing for the Aerospace Defense Command personnel." At the
same time, Air Force headquarters advised the DSP program office to "begin initial development of long lead items necessary to protect the option of
satellite-to-satellite data relay."[27]

Another solution to the housing problem appeared at least momentarily
attractive: to use more Australian employees to man the station. The command dispatched an Air Force/SAMSO manning team overseas to discuss
the proposal with Australian government representatives.[28] The discussion led
to a tentative Australian commitment to assign fifty-two civilian employees
and ten additional military personnel to station operation. The Air Force
pressed for increased Australian manning, although it was apprehensive that
there would be an increased opportunity for labor disputes and strikes that
might seriously, if not disastrously, interrupt the planned around-the-clock
operation.[29]

On December 23, 1971, the Australian government approved construction of eighty-one houses beginning on March 1, 1972. As of December 31,
1971, there were 519 station personnel: 234 U.S. military, 136 U.S. contrac-

tor, and 149 Australian. Another 276 U.S. dependents and 30 Australian dependents pushed the station population to 825.[30]

The second problem concerned station security, including the broader problem of avoiding program identification and public disclosure of the program's objectives. The United States felt strongly about the importance of protecting the station's physical security and concealing the program's mission and technical operations. On the other hand, according to a U.S. Air Force history, the Australian government

> tended to think of station security in terms of its political consequences; as the ruling party it confronted certain political realities in trying to conceal the purposes of a mysterious installation within its boundaries. Yet this concealment effort became increasingly tenuous as the American press with increasing frequency seemed to violate the security restrictions the Air Force had mandated for the U.S. and the Australians.[31]

The history pointed to repeated publication of articles in *Aerospace Daily* and *Aviation Week* that, it said, revealed the broad missions and the location of the overseas ground station. In addition, congressional hearings and open preparation for a Titan III launch at Cape Kennedy "stimulated the alert minds of perceptive newsmen to link the program's mission and launch preparations in progress." In effect, the Air Force history said that, at least with respect to Woomera, Whitlam and the other Australian parliamentarians who said that one could read "all about it" in the American press and congressional hearings were right.[32]

The matter came to a head on October 16, 1970, when *Aerospace Daily* referred to "a mid-November launch of [the new] integrated spacecraft." The news story went on to reveal details about the satellite and the contractors involved in the program, and quoted the Australian prime minister's announcement to Parliament. It also made quite clear where the ground station was located, noting that "the facility is situated about 300 miles northwest of Adelaide in the south central part of the continent. It is known by the aborginal name of Nurrungar and is located in some of the most desolate country in Australia."[33] All that was missing were the geographic coordinates.

The stories led the Air Force to a reconsideration of its secrecy policy, which had clearly failed to prevent discovery of the facility. However, the Australian government had not been concerned with keeping the information from the Soviets and other foreign intelligence services but from its own citizens. Thus, a policy of continued secrecy was preferable. According to the official Air Force history:

In November 1970, at the time of the first satellite launch and when final negotiations were proceeding, the American press revealed the station's purpose, much to the embarrassment of the Australian government, which had not informed its citizens about Woomera. Hence, while U.S. officials wanted to relax security restrictions in order to facilitate site operation and construction, the Australians were anxious to retain secrecy so as to avoid public debate. Although the U.S. eased program security at home in May 1971, it maintained restrictions in Australia for about a year afterwards.[34]*

A third problem, at least in the Air Force's view, was Australian attitudes. A SAMSO history observed: "Perhaps Australians were more relaxed and their workers and managers may not have felt the same sense of urgency that characterized U.S. interest in building and operating the station. . . . To the credit of the Australian authorities they listened and agreed to cooperate fully in solving these problems."[35]

Acquiring a CONUS (Continental U.S.) Ground Station (CGS) proved more troublesome than finding an overseas station. During the first part of 1968, it appeared that a U.S. facility would be required only to receive and process the data relayed from Australia. Just as cost concerns had been a factor in the decision to watch Soviet missile launches from geosynchronous rather than Molniya orbit, they had also led to a decision to deploy a one-satellite constellation.

The single satellite would focus on Soviet and Chinese ICBM launches, leaving SLBM detection to the seven FSS-7 land-based radars located on the East and West Coasts and the Gulf of Mexico. That decision, apparently made in 1967 or early 1968, had been the result of two concerns. Cost was one, with the Defense Department's comptroller's office arguing that there was room for only a one-satellite constellation in the budget. The other was that the infrared sensor would be operating at a far higher altitude than the MIDAS/Program 461 sensors; thus some felt it necessary to first establish that it would work properly before planning on a larger constellation.[36]

---

*The history is not referring to easing security in the United States concerning the location of the U.S. ground station, but with respect to the existence and purpose of the program. Secretary of Defense Melvin Laird had become convinced that if the program was to serve as a credible element of deterrence, its existence and purpose had to be acknowledged. Thus, reference to a "Satellite Early Warning System," or SEWS, was added to the unclassified Defense Posture statement.

Challenging the decision was Daniel Brockway, who joined the Defense research and engineering organization in mid-1968 and was given the 949 portfolio. To substitute discussion for interoffice memos and get everyone on the same track, he briefly established a "C3 group"—coffee, cigars, and conversation—which included representatives of the systems analysis, comptroller's, and logistics offices. One of the first subjects he took up was the plan to forgo satellite coverage of the SLBM launches. He objected that the threat from Soviet SLBMs was increasing with the deployment of Yankee-class subs, which began in 1968, each armed with sixteen SS-N-6 ballistic missiles. While the missile's low accuracy and moderate yield made it not much better for counterforce purposes than earlier-generation SLBMs it did have one significant advantage. Its range of 1,300 nautical miles was considerably greater than that of earlier SLBMs; even its direct predecessor, the SS-N-5 SERB, could travel only 750 nautical miles. In 1968 the CIA noted that "strategic targets in the middle of the US are beyond reach of the 650-mile SS-N-5 carried on the older ballistic-missile submarines, but are within range of the SS-N-6."[37]

He argued that reliance on the FSS-7s, whose "gaping holes and worthlessness" had been pointed out to him by a staff member of MIT's Lincoln Labs, threatened the survivability of the B-52 leg of the strategic triad. B-52 bases in the interior, such as those in Illinois and Kansas, would be subject to attack from missiles fired just offshore, drastically reducing warning time. Indeed, Brockway argued that if the United States had to choose between using a space-based system to cover ICBM launches or SLBM launches it was preferable, given the existing BMEWS and over-the-horizon ICBM detection radar systems, to first address the deficiency in SLBM detection. The result was a decision to deploy three satellites: one to cover ICBM launches, two to cover SLBM launches (with one satellite monitoring the Pacific and the other the Atlantic).[38]

Thus, on January 3, 1969, the Secretary of Defense issued a directive that scheduled the first 949 research and development Phase I launch for June 1970, provided for an improved system (Phase II), and kept open the option of developing an operational series of satellites (Phase III). It also approved procurement of an overseas ground station in Australia and a U.S.-based data-processing facility in the 1969 and 1970 fiscal years, followed by procurement of a U.S. ground station in the 1970 and 1971 fiscal years.[39]

In April 1969 the ADCOM recommended to SAMSO that Waverly Air Force Station, Iowa, where a SAGE radar was being phased out, be considered as the Program 949 ground station for the continental United States.[40] However, Waverly was not the only candidate. On April 27, 1969, the *Denver*

Post headline, in large type that consumed most of the top of the page, read "Buckley Will Help Track Super-Secret Sky Spies." The article reported that construction of an aerospace data facility would begin in June 1969 at Buckley Air National Guard Base (ANGB), located in the Denver suburb of Aurora, that it would cost over $20 million, and that approximately 300 personnel would be assigned to the station. As was the case with Nurrungar and Waverly, Buckley was located far enough inland to be immune to Soviet ship-based jamming.[41]

That story has generally been taken to represent the first public revelation that Buckley ANGB would be the CONUS Ground Station. However, the story stated that the data facility would employ radars to track U.S. reconnaissance satellites—not that it would serve as a downlink for U.S. surveillance or reconnaissance satellites. The story was not only confused but also premature. SAMSO surveyed the site in detail in May 1969 and forwarded its recommendation to the Department of Defense on September 9, 1969. However, the department objected that "the Air Force has not shown conclusively that the choice of [Buckley ANGB] is cost effective." Among its objections was that establishing the CONUS Ground Station at Buckley would run counter to Defense Department policy of closing single-mission installations wherever possible. To make its point clear, the department withheld CONUS Ground Station construction funds until the Air Force made a more convincing case.[42]

It was suggested that hardened facilities should be considered, such as the NORAD Cheyenne Mountain Complex (NCMC). But NORAD/ADCOM strenuously objected, claiming a lack of space and extensive air-conditioning and power requirements. Shortly afterwards, another element of the Air Force suggested placing the station in the much smaller Airborne Command Post. That proposal was also shot down. Sunnyvale, California, the headquarters of the Air Force Satellite Control Facility, may also have been a candidate.[43]

Those objections were followed by a detailed survey of Buckley on November 21 through 26, 1969. According to one Air Force account, the survey "revealed inadequacies the current budget could not correct nor the development schedule allow." One factor that certainly worked in Buckley's favor was the decision that the site should serve as the downlink for an upcoming National Reconnaissance Office (NRO) program, designated JUMPSEAT. JUMPSEAT satellites, the first of which would be launched in 1971, operated in a highly elliptical orbit, the same type of orbit rejected for DSP, and intercepted a variety of Soviet electronic signals. In June 1970 Buckley was formally chosen as the site for the U.S. ground station for the early warning satellite program, allowing construction to proceed. Until sufficient facilities were completed, control of the satellites from the United States would be conducted from Sunnyvale.[44]

Through the last half of 1970 calendar year, architects and engineers designed station installations including the main technical building, which was delivered to the construction site by January 11, 1971. Meanwhile, building sites and foundations had been prepared; construction, begun in January, progressed to a completion date of September 1, 1971.[45] Contractors had also completed the antenna pedestal and on July 1, 1971 began working on the sixty-foot antenna. Nearly completed construction of the technical and administration buildings by the end of June permitted their occupancy by the end of July 1971.[46]

That the construction taking place was to produce the CONUS Ground Station for the early warning satellite program would remain an official secret for over twenty years. Air Force officials in 1971 were apparently afraid that drawing attention to Buckley would only encourage antiwar demonstrators in the area to turn their attention to the new facility. Some also hoped, perhaps *dreamed* would be a better word, that the Soviets would not figure out Buckley's true function and it would avoid becoming an entry on the Soviet target list.[47]

As construction progressed, Washington decision makers became increasingly concerned about the vulnerability introduced into the DSP system by relying on a single U.S. ground station. That concern was triggered by reports from intelligence sources, apparently concerning the threat from Soviet SLBMs. One idea was to build a backup station so that loss of the main station would not eliminate surveillance coverage either in war or during inevitable downtime for repairs and normal maintenance.[48]

The Air Staff and other Washington authorities debated the proposal for some months. Frank Ross, Assistant Secretary of the Air Force for Research and Development, proposed a training-only facility to be located in Mississippi. Dan Brockway recalls that proposal as the subject of "our strongest disagreement." Brockway, the program monitor for Defense Research and Engineering, argued for a facility that not only would have training and software development functions but also would serve as a backup. Given the location of the main station at Buckley, it seemed appropriate to locate the backup nearby, specifically at Lowry Air Force Base, Colorado. According to an Air Force history, "building another ground station simply for its value as a training facility or as a means of improving software and performing system analyses appeared to be unsound since those functions could be performed just as well at existing facilities." The most convincing argument to support a dual system was its backup role and the fact that it would "eventually provide the capability to accept the full data stream for the eastern hemisphere through DSCS (Defense Satellite Communications System) Phase II, or satellite data relay."[49]

As of early February 1971, there was general agreement that "the issue of an additional ground station has been unresolved for too long," and that Lowry Air Force Base, Colorado, possessed certain advantages. On April 15, 1971, Air Force headquarters approved establishment of a backup facility there. The facility would also provide training for operators and would function as a data-processing and software development laboratory. By July 30, 1971, architects and engineers were at work on design of the Lowry Multi-Purpose Facility, attempting to meet a deadline of August 31, 1971.[50]

FIVE

DSP's First Decade

While construction continued at Buckley and Nurrungar during 1970, the first launch of a DSP satellite approached. On June 14, 1969, as the result of a security breach that spring, Program 949 became Program 647, with the unclassified and undescriptive cover name Defense Support Program. It was a name that one individual involved in the program recalls "just showed up out of nowhere."[1]

The following month the program received a new Air Force program director, Colonel Frederick S. Porter Jr. of the Space and Missile Systems Organization. Many involved in the DSP program consider Porter to have been an exceptional officer and a key factor in the program's success. Rather than use normal bureaucratic channels, which would have required him to take his problems and concerns to the Systems Command and Air Staff, Porter dealt directly with the Secretary of the Air Force. The secretary's clear interest in the program made it easier for Porter to convince other elements of the Air Force that problems should be solved before he had to report them, and presumably who was an impediment to their solution, to the secretary. In addition, Porter maintained tight control over information about the program, in order to prevent involvement by other agencies and commands.[2]

The DSP program that Porter was trying to guide to a successful beginning would be responsible for detecting any launches from the continually growing Soviet ICBM force, including the SS-9 SCARP, of which 288 would be deployed by October 1. The liquid-fueled missile, with a throw weight of over 9,000 pounds, could carry a single twenty-five-megaton warhead or three separate five-megaton warheads. Some, including Secretary of Defense Melvin Laird, argued that in its multiple-warhead form the SS-9 constituted a threat to the survival of the U.S. ICBM force and required the United States to deploy an antiballistic missile (ABM) system.[3]

The SS-11 SEGO and SS-13 SAVAGE missiles had also been added to the Soviet arsenal. In contrast to the SS-9, the SS-11 had a throw weight of only 1,000 to 2,000 pounds and limited accuracy. Thus, in November 1970 the intelligence community assessed it as "a weapon best suited against soft targets—cities, industrial installations, and some military targets." While the intelligence community estimated that by October 1, 810 SS-11s had been installed in their silos, the SS-13 contingent was judged to number only twenty. No more than sixty would ever be deployed, but the SS-13 did have the distinction of being the Soviet Union's first solid-propellant ICBM.[4]

Defense planners continued to worry about possible Soviet use of a Fractional Orbital Bombardment System (FOBS) or depressed trajectories. The Mod 3 version of the SS-9 had, by 1970, apparently been tested in both modes. An SS-9 Mod 3, launched northward toward the United States on a depressed trajectory, would be detected much later by BMEWS radars than an ICBM traveling along a conventional trajectory, with warning time cut from fifteen minutes to ten minutes or less. A southward launch, the "wrong way around," would avoid BMEWS altogether. But as one intelligence estimate explained, referring to DSP, "US sensors now being deployed promise to provide early detection of launches regardless of their firing direction."[5]

Launch of the first DSP satellite, scheduled for November 3 from Cape Canaveral, actually took place on November 6, 1970. Early that morning a Titan III blasted off from the Cape carrying the new satellite. As would be the case for all DSP launches for the next two decades, the payload of the Titan III booster launched that day was officially secret. However, the secrecy existed only in theory. The next day's *New York Times* reported that "while the Defense Department refused to discuss the nature of the satellite officially, the mission of the spacecraft together with details about it, including its launching time, were an open secret in the nearby town of Cocoa, Fla."[6]

Indeed, days before the launch the message on a sign above a local bank read:

SPY SATELLITE

GOES 5 47 AM

FRIDAY

TITAN 3 ROCKET

The bank was not the only institution in the area to breach security, with a number of motels also announcing the forthcoming launch.[7] One former program executive recalls that "the townspeople really didn't need the sign posted in front of the bank. All one had to do was to go to dinner at the restaurant down the road . . . and see all the visitors' faces, uniforms, and especially, raucous talk. The crowds on the beach on the night of the launch were quite large."[8]

The bank's sign proved off by only twelve minutes, with the actual launch occurring at 5:35 A.M. But as late as the night before, the Air Force had denied that any launch was planned. "It was a joke," said one reporter. "We knew and they knew we knew."[9]

Although the Air Force would only say the rocket had been launched with an experimental payload, the *New York Times* was more informative. Its story reported: "Reliable sources disclosed that the 1,800-pound satellite, codenamed 647, was destined to be placed in orbit about 22,300 miles above the Indian Ocean. From there is would provide roughly 30 minutes of warning of approaching intercontinental missiles."[10]

The satellite was the first of four DSP Phase I satellites to be launched between 1970 and 1973. The satellites actually weighed about 2,000 lbs and had a sensor package weighing 700 pounds and a design life of eighteen months (which would be exceeded by a substantial margin). The primary sensor, built by Aerojet, was a 3.3-foot-diameter, 12-foot-long Schimdt infrared telescope with an array of 2,048 lead sulfide detectors, which could detect infrared signals in the 2.69- to 2.95-micron range* and provide below-the-horizon coverage—spotting the missile's plume against the background of the earth.[11]

The twenty-three-foot-long, ten-foot-wide satellite would rotate six times a minute, with the rotation sweeping the sensor field of view around the earth so that almost all of the hemisphere was covered. Thus, any sufficiently intense infrared source would, in the absence of cloud cover, be viewed once each ten seconds. In the presence of cloud cover, a missile would be detected once it broke through the clouds. Repeated observations prior to missile burnout would help analysts determine that the infrared source was a missile and estimate its trajectory.[12]

Radiation would register on the infrared detectors and cause them to

---

*Choice of a system response in this range reflected the need to prevent the returns from missile plumes from being lost in a sea of infrared signals from earth. Since the atmosphere absorbs about 95 percent of all infrared signals in that range, most of the emissions generated at that wavelength would not make it through the atmosphere. At the same time, the fact that some signals, of various wavelengths would leak through the atmosphere and register on the satellites' infrared detectors would allow the satellites to be used to collect terrestrial intelligence as well as to detect missiles prior to their passage out of the atmosphere.

produce signals whose amplitude was proportional to the brightness of the radiation. It was then necessary to discriminate between signals representing missile launches and signals representing other phenomena, a task that would be initiated by signal- and data-processing electronics within the sensor and later completed by computers at the ground stations.[13]

Detection of missile launches, the primary objective, was designated "Mission A." The satellites' nuclear detonation detection mission was "Mission B." The primary means of detection were the RADEC-I and RADEC-II radiation detection packages, developed by the Sandia National Laboratory. The RADEC-I sensors were dedicated to detecting atmospheric and near-earth explosions, while the RADEC-II sensors were targeted on explosions in outer space.[14]

RADEC-I sensors included large- and small-aperture bhangmeters, an atmospheric fluoresence detector, and an X-ray locator system. Bhangmeters are optical sensors that detect the bright flash from the fireball of a nuclear explosion. The atmospheric fluoresence detector would detect nuclear detonations that took place between 31 and 1,240 miles. It recorded the optical time history of the nitrogen fluoresence signals produced when X rays from an exoatmospheric nuclear explosion excited air molecules at the top of the atmosphere. The X-ray locator consisted of several detectors, including an X-ray spectrometer, that measured the direction and arrival time of X rays from near-earth exoatmospheric detonations.[15]

The RADEC-II sensors, which were designed to focus on explosions above 1,240 miles, included an omnidirectional X-ray spectrometer, a neutron detector, and a gamma detector. The neutron detector would measure the neutron flux from a detonation, while the gamma detector would sense the gamma radiation associated with the detonation.[16]

Any data collected by the RADEC sensors would be transmitted to the DSP stations at Buckley and Nurrungar. The data would be simultaneously relayed to the key customer, the Air Force Technical Applications Center's Advanced RADEC Analysis Monitor (ARAM), a part of AFTAC's Satellite Operations Center (SOC), and the Sandia RADEC Analysis Monitor (SRAM) located at the DSP ground stations.[17]

In one way the *New York Times* story proved incorrect. Despite Air Force intentions, DSP Flight 1 (F-1) did not go into the planned geostationary orbit over the Indian Ocean. During the launch, the upper stage of the Titan IIIC booster malfunctioned and the satellite was propelled into an elliptical orbit with an apogee of 19,378 nautical miles, a perigee of 14,038 nautical miles, and an inclination of 7.8 degrees.[18]

In such an orbit the satellite would not be able to cover one portion of the earth on a continuous basis. However, its rather leisurely journey around the earth—twenty hours, compared with ninety minutes for low-earth-orbiting reconnaissance satellites—meant there was substantial "dwell time" above the Northern Hemisphere, which was put to good use. The satellite beamed its observations of U.S. and Soviet missile launches as it passed within range of various tracking stations (in Guam, Hawaii, Vandenberg Air Force Base, New Hampshire, and the Indian Ocean). While Nurrungar monitored the satellite as well as circumstances permitted, the initial responsibility of controlling F-1 (and subsequent satellites) rested with the Air Force Satellite Control Facility and its remote tracking stations. By April 10, with most of its equipment installed, the station was able to resume responsibility for satellite command and control during the times the satellite was in view.[19]

Because the satellite furnished data from both U.S. and overseas areas, it was called "the best mistake that ever happened" by those in ADCOM headquarters who worked on the project. Engineers gathered information that was useful for designing follow-on satellites. The launch turned out to be 99 percent useful in meeting its research and development objectives, in part because the SAC and the Navy modified their launch schedules so that their missile tests would take place while Flight 1 was in position to detect them. The satellite was also considered to have satisfied about 30 percent of operational requirements.[20]

Returns from Flight 1 confirmed the 1967 data on the SS-N-6 SLBM. And, because DSP's pointing accuracy was better than had been projected, it could produce highly accurate estimates of the launch point. That, in turn, made it possible to devise a special algorithm so that SS-N-6 signals would be recognized as coming from SS-N-6s.[21]

Beyond discoveries relating to Soviet missile capabilities, Flight 1 produced other information that illustrated the new systems capabilities and utility for a variety of missions. The very first "interesting IR returns," according to a program official, were from a forest fire raging near Santa Barbara. Along with other early DSP satellites, Flight 1 would contribute significantly to the cataloging of infra-red stars.[22]

In light of the failure of the first satellite to reach its proper station, TRW was requested to bring the single test satellite to flight standard just in case it was needed. It was not, as two fully successful launches, in May 1971 (F-2) and March 1972 (F-3), followed. The first countdown began on the evening of May 4, 1971, and continued with only minor problems until T-32 seconds, when the mission was aborted. The second countdown proved uneventful, and the Titan IIIC lifted off early on May 5.[23] The F-2 spacecraft went into geostationary orbit approximately 22,300 nautical miles above the earth in the Eastern Hemisphere at 65E longitude. It weighed 2,100 pounds and carried the standard payload of infrared detectors, star trackers, and nuclear blast detectors in the form of small horns on the spacecraft.[24]

On May 19, 1971, Nurrungar assumed command and control of the new satellite. But whereas the Nurrungar facility had no problem communicating with the spacecraft, communications between Nurrungar and the United States "proved slow and unreliable." Two months earlier, Air Force Under Secretary John McLucas had asked SAMSO to curtail the Nurrungar facility and eventually phase it out entirely. However, it soon became clear such plans would be premature, since there was no alternative means, such as a relay satellite, for transmitting the data.[25]

If McLucas's plan had been carried out, that would have provided at least partial satisfaction to those in Australia who were already calling for the elimination of all American military facilities there—particularly Nurrungar, the Pine Gap ground control station for the RHYOLITE signals intelligence satellite, and the North West Cape naval facility. The Australian Labor Party, at a conference that year, adopted a platform that called for the elimination of "foreign-owned, -controlled or -operated bases and facilities in Australian territory, especially if such bases involve a derogation from Australian sovereignty."[26] The platform also called for the federal Parliament and the public to be informed of the "general purpose and possible consequences of the joint defence installations and facilities at Pine Gap and Woomera."[27]

A third DSP satellite was orbited on March 1, 1972, only a day later than scheduled, and placed into geosynchronous orbit over the Atlantic at 80W. Four months later the Air Force announced, in a classified document, that the third geostationary satellite would be added to the DSP constellation by October 1973. It also authorized development of a capability at the Buckley station to provide sixty-five-second warning of an SLBM launch if it occurred less than 1,700 nautical miles from the U.S. coast, and specified that failed satellites should be replaced within thirty days.[28]

Implementation of the three-satellite constellation would require another launch, which occurred on June 12, when the fourth DSP satellite was launched on a Titan IIIC booster from Cape Canaveral and successfully placed in its proper orbit, apparently over the western Pacific at approximately 134W.[29]

With the successful launch of F-4, the Air Force had in place a set of satellites with extensive detection and reporting capabilities. According to a then-classified assessment, in 1975 the probability of launch detection was greater than .96; when it occurred, warnings of ICBM launches took place within two minutes of launch. Analysis of the data at the ground sites provided an estimate of the launch point within two to ten miles, a launch heading within 5 to 25 degrees, and launch time within three seconds. The false report rate was "virtually nil" for ICBM launches and several per week for SLBM launches under unfavorable solar conditions (the Northern Hemisphere summer).[30]

The use of two satellites to cover the Western Hemisphere—one from the Pacific, the other from the Atlantic—also made it possible to take further action to improve the probability of detecting the dim signal from an SS-N-6. Since satellite coverage overlapped extensively, as shown in Figure 5.1, it was possible to subject the data from the satellites to a rudimentary stereo processing technique known as DUAL. The technique further improved the ability to detect the SS-N-6 and pinpoint the launch area.[31]

The data from the DSP-W satellites, of course, were transmitted directly to the CONUS Ground Station at Buckley. From there they passed through the Data Distribution Center (DDC) on their way to their users, which included the North American Aerospace Defense Command Cheyenne Mountain Complex, the SAC, the National Military Command Center, and the Defense Special Missile and Astronautics Center (DEFSMAC)—the latter being a joint NSA-DIA organization that receives data related to imminent foreign missile and space launches and, in turn, cues U.S. intelligence collection systems and organizations.[32]

At least two routes were available for the relay of data from the DSP-E satellite. One was via a submarine cable that passed through New Zealand, Hawaii, and Vancouver and a landline from Vancouver to the DDC. Installation of a satellite communications terminal at Nurrungar permitted the station to transmit mission data to the DDC via the Defense Satellite Communications System (DSCS) rather than through leased commercial circuits.[33]

By the end of June 1973, the DSP satellites had detected 1,014 missile launches, including several in December 1972. In that month, twelve Soviet land-based ballistic missiles were fired. Operational tests involved

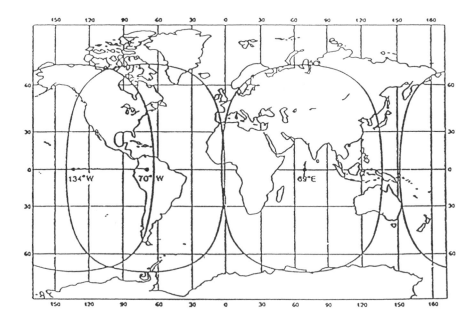

5.1. Approximate Earth coverage of DSP satellites. (From Anthony Kenden, "U.S. Military Satellites, 1983," *Journal of the British Interplanetary Society* 38 (February 1985): 63. Reprinted courtesy *JBIS*.)

three SS-4 medium-range missiles, one SS-7, and six SS-11s, while developmental tests involved two ICBMs launched from Tyuratam, designated TT-6 and TT-7. Several SS-N-8s and SS-N-6s were also tested in December, including what appeared to be a modified SS-N-6. The European Command's January 17, 1973 intelligence appraisal also noted, "What is believed to be the first firing from a seaborne platform of a new ballistic missile . . . occurred on 9 December in the White Sea."[34]

The satellites also proved flawless in performing Mission B—the detection of nuclear explosions in the atmosphere. Although the United States and the Soviet Union had conducted only underground tests as result of the 1963 Partial Test Ban Treaty, both France and China continued aboveground tests for many years afterward. The Chinese conducted four such tests between the launch of F-2 and June 30, 1973. The French conducted nine atmospheric tests at its Mururoa test site in the Pacific during the same interval. Then, in July and August 1973, it detonated five nuclear devices at the site— with all the detonations being detected by F-2. In each case the detonation was detected not only by the dedicated nuclear detonation detection sensors but also by the infrared sensor, which could both detect the initial explosion and also follow the rising mushroom cloud.[35]

Reports of nuclear detonations were made within minutes of the event. In addition to an estimate of the geographic location, the nuclear detonation sensors permitted estimation of the time the device was detonated, its yield and yield-to-mass ratio, and additional characteristics of the explosion.[36]

The French explosions of the summer of 1973 were followed by explosions of another kind—those produced by the October Yom Kippur War. To Ellis Lapin of the Aerospace Corporation, the dry, relatively cloud-free conditions in the Middle East offered an opportunity to clarify the extent of DSP capability. He hoped that the infrared returns generated by the Middle East war would help explain some of the returns picked up during the Vietnam War that analysts were not, at the time, able to associate with specific events. And since he was the DSP principal manager at the Aerospace Corporation, which provided technical support to the Air Force on that and a number of other satellite programs, he was in a position to have his theory tested.[37]

With the concurrence of the DSP program office at SAMSO, and in cooperation with ADCOM, a six-man team composed of ADC, SAMSO, Aerojet, and Aerospace Corporation personnel left for Nurrungar virtually immediately. The team was led by Ed Taylor of the Aerospace Corporation and included Bob Richards, who had joined Aerojet in 1956 and would become the head of its Electronic Systems Division. Once there, they performed initial evaluation of data as they were being collected and established procedures for data collection and reporting. They compared the infrared returns with the location of possible target areas and were able to report the most interesting data to the United States in near real time.[38]

Daily messages were sent, via the AFSCF or Los Angeles Air Force Station, to Washington, enumerating daily sightings. It soon became clear that some of the returns from Vietnam had been of antiaircraft missiles. With a burn time of only thirty seconds for an antiaircraft missile, such as the Syrian and Egyptian SA-2s and SA-3s, a DSP satellite could produce a maximum of three infrared returns. Thus, according to a program executive, "It was touch and go as [to] whether we could get enough returns to verify that it was a launch and not noise."[39]

After the war, an extensive analysis, designated Project HOT SPOT, was conducted by Aerojet, SAMSO, and the Air Force Foreign Technology Division (FTD), the Air Force's scientific and technical intelligence organization.

In addition to confirming DSP's ability to detect and discriminate surface-to-air missiles, the analysis also revealed detection of warhead detonations and aircraft fires. More important, it highlighted the capability demonstrated by the Program 461 satellites to detect IRBMs. A December 7, 1973, report by the FTD and SAMSO, entitled "Preliminary Analysis of Project Hot Spot IR Signals," apparently concluded that certain infrared data gathered by DSP satellites conformed to the predicted intensity signature and trajectory for a Scud. A second study, by Aerojet, entitled "Application of Infrared Tactical Surveillance," further substantiated the conclusion that DSP satellites could and should be employed to monitor tactical missile launches.[40]

In the ensuing years, DSP satellites would provide important data on both the testing and the wartime use of such missiles. In less than a decade they would assist the U.S. intelligence community in monitoring developments in a Middle East war. In 1991 they would be a significant element in the effort to drive Iraq out of Kuwait.

While the successful orbiting of F-4 established a three-satellite constellation, it was a constellation that did not satisfy everyone. There had been sentiment within the Air Force for a considerable time for launches of additional DSP satellites into highly elliptical orbit.[41]

The desire to place DSP spacecraft in elliptical orbit was, in part, due to concern over Soviet SSBN operations in the Arctic as well as increases in Soviet SLBM range. In 1969 the Soviet Navy began testing its single-warhead SS-N-8 Mod 1 SLBM. As mentioned earlier, its SS-N-6 predecessor had a range of approximately 1,300 nautical miles. In October 1972 the intelligence community estimated that the maximum range for the SS-N-8 would be between 3,100 and 3,500 nautical miles, which would substantially increase the patrol areas of Soviet subs. However, in November and December 1972 tests, the SS-N-8 demonstrated an even greater range—4,200 nautical miles.[42]

When the missiles were deployed on Delta I and II submarines, which was initially expected to take place in 1973 but actually occurred in April 1974, they could be fired from the Arctic seas, quite near naval bases such as Murmansk, and still hit an array of targets in the United States. Close to their home ports, those submarines would be less vulnerable to U.S. anti-submarine warfare forces. At the same time, the geostationary DSPs were unable to detect Arctic launches, those emanating from DSP's blind spot: 82 to 90N latitude. Nor did land-based systems provide complete coverage. FSS-7 SLBM radars, which were oriented toward the surrounding oceans, provided no detection capability at all against SLBMs launched from the Arctic.[43]

It was argued that adding an elliptical component to the constellation would immediately improve SLBM launch capability, close gaps in coverage caused by adverse angles and the hole in DSP coverage, and provide backup in the event that one of the satellites failed. The Air Force argued that the capability could be available by mid-1973.[44]

In September 1971 Deputy Secretary of Defense Kenneth Rush had approved the elliptical satellite concept, and in November he authorized a second computer for the CONUS Ground Station. In December 1972 Rush authorized the Air Force to continue development of an elliptical option. But in August 1973 Deputy Secretary David Packard indicated he would postpone certain software modifications and the elliptical orbiting satellites until after 1980. However, the DSP spacecraft would be modified to permit them to be placed in elliptical orbit.[45]

While dedicated DSP satellites in elliptical orbit were ruled out by the deputy secretary for at least the remainder of the decade, there were alternatives. One proposal, to give future satellites an above-the-horizon capability, had come from Defense Research and Engineering and could be accomplished at significantly less cost than adding elliptical satellites or designing a completely new system. Adding an above-the-horizon capability meant orienting some of the infrared detectors to detect not the hot, first-stage burn of missiles against the earth's surface (below-the-horizon) but the cooler, second-stage burn against the background of space. Such detectors would be capable of seeing missiles fired from Arctic waters.[46]

Once the proposal had been accepted by Secretary of Defense Melvin Laird, it was necessary to move one of the DSP satellites in orbit in order to collect the data necessary to design the ATH sensor. That happened only after a directive from Laird to a reluctant ADCOM, which then ordered the satellite in question to be moved for a brief period. Not long before the satellite was to be returned to its normal station, "a target of opportunity" occurred, permitting the necessary infrared data to be collected.[47]*

There was also another means of detecting Arctic waters launches from space, although it would leave some in the early warning community dissatisfied. Since 1971 the Air Force and the NRO, under a program desig-

---

*There were some limitations and special requirements associated with ATH detection. Because the second and third stages of flight would be detected but not the first, it was not possible to determine launch points. In addition, special stereo processing is required to avoid "azimuth flip," a phenomenon that occurred because the above-the-horizon sensors were viewing the missile against empty space and not the background of the hard earth. As a result, a missile fired within the Soviet Union for test purposes could be mistaken for a missile fired toward the United States.

nated JUMPSEAT, had launched several signals intelligence spacecraft into Molniya orbit. The satellites intercepted communications emanating from Soviet Molniya satellites as well as from ground-based targets.[48]

The Air Force and NRO decided to equip the spacecraft with an infrared sensor, different in type than those on DSP satellites. The primary purpose of the sensor, which was originally or would eventually be code-named HERITAGE, was to collect technical intelligence on any short-burning, high-acceleration ABMs, similar to the U.S. SPRINT, that the Soviets might have in development—missiles that DSP would not be able to detect. The HERITAGE sensor was a modified staring sensor that had a much shorter time frame (revisit time), allowing it to produce more observations of a particular region in the same amount of time. Because of its orbit, such a sensor could also be used to provide warning of SLBM launches from the Arctic area. Eventually, the use of these sensors to supplement the DSP geostationary satellites would be designated DSP-Augmentation (DSP-A).[49]

But not everyone was happy with the ultimate result. One DSP executive recalls that he argued "long and hard" over the sensor design because he did not like certain mechanical aspects. But while DSP proper belonged to ADCOM and the early warning community, and the infrared intelligence derived from missile launches was a subsidiary function, HERITAGE was an NRO program. Warning was "way down on the list of priorities," according to one individual who was knowledgeable about the program. In the view of one former executive involved in DSP, it was "not on the list at all," with its warning value being used after the program began as a means of saving the program at budget time. As he viewed it, DSP-A was "a sordid story from beginning to end"—in terms of the requirements process, initial contractor selection, and a number of other factors.[50]

By the end of 1974 the DSP satellites had added eighteen months of detections to the June 1973 total of 1,014—without having missed a single ICBM launch. Among the over 966 Soviet and 16 Chinese test launches detected were those of the liquid-fueled SS-17 SPANKER and SS-18 SATAN ICBMs, which began in 1972, and those of the similarly fueled SS-19 STILLETO, which began in 1973.[51]

In addition to performing the assigned mission, DSP also exhibited greater-than-expected capabilities in a number of areas. As noted earlier, pointing accuracy was better than expected, permitting estimation of a missile's trajectory as well as prediction of its impact area. The satellites also had exhibited far greater life spans than had been predicted.[52]

As would be expected, particularly of satellites operating past their design lifetime, various problems developed. In March 1974 the attitude control system of F-1, which kept the satellite pointed in the right direction, ran out of fuel, and the satellite had to be deactivated after forty months of operation—twenty-two months after it had been expected to end operations.[53]

In May the AFSC, partly in anticipation of future problems, recommended the launch of a replacement satellite. By the beginning of 1975 there were problems with F-2, F-3, and F-4. F-4's problem was the most serious. A steady leak from the gas generator assembly, which fueled the low-level thrusters of the satellite's attitude control system, developed in January 1975. Since the thrusters controlled the satellite's rate of spin and its pointing accuracy, the satellite's continued viability was called into question.[54]

That problems emerged with Flights 2 and 3 were not surprising, since the satellites were twenty-four and fifteen months respectively, beyond their design life. According to former MIDAS/461 program manager Marvin Boatright, the "Air Force was reluctant to let go of a bird and wouldn't call for a replacement until someone became very nervous." If the satellite could produce substantial data, the Air Force wanted to keep it operating. When problems emerged, the Air Force would ask for "work arounds" to keep it functioning.[55]

That was amply illustrated by Flight 2, which had one star sensor unit and one digital telemetry unit fail. Also, two of the three atmospheric burst locator lamps had failed. Failure of the backup star sensor unit would mean the ability to produce accurate data associated with a missile launch or nuclear explosion would be lost. Failure of the third lamp would cause loss of the satellite's nuclear detection capability. These problems were corrected through the use of modified software, although not without reducing launch-point accuracy and the probability of detection, and increasing the false report rate. Tests and further analysis of the problems were completed in March 1975, with a follow-on fix that was expected to return the system to near-normal operational capability. But, clearly, the fragile constellation required some fresh blood. When the Air Force approved another satellite, ADCOM wanted to launch it in March or April, but a full schedule for the launch area initially delayed the date until June. Then two Titan booster failures further delayed the launch, which was rescheduled for mid-September 1975.[56]

Additional delays postponed the launch into the winter of 1975. Finally, on December 14, 1975, DSP Flight 5 blasted off from the launch pad at Cape Canaveral to replace Flight 3. Flight 5 was the first of five DSP Phase II

Multi-Orbit, Performance Improvement (MOS/PIM) spacecraft, capable of operating in elliptical orbit, which had been been designed in response to concerns over the coverage limitations of the geosynchronous DSP satellites with respect to the new Soviet long-range SLBMs. They carried an enhanced nuclear diagnostic capability, improved communications and sensor capabilities, and a greater, two-year, design life.[57]

Five days after launch, however, the explosion of the hydrazine fuel line, the result of the line's repeatedly freezing and thawing, resulted in the satellite spinning and tumbling uncontrollably. Its spin rate was increased even further after a collision with an object of unknown source and physical dimensions. As a result, no mission data were available to the ground stations. And the failure was of such a nature that there was no possibility of restoring the satellite to operational status.[58]

A day after the loss of F-5, James Wade, director of the Defense Department's SALT Task Force, addressed allegations that another DSP satellite had been the target of hostile action by the Soviet Union. Over the preceding two months, it had been suggested that the DSP-E satellite might have been the target of Soviet lasers in an attempt to test their ability to "blind" the warning satellite. A 1973 intelligence study had noted a Soviet interest in laser blinding and that a laser system capable of interfering with the optical train components of a photoreconnaissance satellite, which operated in much lower orbits, "could become operational at any time."[59]

On five occasions in October and November 1975, DSP-E picked up intense returns at 2.7 microns from sources in the Soviet Union. The first instance occurred on October 18, 1975; two others took place on November 17 and 18, 1975; and on one occasion the illumination persisted for more than four hours. According to a report in *Aviation Week & Space Technology* on December 8, 1975, the Defense Department was considering the possibility that the satellite had been deliberately dazzled by a hydrogen-fluoride chemical laser.[60] However, subsequent analyses suggested that the illumination was less likely to have been caused by laser radiation than by natural phenomena, particularly fires from decrepit Soviet natural gas pipelines. On December 15, 1975, Wade testified about the incidents:

The facts are not yet in.

The issue is that a [DSP satellite] may have been interfered with, possibly [by] a laser beam. There is, however, a high probability that it could be nothing more than a gas flare, or a gas fire. Such a signal is sufficiently large, that it would cause a perceptible noise in the sensor system. . . .

This really could be nothing more than a large transmission line, blowing off gas. The satellite sensors are sensitive enough for the type of signal. . . . Certainly some of the gas fires that have occurred in the Mid-east and the Soviet Union may have radiant intensity higher than that of a SS-11, for example. Therefore, the signal presented to the satellite by such a gas fire could be on a level approximating stages of a missile. I suggest we wait and see what the facts are. This activity has not interfered with the ability of our satellites to do their job.[61]

Following this testimony, the Department of Defense informed the House Armed Services Committee:

A US early warning satellite system recently detected several events in remote areas of the Soviet Union which contain no military installations. Analysis of these events essentially eliminates initial concerns that lasers were being used against the satellite system. Instead the events are believed to have occurred as a result of several large fires along Soviet natural gas pipelines. These events were local in nature and did not reduce the satellite system's ability to provide early warning of ballistic missile launches.[62]

At his first press conference on December 22, 1975, the then Secretary of Defense, Donald H. Rumsfeld, stated that the initial "speculation" about the use of lasers to blind the satellites was not confirmed by examination within the Defense Department.[63]

While Wade and Rumsfeld dealt with the laser-blinding issue, ADCOM determined that F-5 was irretrievably lost and asked that a replacement satellite be launched as soon as possible. DSP Flight 6 was launched on June 25, 1976, deployed between 65W and 69W longitude, and turned over to ADCOM as an operational satellite on July 28, 1976. Flight 6 thus became the first Phase II vehicle to be used for operations. In addition to the Phase II modifications, a modified sensor which permitted an above-the-horizon experiment was carried. The modification consisted of including more sensitive infrared detectors to allow the satellite to detect infrared returns of missile launches from beyond the horizon and observe lower-intensity tracks from locations not normally in the field of view. Such a capability promised to give

DSP the ability to detect SS-N-8 launches from the polar region.[64] A new sensor, referred to as the Impact Sensor, was added to the satellite to detect impacts of external objects and/or forces on the satellite—such as the impact that had increased F-5's spin.[65]

As a result of the launch there were six DSP satellites in orbit, four of which were operational: three primaries and one backup. Flights 2, 3, and 4 were operational, but, as noted earlier, each had problems that degraded their mission capability.* The continued deterioration of in-orbit satellites prompted General Daniel James Jr., the Commander in Chief, Aerospace Defense Command, (CINCAD), to request that Flight 7 be launched. On February 6, 1977, the seventh DSP satellite was launched into the standard geostationary orbit and positioned over the Pacific at 135W. After testing of its sensors and subsystems Flight 7 was handed over to the ADCOM on February 18 and was accepted for operation on February 26.[66]

The successful launch of Flight 7 meant that for the first time the United States would have a complete network of primary satellites and Limited Reserve Satellites (LRSs), or backup satellites, with one backup satellite for each hemisphere. It also signaled the beginning of the practice of shifting the orbital locations of still-operational satellites in the wake of a successful launch.

Thus, after the checkout of Flight 7, Flight 3, which had been serving as the Pacific satellite, was moved to the Eastern Hemisphere as the backup satellite. An operational test for Flight 3 was intended to demonstrate the ability of the overseas ground station to recall and operate the satellite, and to provide an indication of how long it would take to recall a satellite to active duty should an emergency recall ever be necessary.[67]

In 1979 the DSP constellation again needed replenishment. On February 15, 1979, a memo from the CINCAD directed that a new satellite be launched to replace Flight 7, then located over the Pacific. Flight 7 would then be moved to the Eastern Hemisphere to replace Flight 2, whose troubles had continued and worsened. On June 10 a Titan IIIC successfully injected the satellite into the required geostationary orbit; once there, the satellite was designated Flight 8. It was placed in operation on June 22 and formally turned over to the ADCOM on July 11.[68]

---

*The criteria for satellite replacement in terms of mission capability was a sufficient degradation of the sensor's ability to detect missile launches, specifically the minimum-intensity threat of the time, the SS-N-6 SLBM. Such a degradation could result from the normal long-term temperature rise from extended time in orbit, which would produce a long-term decrease in sensitivity and thus increased background noise within the sensor.

TABLE 5.1. *Satellite Stations After Launches F-4 Through F-8*

	F-4 June 12, 1973	F-6 June 25, 1976	F-7 Feb. 6, 1977	F-8 June 10, 1979
F-1	NONE	NONE	NONE	NONE
F-2	EH	EH	EH	EHLRS
F-3	ATL	PAC	EHLRS	WHLRS?
F-4	PAC	EH or WHLRS	WHLRS	RETIRED?*
F-5**	NYL	NONE	NONE	NONE
F-6	NYL	ATL	ATL	ATL (65W)
F-7	NYL	NYL	PAC	EH (65E)
F-8	NYL	NYL	NYL	PAC (135W)

*Possibly retired after malfunction.
**Failure of F-5 five days after launch precluded it from ever occupying a station.
Key: EH = Eastern Hemisphere; WH = Western Hemisphere; LRS = Limited Reserve Satellite;
NYL = not yet launched.
Sources: Aerospace Defense Command, *The History of the Aerospace Defense Command Fiscal Year
1972*, pp. 114–15; ADCOM, "DSP Flight 3 Operations Test," August 12, 1977; ADCOM,
"Flight 8 Turnover," July 10, 1979; ADCOM, "DSP Operational Satellites," August 8, 1979;
ADCOM, "Flight 9 Justification Briefing," July 29, 1980; ADCOM, "DSP Launch Initiation,"
October 21, 1980; ADCOM, "Defense Support Program (DSP) Improvement Status," April 24,
1979.

Table 5.1 lists the stations for the DSP satellites that resulted from the
launch of F-4 to that which existed weeks after the launch of F-8 on June 10,
1979. Figure 5.2 presents a view of Earth from DSP Flight 7 in its position
as the Eastern Hemisphere's primary satellite.

It had been anticipated, during the MIDAS era, that piggybacking a nuclear detonation detection system on launch detection satellites would
allow termination of the VELA satellite program. The first two of the initial
generation of VELA satellites, which were launched in pairs, had been placed
in an approximately 70,000-mile orbit in October 1963. Another two sets of
the first generation VELAs followed in July 1964 and July 1965. The satellites
were faceted spheres, which carried X-ray, gamma-ray, and neutron detectors.[69]

Three pairs of advanced VELA satellites were launched in April 1967,
May 1969, and April 1970. That the April 1970 launch proved to be the final
one was expected, given that the first DSP launches would take place well
before the lifetime of the last VELAs would come to an end. What was not
anticipated was that one of the final VELAs, satellite 6911, with an expected
lifetime of eighteen months, would operate for *fourteen* years.[70] Thus, 6911
was still in service on September 22, 1979.

At 3:00 A.M. that day, the bangmeters on VELA 6911 detected a brief,

5.2. View of Earth from F-7

intense double flash of light, which analysts concluded had come from near the southern tip of South Africa. In many ways that double flash was characteristic of the forty-one previous double flashes picked up by VELA spacecraft when they had detected nuclear detonations. It was reportedly different, however, in that the two bhangmeters did not record the usual roughly parallel intensities of light, as was expected from a distant event.[71]

Whether a nuclear explosion had taken place, and who might be responsible if one had occurred, became a subject for intense study and debate by the U.S. intelligence community and a number of other organizations—particularly the Energy Department's national laboratories and the Naval Research Laboratory. The easier part of the problem was to identify those most likely to be responsible for a test. An interagency intelligence memorandum completed in December 1979 identified South Africa and Israel, either separately or jointly, as the key possibilities. The memorandum did note the DIA's belief that "if an atmospheric test were in the technical interest of the USSR, an anonymous test near an unwitting proxy state such as South Africa could have provided an attractive evasion method."[72]

To try to settle the more difficult question, the United States sought to assemble data that might have been recorded by a variety of intelligence and scientific sensors throughout the world. It also engaged in an active campaign to collect, using both technical and human resources, any evidence of nuclear debris that might have resulted from such a test. Thus, the Air Force conducted aerial sampling missions, while the CIA deployed personnel to West Africa to gather leaves that might be the host to nuclear residue.[73]

In addition to detailed analysis of the VELA data, the data from two DSP satellites, which were viewing some of the area viewed by the VELA satellite, were examined. A cross in the lower corner of Figure 5.2, which shows Flight 7's view of the world, indicates the location of Prince Edward Island, often suggested as the site of the alleged test. However, according to a Los Alamos report:

> The bhangmeters on Flights 6 and 7 did not trigger. The weather in the South Atlantic was overcast to broken with low and high clouds, so it is necessary to conclude that either the event didn't take place within their fields of view, or the event signal was weakened by transmission through clouds and wasn't strong enough to attain the necessary threshold irradiance.[74]

Such ambiguity plagued the entire effort to determine whether a nuclear test had, in fact, occurred. Different elements of the intelligence and scientific communities would reach different conclusions. No consensus was ever reached, and the controversy continues to this day.[75]

In early 1980 those responsible for the operation of the DSP satellite system were more concerned with problems at one of the ground stations than with interpretation of DSP data with respect to the September 22 event. In a March 8, 1980, visit to ADCOM commander Lt. Gen. James V. Hartinger, Assistant Secretary of Defense for Command, Control, Communications, and Intelligence (C^3I) Gerald Dineen addressed the issue of an apparent increase in outages at Nurrungar.[76] The outages were still an issue at the end of the year. From November 1, 1980, through December 30, 1980, the station had experienced an unusually high number of unscheduled outages. On November 20 and 21, 1980, it had nine outages directly related to a power supply problem, which was corrected by the end of December.[77]

Unscheduled outages also occurred on December 10 and 12, followed by an extended outage on December 28. The DSP overseas ground station completed corrective maintenance on both computers. The outages required an increased state of readiness on the part of ground-based warning assets, including the FPS-17 and FPS-79 radars in Turkey, the PAVE PAWS SLBM radar at Otis Air Force Base, Massachusetts, the COBRA DANE radar on Shemya Island, Alaska, and the BMEWS. On November 22, 1980, the CINCAD thanked all agencies involved for their support during this period, noting, "Your prompt and energetic efforts insured that the missile warning degradation resulting from the loss of DSP-E was held to a minimum."[78]

A few weeks after the DSP-E system was restored, the Western Hemisphere capability needed attention. Strategic Air Command and ADCOM representatives attended a December 12, 1980, meeting to determine whether there was a requirement to launch Flight 9. They presented performance data that was felt to urgently justify the replacement of Flight 6. An earlier decision to launch in light of declining capability was upheld, and a revised launch date of February 19–26, 1981, for Flight 9 was established.[79]

Plans called for placing Flight 6 at 65W approximately seven days after launch, at which time most of the Flight 9 prehandover testing would be completed. Flight 9 would arrive at 70W ten days after launch. Flight 6 would then remain at 65W until thirty days after launch; at that time it would be moved toward 100W, where it would become the Western Hemisphere Limited Reserve Satellite. In that position it could be moved to replace either the DSP-W (Atlantic) or DSP-W (Pacific) satellites should problems arise with either.[80]

The launch of DSP Flight 2 into its proper orbit on May 5, 1971, occurred almost exactly eight years after Harold Brown's downbeat presentation to Congress on May 6, 1963. That testimony was followed almost instantaneously by the success of MIDAS 7 on May 9, which demonstrated the feasibility of infrared launch detection from space. In the interval, the data from that and subsequent tests in 1963 and 1966 had been used to create a system that had fulfilled the original fundamental objective of the program: to provide the United States with the earliest possible warning of a Soviet missile launch.

In addition, DSP proved far more capable than had been expected or demanded. It showed that it could perform its two basic missions: warning of ICBM or SLBM launches and detecting nuclear detonations. But DSP's first decade also established that the endurance of the spacecraft and its sensors and sensor performance exceeded expectations. In addition to remaining operational for a longer time, the infrared sensor's accuracy allowed it to serve as more than a "bell-ringer," also providing valuable data on trajectories and impact areas. DSP's ability to detect intermediate-range missiles had also been verified, as was another bonus capability (discussed in chapter 7). Thus,

DSP proved capable of performing some of the missions that had been suggested for the MIDAS program but had not been demanded of DSP.

As one observer has noted: "If ever serious thought had been given to canceling MIDAS outright in the 1960s, such thoughts disappeared in the 1970s. The space system was an integral part of the nation's early warning system."[81]

SIX

Surviving World War III

On May 26, 1972, approximately eighteen months after the launch of the first DSP satellite, Richard Nixon and Leonid Brezhnev met in Moscow to sign the two treaties produced by the Strategic Arms Limitation Talks (SALT), which had commenced in Helsinki, Finland, in November 1969. The ABM Treaty limited each nation to a maximum of two sites and 100 launchers at each site for launching ABMs. The SALT Treaty, whose formal title was the unwieldy Interim Agreement Between the United States of America and the Union of Soviet Socialist Republics on Certain Measures with Respect to the Limitation of Strategic Offensive Arms, imposed ceilings on the number of ICBMs and SLBMs each side could deploy. The United States was permitted up to 1,710 strategic missiles, while the Soviets were allowed 2,358.[1]

The U.S. Arms Control and Disarmament Agency (ACDA) characterized the treaty as a "holding action," limiting strategic competition and providing "time for further negotiations." Very early in his presidency, Jimmy Carter asked the JCS about going far beyond such a holding action. Carter wondered about negotiating drastic reductions in American (and Soviet) strategic ballistic missile forces—possibly down to 200 missiles—a question that implied a nuclear strategy of pure retaliation against a limited number of targets.[2]

Ultimately, the Carter administration would produce a major revision in U.S. nuclear strategy, but not the one implied by Carter's question. The process that produced this revision began with the signing of Presidential Review Memorandum 10 (PRM-10), "Comprehensive Net Assessment and Military Force Posture Review," on February 18, 1977. The memorandum called for a two-part study, with one part focusing on issues such as technological developments with regard to new weaponry, deterrence at reciprocally lowered strategic levels, and the viability and desirability of the "triad" posture. The second part of the study was to focus on "the overall trends in the political, diplomatic, economic, technological, and military capabilities of the United States, its allies, and potential adversaries."[3]

The resulting effort involved no less than eleven task forces, employing 175 people from every agency in the national security bureaucracy, who spent some six months of full-time work on the project. PRM-10 contained studies of regional issues, intelligence, and military matters. The eleven task force reports were summarized, expanded upon, and analyzed by a four-person executive staff, which produced an executive summary of over 300 pages, *Military Strategy and Force Posture Review: Final Report.*[4]

Annex C of the PRM-10 study treated the subject of targeting Soviet strategic forces; other Soviet military forces (which included naval bases and shipyards, airfields, ammunition depots, and theater nuclear forces); command, control, communications, and intelligence (C^3I) headquarters; and economic recovery sets (including oil refining, electric power, steel, coal). Such classes of targets had been spelled out in the Nixon administration's National Security Decision Memorandum 242 (NSDM-242), "Planning the Employment of Nuclear Weapons," when the idea of limited options had first been raised. PRM-10 followed those standard lines.[5]

In response to the policy review, Carter signed the top secret Presidential Directive 18 (PD-18), "U.S. National Strategy," on August 24, 1977. PD-18 announced that "the United States will maintain an overall balance of military power between the United States and its allies on the one hand and the Soviet Union and its allies on the other at least as favorable as that that now exists."[6] PD-18 also directed the initiation of three major studies: the Nuclear Targeting Policy Review (NTPR), the ICBM Force Modernization Study, and the Secure Reserve Force Study.[7] The first of these studies would lead to significant revisions in U.S. nuclear strategy—revisions that would impose new requirements on DSP and its successors-to-be.

By 1979 the NTPR had produced two reports. Its top secret Phase I report discussed basic options, problems, and dilemmas of nuclear strategy as applied to the United States and Soviet Union. The options it examined ranged from punishing Soviet society as a whole (including targeting of urban areas) to focusing U.S. nuclear retaliation on the approximately 110,000 individuals who constituted the Soviet leadership.[8]

The Phase II report that followed contained more specific recommendations. The result of the review was a significant shift in U.S. nuclear strategy. A key, although not the only, U.S. objective in a large-scale nuclear exchange, under the doctrine enunciated in NSDM-242, was delaying Soviet recovery to major power status for as long as possible. That objective was to be achieved, in significant part, by inflicting severe damage on key Soviet industrial sectors, such as oil refining, coal, steel, and electric power.[9]

Under the new doctrine, initially referred to as the "countervailing strategy," the United States' prime objective in a large-scale nuclear conflict would be the destruction of Soviet political and military assets: the leadership itself (estimated at 110,000 people); command, control, and communications capabilities; strategic nuclear forces; and other military forces. The rationale for the new doctrine was that the best means of deterring the Soviet leadership from engaging in or risking a nuclear conflict would be to directly threaten their survival and their ability to achieve offensive objectives.[10]

Additionally, in another extension of NSDM-242, the new doctrine required the United States to develop a capability to conduct a protracted nuclear conflict of limited strikes and counterstrikes that might last anywhere from days to six months. Such a strategy required military forces, intelligence assets, and tactical warning capabilities that could survive in a nuclear war environment.[11]

The new doctrine was codified in PD/NSC-59, "Nuclear Weapons Employment Policy," of July 25, 1980—a three-page, single-spaced top secret presidential directive. However, implementation of that policy had started considerably earlier in the Carter administration. It was a policy shift that would affect the content of the Single Integrated Operational Plan (SIOP), the set of potential U.S. nuclear war plans, as well as plans for the procurement of weapons, C³I, and space systems. Among the space systems whose future would be most significantly influenced was DSP.

Thus, in 1979 Seymour L. Zeiberg, Deputy Under Secretary of Defense, Strategic and Space Systems, observed:

> We have a variety of responses to a nuclear attack and are not constrained to a doctrine whereby any level of attack, regardless of size, would draw a massive response. This requires that in the event of a nuclear war, whether limited or general, timely and reliable warning of a ballistic missile attack, including the prediction of warhead impact points, be available to support the National Command Authorities (NCA) in their selection of appropriate Single Integrated Operational Plan (SIOP) responses. *Therefore, a survivable system that will provide initial, trans- and post-attack information is needed to detect the missiles and to track them. . . . Details of the missile launches and predicted impact points must be transmitted quickly, via a survivable communications network, to the NCA for rapid attack assessment to assure adequate decision time for selecting the proper responses.*[12] (emphasis added)

Satisfying such requirements would require a major upgrade of both the space and ground segments of the DSP system. It would be necessary to consider an assortment of options that might allow attainment of the goals spelled out by Zeiberg and to harness available technology to realize the options chosen—or in the cases where the technology was not yet developed, to develop it.

On November 19, 1979, the Air Force issued its "Mission Element Need Statement (MENS) for Improved Missile Warning and Attack Assessment," which echoed Zeiberg's sentiments. A month later, on December 20, the Defense Systems Acquisition Review Council (DSARC) met to discuss how to improve DSP survivability and performance. That meeting had been convened at the request of the House Armed Services Committee, in light of Air Force intentions to put more funds into its Mosaic Sensor Program (MSP), in anticipation of eventually replacing the DSP spinning sensor with a mosaic staring sensor. The DSARC was to consider the various alternatives and prepare a report for Congress.[13]

At the DSARC meeting the Air Force presented five options to improve the survivability of the DSP spacecraft themselves. The first was simply continuing the existing development and procurement programs. A second option was to add a 4.3-micron sensor element to offset the threat of Soviet laser jamming of 2.7-micron detectors. A more costly alternative would involve adding two satellites, in inclined elliptical orbit, with a capability to retransmit mission data, thus eliminating the dependency on communications satellites to relay warning data. Even more ambitious was the fourth alternative—development of the Advanced Warning System (AWS) with a first-generation mosaic staring sensor, which would provide an improved attack assessment capability. The final option would include both deployment of a first-generation mosaic staring sensor and the short-term improvements that were part of the second option.[14]

The evaluations the DSARC received from a number of Defense Department offices of the Air Force's preferred option, the development of an AWS with a mosaic sensor, were not favorable. While the potential of the mosaic staring sensor approach was considered promising, the technology did not appear to be sufficiently developed and would yield only marginal improvements in survivability over the second option, but at substantially greater cost and technical risk. In addition, it would not, as had been hoped, meet the requirements for tracking and detection of post-boost vehicles in support of predicting impact points. Precise prediction of where enemy warheads would land was considered essential for conducting any limited nuclear war, for in order to select the appropriate responses it was necessary to know what areas were (or were not) being targeted. In mid-February 1980 the Air Force received the bad news in the form of a DSARC impelementing memo, which announced that the second option had been selected as the best means of ensuring quick, significant improvements in survivability within fiscal constraints.[15]

For the time being, the decision was made to proceed with Option 2, and it was directed that the ongoing MSP be terminated. In addition, the Air Force and the DARPA were to jointly formulate and present to the Under Secretary of Defense for Research and Engineering a coordinated plan for technology developments, experiments, and demonstrations in support of eventual mosaic sensor missile surveillance from space. The result would be a contract awarded to Aerojet and Grumman for explorations of mosaic sensor technology, a contract that would run for two years (1981–82) before it was canceled by Assistant Secretary of Defense for C^3I, Thomas Cooper.[16]

The emphasis on being able to fight a prolonged nuclear war required not only survivable spacecraft but also ground stations that could operate during a nuclear exchange, for data stored in the satellite that could not be transmitted back to earth would be of little use to the war fighters. Even before the Carter administration's revision of strategic doctrine, there were plans to provide for simplified processing stations that could be used during war to receive data from the DSP satellites. The new strategic doctrine gave those plans additional urgency.

In December 1974 it had been reported that SAMSO would initiate an effort to proliferate ground stations, with the selection of IBM to develop a Simplified Processing Station (SPS) over 1975 and 1976. Earlier in the year the Air Force had told Congress that development of a prototype was necessary to evaluate the military utility of the system and to determine whether to proceed with the production of several such stations, which could be transported by aircraft. Proliferated sites would allow for the dispersal of receiving capability, and thus increase the probability that data from the sensors would reach users. They would also provide emergency backup for the ground stations.[17]

The SPS, illustrated in Figure 6.1, was to consist of a Satellite Communications Module that would receive data from the satellites and a Data-Processing Module that would process the data and transmit it to users by way of the Ground Communications Network (GCN).[18]

It would be a truncated version of the Large Processing Stations at Buckley and Nurrungar, and was to have roughly the same capacity for receiving and processing information and transmitting it to users. At the same time,

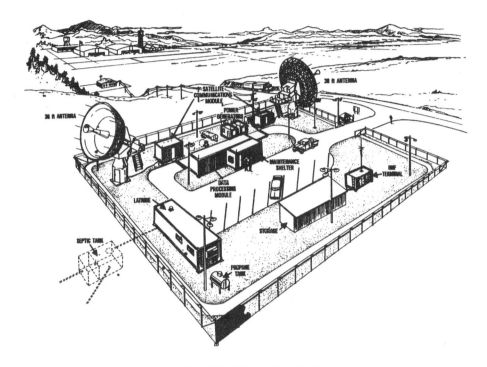

6.1. Simplified Processing Station

it would be automatic in its operation and would have lower operating and maintenance costs than the large stations. Finally, it would be a small, modular system that would be easy to transport and deploy. By building a number of SPSs and collocating them with users, it would be possible to provide users with direct readout of sensor data and increase the survivability of the DSP ground segment.[19]

ADCOM envisioned the stations as located within the ultrahigh-frequency "footprint" of the airborne command posts for CINCNORAD; commander in chief, Pacific Command (CINCPAC); commander in chief, European Command (CINCEUR); and the National Emergency Airborne Command Post (NEACP).[20]

During late 1975, the design reviews for the SPS hardware and software were completed, followed by construction of a Data-Processing Module Shelter and a Satellite Communications Module and preliminary testing. Initial operational testing and evaluation were originally scheduled to begin in 1977, but a program slip of nine months, caused primarily by technical problems with the data-processing equipment, had moved the starting date to

April 1978. Upon completion of testing and evaluation, a one-year operations and maintenance period would begin.[21]

On June 10, 1977, ADCOM published Required Operational Capability (ROC) 3-77 for an SPS. The optimum solution, according to ADCOM, was five SPS sites—two in the Eastern and three in the Western Hemisphere. Each site would have to be able to simultaneously process launch detection and nuclear detonation data from two satellites. An acceptable solution involved four SPSs, two in each hemisphere, with only the Western Hemisphere sites having a dual capability, allowing them to simultaneously process data from two satellites, while each Eastern Hemisphere SPS would be able to simultaneously process launch detection and nuclear detonation detection data from a single satellite.[22]

Although the Air Force believed there was an urgent need for the SPS program, it had met a series of delays, with cost being a key problem. SPS production funds were eliminated during planning for the 1978 fiscal year, while military site preparation construction funds were frozen because of a change in location for the prototype. The Requirements Review Group (RRG) reviewed the Air Force proposal on December 19, 1977, and approved the requirement for DSP data survivability but not the number of sites. The group also approved the pre-attack data survivability requirement but not the "transattack" and "postattack" periods (it was confident that SPS sites could help guarantee the receipt of DSP data prior to war by serving as backups to the fixed stations, but not once war began) and called for the investigation of other solutions. As a result, the Air Staff directed AFSC and ADCOM to consider alternatives. In April 1978 ADCOM offered one: the Mobile Ground Terminal (MGT), or "Midget."[23]

As its name indicated, rather than just being transportable, as the SPS was to be, the terminals would be mobile. Specifically, the MGTs and associated communications terminals would be deployed in truck-mounted vans that could change locations frequently. SPS hardware and software technology would be used for the MGTs.[24]

In May 1978 Air Force headquarters had directed the AFSC to prepare a memorandum on the need for the mobile terminals and to analyze the concept's feasibility. In July the Air Force told the AFSC to prepare a plan for the acquisition of six MGTs—one preproduction model and five production models.[25]

But, in the absence of a definitive policy decision canceling the SPS program, the concept of multiple SPSs was still being pushed. In June 1978 the prototype SPS was shipped to Vandenberg Air Force Base for further testing. Plans to deliver the SPS for operational use in April 1978 were altered to call for delivery in December 1978.[26]

The commander in chief of the Aerospace Defense Command approved a briefing given to the Air Staff, the JCS, and the Assistant Secretary of Defense for C³I on August 1–2, 1978, stating that ADCOM recommended procurement of six SPSs. The ultimate result was a definitive decision, although not the one ADCOM would have liked.[27]

On August 2, 1978, ADCOM briefed the Air Staff, asserting a need for six SPSs to continue the DSP data-processing mission and provide for some increased data survivability through proliferation. The Air Force agreed that ensuring the flow of DSP data was critical, but it rejected ADCOM's solution. Air Force headquarters gave a number of interrelated reasons for its negative reaction: the SPSs were at fixed locations and were targetable, and therefore would not survive the transattack or postattack periods.[28]

Air Force headquarters felt that a more mobile, more survivable processing station—such as the MGTs—might be required. The view was shared by higher-level Defense Department officials. Thus, in October 1978 the Principal Deputy Under Secretary of Defense directed the Air Force to prepare, by March 1, 1979, a plan for initial deployment of such mobile processing stations.[29] In turn, the DSP Program Management Directive (PMD) of November 22, 1978, directed AFSC and ADCOM to "conduct an analysis to determine the feasibility of a mobile receiving terminal to enhance data survivability." The result of the effort was the replacement of the two organizations' previous preferred alternative—procurement of several SPSs—with deployment of MGTs and associated communications terminals in truck-mounted vans that could change locations frequently.[30]

A December 27, 1978, Air Force letter to the Defense Communications Agency on "Mobile Terminals for the Defense Support Program" stated that six MGTs would be bought instead of SPS. The SPS that had been built would not be discarded but moved to a fixed location overseas. Pending its move, the SPS was transported to the Cornhusker Army Ammunition Plant at Grand Island, Nebraska, approximately 125 miles west of SAC headquarters at Offutt Air Force Base, Nebraska.[31]

The following April, William J. Perry, then the Pentagon's research and engineering chief, described why he believed MGTs were needed. He argued:

Our [missile warning] satellite system consists of three satellites deployed in geostationary orbit over the eastern and western hemispheres to cover

Soviet ICBM and SLBM launch areas. While the system has performed admirably, it is nevertheless fragile. We are therefore proceeding with the development of mobile truck-mounted terminals easily proliferated and indistinguishable from other Service vans.[32]

In January 1981 Congress was told that six MGTs would be required. A little over two years later, in March 1983, Donald C. Latham, then Deputy Under Secretary of Defense for C^{3}I, offered further justification for the program, testifying that

> mobile ground terminals (MGTs) . . . will provide an austere backup capable of processing the [DSP] satellite data and provide direct readout of warning information to the users. These MGTs are virtually impossible for the Soviets to target and that frees us from dependence on the very vulnerable fixed ground processing facility used today.[33]

Seymour Zeiberg, as might be expected, was also a strong supporter of the MGTs, although they were not quite what he had in mind. He would tell another Defense Department official to "have them design terminals for deployment in Winnebagos. Now *that's* camouflage and survivability."[34]

The quest to ensure survivability in space and on the ground for the DSP system, and the ability to make maximum use of DSP for warfighting purposes, would have a major impact on the evolution of the DSP system—an impact that lasted throughout and beyond the Cold War, although not always in the ways intended. The development of the SPS and its deployment overseas would increase the flexibility of the DSP system and eventually facilitate the deployment of a four-satellite constellation. Beyond the SPS and addition of the 4.3-micron detection capability, there would be additional modifications of the sensor, programs initiated to protect the satellite and its sensor, and initiatives to allow the transfer of data from satellite to satellite. Some of these programs not only would influence the nature of DSP but also would affect the debate over a successor system.

SEVEN

Slow Walkers, Fast Walkers, and Joggers

On January 20, 1981, Ronald Wilson Reagan became the fortieth president of the United States. He entered office promising a massive strategic modernization program that would focus on both the deployment of new weapons systems and the improvement of command, control, and communications (C^3) systems. By convention, C^3 also included warning systems such as DSP. The new president's plan called for the expenditure of more than $200 billion over five years for bombers and cruise missiles, ICBMs, SLBMs, nuclear defense, and C^3 systems.[1]

The guidance for the program was contained in National Security Decision Directive 12 (NSDD-12), "Strategic Forces Modernization Program," which Reagan signed on October 1, 1981. That was followed by NSDD-13, "Nuclear Weapons Employment Policy," possibly signed the same day. NSDD-13 not only had an identical title to Carter's Presidential Directive 59 (PD-59); it also had an identical message: that in a nuclear war the United States must be prepared to destroy a wide variety of targets in the Soviet Union over a significant period of time, as much as six months, with particular emphasis on military and political leadership targets. Thus, the requirement introduced at the end of the Carter administration for a survivable and enduring DSP system would continue to guide DSP development during the Reagan administration.[2]

While that long-term effort proceeded, the immediate concern of those involved in DSP operations was to maintain a fully operational constellation. On January 13, 1981, a week before Reagan's inauguration, the DSP satellite that was to become Flight 9 was shipped to Cape Canaveral in preparation for a February 16 launch. A problem with the booster forced postponement, but the difficulty was eventually resolved and the launch took place on March 16. The launch was entirely successful, and the satellite entered a virtually geostationary orbit (21,987 by 22,027 miles with a 2-degree inclination) the following day.[3]

Flight 9's operational position was at 70W, where it was to serve as the Atlantic satellite. But while Flight 9's launch and injection into orbit were untroubled, its performance was not. As a result, the Space Division recommended at a prehandover briefing on March 25, and both ADCOM and the SAC agreed, to a delay in F-9 handover until certain problems were resolved.[4] After corrective action it was proposed that formal turnover of the satellite to ADCOM as a fully operational asset take place on April 15, 1981. Flight 6 would be positioned as the Western Hemisphere backup satellite, between 90 and 110W, while Flight 3 was to be evaluated for end-of-life testing following Flight 9 turnover.[5]

But immediately after turnover the CONUS Ground Station began experiencing problems with Flight 9. These problems were generated, at least in part, by ringing cells—a phenomenon common to all satellites throughout their lifetime, particularly during the first several weeks of operations. The ringing cells, which are more active than other infrared detector cells, inject more noise into the system—noise that can produce false returns. Once the "ringers" are identified and properly thresholded, the false-return problem is eliminated.[6]

Unwanted returns can also be generated by an actual physical event, such as the reflection of the sun off ice and snow on high mountains such as the Andes and Himalayas. To deal with that and similar problems, an operator can resort to "blanking" to delete returns from those areas. The operator can choose to eliminate all returns from a certain area or choose a threshold that a return must exceed to be "noticed" by the system. In the case of Flight 9, the CONUS Ground Station resorted to extensive blanking.[7]

The problems with Flight 9 were to come at an inopportune time. On March 30 President Reagan addressed a meeting of the AFL-CIO at the Washington Hilton International to request support for his economic program. After his speech, RAWHIDE, as Reagan was code-named by the Secret Service, left via the VIP exit and headed for the waiting limousine. But while the president was raising his hand to wave to people across the driveway, a bullet fired by twenty-five-year old John Hinckley Jr. of Evergreen, Colorado, tore into his chest. Another of Hinckley's bullets seriously wounded James Brady, Reagan's press secretary, causing permanent brain damage.[8]

At such times fears of a conspiracy, possibly connected to a nuclear attack, are heightened. Thus, a Secret Service agent explained that after hearing of the shooting the "first thing on our minds was . . . if they got the President in Washington, were they waiting for the Vice President in Austin [where his plane landed on the way back from Dallas]?" An immediate concern on November 22, 1963, was whether the Soviet Committee for State Security, the KGB, had been involved in the assassination of President John Kennedy and whether his murder was the prelude to a nuclear attack. And Ronald Reagan had taken office as the president most critical of the Soviet leadership and its policies. Further, if an attack was to be launched, it would likely begin with an SLBM launch from Soviet subs in the Atlantic, against Washington-based command and control targets, in an attempt to cripple the United States ability to respond.[9]

The belief that there was some probability, however small, that the United States might be subject to attack is reflected in the battle over possession of Reagan's coded card, which he would use to authenticate nuclear strike orders in case secure voice communications were not possible. Other than identification numbers to distinguish the particular card holder, the cards carry identical arrangements of numbers and letters, which, when read back, "provide an easy way of verifying that the person you're talking to has a card and therefore is on the list of players involved in the decision-making process." Thus, *Washington Post* writers Patrick Tyler and Bob Woodward reported, "The loss of the card, taken as part of the evidence-gathering after the shooting, caused a serious dispute over its possession between the President's military aides and the FBI."[10]

Five days after Hinckley's attempt, with the president still in the hospital, the first of two serious false returns occurred, apparently generated by DSP Flight 9, which was watching for SLBM launches from the Atlantic. The false signals of April 5 were followed by even more serious false returns on April 13. An ADCOM memo noted that "the false reports of April 13 could have resulted in unacceptable posturing of SAC forces."[11]

Surely, some additional skepticism was introduced by the recent history of false alarms generated by other elements of the U.S. early warning network. On November 9, 1979, monitors at NORAD indicated that a massive Soviet attack, involving both ICBMs and SLBMs, had been launched. The attack depicted on the monitors conformed to Pentagon expectations of how the Soviets would employ their strategic forces in case of an attack. In this case a test tape simulating a missile attack on the United States had been mistakenly played and transmitted outside of NORAD headquarters. A thousand ICBMs were placed on low-level alert, and ten tactical fighters scrambled from a base in Ohio. Fortunately, within six minutes the alert was correctly identified and confirmed as a false report—but not before some of the associated events were detected by Soviet communications intelligence assets. Notification that an attack was under way, consisting of 2,200 missiles, reached as high as national security adviser Zbigniew Brzezinski, who was within one minute of calling President Carter when he was informed that it was a false alarm.[12]

Additional false alarms followed on June 3 and June 6, 1980. On June 3, at approximately 2:26 A.M. Eastern Daylight Time, the SAC command post display system indicated that two SLBMs had been launched toward the United States. Eighteen seconds later, the display showed an increased number of SLBM launches. Alert crews were directed to move to their aircraft and start their engines. After a brief period the SAC display cleared, only to be followed by indications of ICBM and SLBM launches. A similar situation arose on June 6, 1980, while NORAD and the Defense Department were attempting to determine the cause of the June 3 false alarm. It was subsequently determined that the cause was a faulty integrated circuit in a communications multiplexer.[13]

In their report to the Senate Armed Services Committee, Senators Gary Hart and Barry Goldwater observed: "In no way can it be said that the United States was close to unleashing nuclear war as a result of the June 3 and 6 incidents. In a real sense the total system worked properly in that even though the mechanical electronic part produced erroneous information, the human part correctly evaluated it and prevented irrevocable reaction."[14]

In each case the incidents were resolved via telephone conferences between officers at the relevant command posts. The lowest-level conference was the missile display conference, which was handled by duty officers and called whenever returns that might represent missile launches are observed; in the first six months of 1980, there were 2,159 such conferences. More serious was the missile threat assessment conference, which involved higher-ranking officials at the command posts; there were seventy-eight such conferences in all of 1979. Neither of the DSP-related incidents, nor any previous ones, had resulted in convening of a Missile Attack Conference, which would include the president, to consider U.S. courses of action. Nor did the April 5 and 13 false alerts produce such a conference. However, in other circumstances such false returns and the reaction to them, such as alerting SAC forces, could exacerbate a crisis situation, for Soviet electronic intelligence satellites and other systems could detect the heightened state of alert. The Soviets might then respond in turn. In any case, warning assets that appeared to generate false alarms or exhibited other problems needed to be fixed.[15]

On April 14, 1981, an ADCOM memo noted that DSP Flight 9 missile warning performance had been declared unsatisfactory by NORAD. In addition to the false reports, the increase in blanking required to control spurious internal reflections from the sensor played a role in the decision. As a result, ADCOM decided to hand back Flight 9 to the Satellite Control Facility effective 9:00 P.M. Greenwich Mean Time on April 14. A half hour later the CONUS Ground Station would begin operating Flight 6. As soon as practical, F-9 would be moved back to 70W.[16]

In an April 21, 1981, message, SAC informed ADCOM that "we feel that the CGS has satisfactorily corrected those problems associated with ground station support for Flight 9," which were determined to be improperly coded ground station software. SAC proposed that, after receipt of the Space Division's input, "a handover board of senior officers convene to review Flight 9 capabilities and to recommend appropriate dates for handover and turnover."[17]

While Flight 9's problems appeared to be solved, the failure of both star sensors on F-8, the Pacific satellite, also caused great concern. Star sensors are used to determine the position of the satellite, knowledge that was essential to determining the launch point and launch azimuth of missiles viewed by the satellite. By the end of July 1981, both the star sensors on F-8 had failed.[18]

The problem with Flight 8 plus a decline in Flight 6's warning capability had led ADCOM to return Flight 9 to operational status. At the same time F-6 would be handed over to the Satellite Control Facility, which would return it to its backup position. But on April 28 ADCOM requested repositioning of Flight 6 because of the reinstatement of Flight 9 and continued poor performance of Flight 8. On May 6 it was noted that F-8 was exhibiting degraded performance, which could result in satellite failure. F-6 was then being relocated as a temporary replacement for F-8.[19]

The continued problems with Flight 8 led to a shift in military space plans, with early warning requirements taking precedence over communications requirements. The JCS concurred with the Air Force decision to launch DSP Flight 10 in February 1982, while delaying the planned launch of a defense communications satellite.[20]

The DSP satellite was shipped to Cape Canaveral on January 31 and was successfully launched into geostationary orbit (22,022 by 22,070 miles with a 2-degree inclination) on March 6, 1982. On-orbit checkout was completed on March 31, and the satellite was turned over to ADCOM on April 6 when it reached 70W, replacing F-9 as the Atlantic satellite.[21]

At the DSP Flight 10 handover briefing it was noted that infrared sensor

TABLE 7.1. *Stations and Satellites After Launch of F-10*

	Longitude	Satellite
Atlantic	70W	F-10
Pacific	135W	F-9
Eastern Hemisphere	65E	F-7
WHLRS	85W	F-8
EHLRS	75E	F-6

testing on Flight 10 had been successful, that five missile launches had been detected in real time, that sensor-pointing accuracy had been good, and that the internal reflections seen on Flight 9 had not been observed on Flight 10. In addition, the star sensor was working normally, there were no problems commanding the satellite from the Satellite Control Facility at Sunnyvale and Buckley, and the satellite support systems—power, thermal, propulsion— all looked good. However, the nuclear detonation detection system on Flight 10 was not initially fully capable.[22]

The successful launch of Flight 10 led, of course, to the partial restructuring of the DSP constellation. Flight 9, replaced as the Atlantic satellite by F-10, in turn replaced the sometimes shaky Flight 8 as the Pacific satellite. F-7 remained as the primary Eastern Hemisphere satellite. F-6, which had been serving as the Western Hemisphere Limited Reserve Satellite (WHLRS), finally arrived at 75E on December 2, 1982, assuming the role of Eastern Hemisphere backup in place of F-2, which was to be retired. Flight 8 was moved to 85W to become the Western Hemisphere backup satellite. Table 7.1 summarizes the positions of satellites after the launch of F-10.[23]

On April 23, 1982, headquarters noted that "replacement of the degraded DSP-W Pacific satellite [and the deployment of F-10] in the Atlantic restores high confidence in NORAD's ability to do tactical warning."[24]

While 1982 would see no missiles fired at the United States, the DSP satellites would have plenty to observe as a result of repeated Soviet and Chinese missile tests. On June 18 the Soviets conducted what was described as an integrated test, which, according to Secretary of State Alexander Haig Jr., represented an "unusually high level of strategic activity."[25]

The test began with the firing of two SS-11 missiles from operational silos into the Kamchatka impact area, quickly followed by launch of a SS-20 intermediate-range missile from its operational site. In addition to firing a ballistic missile from a Delta submarine in the White Sea, Soviet forces fired two ICBMs from Kapustin Yar as part of a ballistic missile defense test. An antisatellite test against a target in space was also part of the exercise.[26]*

On October 12 a Chinese SLBM was launched from a boat off the mainland, its impact point being approximately 900 nautical miles away, in an area about 300 nautical miles from Okinawa and 180 nautical miles from Taiwan. In the same period the Soviets fired two of their new SS-N-20 SLBMs from a submerged Typhoon submarine in the Barents Sea. Fired about fifteen seconds apart, one SLBM landed on Kamchatka Peninsula, while the other impacted in the northern Pacific.[27]

The capabilities of the DSP constellation that detected these late 1982 launches varied with specific circumstances. In general the satellites could provide a launch count accurate to within 20 percent, locate the launch site to within three nautical miles, and determine the missile's azimuth with sufficient accuracy to determine whether it was headed toward the United States.[28]

The ability to locate the launch site to within three nautical miles meant that DSP data would have allowed the United States to determine from which missile field a Soviet ICBM had been fired—whether it was a SS-17 from Yedrovo or a SS-17 from Kostroma. But it would not have been sufficient to determine from which *silo* in a field the missile had been fired. Thus, it would not permit the United States to retarget its own ICBMs, ignoring the already vacated silos and concentrating on the ones still containing missiles.[29] An attempt to attain that capability would be made in 1983, 1984, and beyond as development of the next generation of DSP satellites continued. During those years and later, additional capabilities of the satellites would be verified and demonstrated.

During the spring of 1983 Rear Admiral John L. Butts, the Director of Naval Intelligence (DNI), made the yearly pilgrimage to Capitol Hill required of all DNIs. He would give the members of the congressional armed services and appropriations committees, who would be allocating funds for the next fiscal year, his assessment of the threat posed by the Soviet

---

*The integrated test led to the first, in what became a series of classified videos concerning DSP operations: "The Event of 18 June 1982—Executive Summary." Subsequent videos included "War of the Cities," "2 December 1990," "Events of 25 February 1991," a history of DSP, and a video on the use of DSP to detect aircraft (a subject discussed below). The integrated test video was shown to President Reagan.

Navy. In testifying before the Senate Armed Services Committee, Butts told the legislators that "the Soviet Navy . . . poses a clear and present challenge to our national interests."[30]

That was a statement that many previous DNIs could not have made. For the Soviet Navy that existed in the early Cold War years had served as a protector of Soviet home waters, not as a means of power projection. But Admiral Sergei G. Gorshkov would change that, after being summoned in 1956 from the Black Sea Fleet, at the age of forty-five, to become the Navy's commander in chief.[31]

Beginning in 1972, the magazine *Naval Collection* published a series of eleven articles attributed to Gorshkov, with the general title of "Navies in Peace and War." The articles, which would form the basis for Gorshkov's 1976 book, *Sea Power of the State*, argued that navies were of increasing importance in affecting the outcome of wars of all varieties, as well as important tools of national policy in peacetime.[32] The articles more than pointed toward a new Soviet Navy in the distant future, for the Soviet Navy already was in the process of changing in accord with Gorshkov's philosophy. Thus, a March 1983 NIE on Soviet naval strategy noted that "by the mid-1970s . . . the Soviet Navy had evolved from a force primarily oriented to close-in defense of maritime frontiers to one designed to undertake a wide variety of naval tasks, ranging from strategic nuclear strikes to worldwide peacetime naval diplomacy."[33]

In the event of war, the naval tasks would also include "sea control/denial efforts against Western surface forces in . . . the Norwegian, North, and Mediterranean Sea and the Northwest Pacific Basin."[34] The target of those efforts would be U.S. and allied aircraft carriers and destroyers that could strike Soviet land and sea targets and protect merchant ships at sea. But what worried the U.S. Navy in the mid-1970s was not just Gorshkov's growing fleet at sea but the land-based aircraft of Soviet Naval Aviation (SNA).

Admiral James L. Holloway III told the 1976 meeting of the U.S. Naval Institute that "our deployed fleets must have the defensive strength to defend themselves against attacks of land-based air, because we are seeing more and more the development of long-range aircraft with anti-ship missiles as a threat which can develop rapidly and can extend to almost any spot on the globe."[35] The aircraft that particularly concerned the admiral and others was known in the West by the designation "Backfire." To the Soviets it was the Tu-22M. In 1969 U.S. intelligence first noticed the existence of the Backfire A—the design verification model used for testing and training. In May 1973 the Backfire B, the first operational model, was discovered. Then, in 1974 the new planes began showing up in satellite photographs of Soviet airfields.[36]

In addition, the photographs showed a break with past patterns of deployment. Previously, the needs of Soviet Long-Range Aviation (SLRA) were filled before those of the Soviet naval aviators. But U.S. intelligence saw approximately equal numbers of Backfires being assigned to each air arm. In early 1978 the intelligence community estimated that more than 500 might be built by mid-1987, with deliveries evenly divided between the SLRA and SNA. Later that year it estimated that about 45 of the new aircraft were in operation for the LRA and 40 for the SNA. Those 40 had gone to the Northern and Black Sea Fleets.[37]

The Backfire was a significant improvement on the plane it was replacing—the Tu-16 Badger. In 1980 congressional testimony, Vice Admiral James H. Doyle Jr. would characterize it as an "order of magnitude upgrading of Soviet Naval Aviation." In March 1983, by which time ninety planes had been assigned to SNA regiments (two in the Baltic Fleet, one in the Black Sea, and one in the Pacific), the CIA assessed the Backfire's maximum speed when flying at optimal altitude as 1,050 knots, in contrast to 535 knots for the speediest version of the Badger (the Tu-16C). The Backfire, according to the CIA and other intelligence agencies, also possessed significantly greater range than the Badger.[38] According to Sumner Shapiro, the DNI from 1978 to 1982, this meant the Backfires would "materially extend the range and scope of Soviet capabilities to strike surface forces, mine and bomb ports and harbors, and provide support to theater operations." Backfires based with the Baltic and Black Sea Fleets could carry out missions over any part of the Mediterranean and deep into the Atlantic, with control over the latter being particularly crucial to any U.S.–NATO war effort. To the Soviets, the Backfire would be "a vital part of [Soviet] strategic defense forces to keep Western carrier battle groups from striking important targets within the Soviet landmass." In a September 1982 exercise Backfires practiced doing that, conducting a simulated strike against a U.S. carrier battle group, as part of a Pacific Fleet exercise targeted against two U.S. carriers east of the Kuril Islands. In a real war the Soviets were expected to send one or two regiments (twenty to forty aircraft) against the battle groups.[39]

A key factor in making the SNA Backfires a serious concern was the ordnance they would be carrying. It was a matter of not just how far the planes could go or how fast they could get there but also what they would be bringing with them. Rather than the bombs that its air force counterpart would deliver on their targets, the naval Backfire would be armed with one or two AS-4 antishipping missiles. A single missile could be carried semirecessed under the fuselage—in which case the plane could fly on afterburner at any time during the mission, including on takeoff. If two missiles were carried, one under each wing, the plane could go to afterburner only after firing its missiles, when it was in its escape mode.

The missile, whose Soviet/Russian designation is the Kh-22 and whose U.S. code name is KITCHEN, is liquid-fueled and weighs approximately 14,000 pounds. It can carry a conventional or a nuclear warhead, and can travel at 1,850 to 2,590 miles per hour. At high altitude its range is approximately 250 nautical miles, at low altitude, 150 nautical miles. In a combat situation the AS-4 would be launched at about 36,000 feet and then climb to about 80,000 feet. At about 80 nautical miles from its target it would begin its steep 3-degree dive, with a velocity of about Mach 1.1. Flight time from its typical operational range of 150 nautical miles would be 9.4 minutes.[40]

At the same time that the DNI was telling Congress of such concerns, the U.S. Navy was trying to determine whether data from DSP satellites could be used to provide earlier warning of the SNA threat than could be obtained from terrestrial assets. The general idea that an infrared sensor in space could be used to spot jet aircraft flying on afterburner had originated with Joe Knopow in 1956.

Verification of DSP's ability to accomplish that mission had occurred around 1974. At the time Kenneth Horn, who had received his Ph.D. in aeronautics and engineering from Stanford in 1966, was head of the advanced mission requirements section in the systems analysis office of the DSP program office at the Aerospace Corporation. Horn and others would pore over the data produced by the DSP satellites, primarily with regard to missile signatures. But there were also infrared returns from Soviet territory that occurred at regular intervals and formed a relatively straight line, quite different from those associated with a missile. Horn and his colleagues were alerted to this by the overseas ground station operating crew and then examined the data in an attempt to determine what had produced the anomalous returns. They concluded, based on the direction of the returns and the speed of the object (indicated by the distance between returns), that the returns correlated with an aircraft flying on afterburner. What type of aircraft was not apparent simply from the returns, but a check of intelligence data on the characteristics of Soviet aircraft pinpointed the Backfire as the prime candidate.[41]

Although the discovery was an interesting one, it did not immediately result in DSP's employment as a warning asset for U.S. naval forces. Indeed, by 1982 Aerojet-General had spent about eight years trying to interest the Air Force in such an application and had been getting nowhere. The Air Force's lack of interest was not surprising, and was probably due more to fear of compromising the primary DSP mission of warning of strategic attack than to reluctance to share a key Air Force asset with a rival service.* Not surprisingly, Aerojet had more luck with the Navy. As a result, a Navy contingent spent several months at Nurrungar in 1983 to determine if earlier warning from DSP satellites was a possible means of reducing the Backfire threat—either by detecting Backfires using their afterburners on takeoff or by detecting the launch of the AS-4 itself.[42]

On May 3, 1983, the Navy briefed General James V. Hartinger, commander of the Air Force Space Command (which was formed in 1982 and inherited responsibility for DSP from ADCOM), on its recommendations for the future of the project code-named SLOW WALKER—a term that arose from the nomenclature "walking dots," used to describe the typical signature of a rocket plume. The signature consisted of repeated observations, every ten seconds, which gave a new dot-return in the sequence that corresponded to the rocket plume observations. The "slow walker" aircraft returns went on for several minutes, in contrast to the relatively short duration of the rocket burn. The dots "walked" slowly across the display screens of the observers at Nurrungar. The returns from missile launches also differed in that they depicted the ascension of the missile, which then terminated its powered flight and sent its payload toward its target(s), in contrast to the relatively straight sequence that would be obtained from the aircraft. In addition, the dots from a missile return would become progressively more distant from each other as the missile accelerated, while the aircraft returns would be evenly spaced.[43]

Hartinger approved the Navy proposal to develop a limited operational capability, which would include "rerouting communications to tactical Navy users" and work on detection algorithms. In the first half of 1984 two mini-computers were installed and data connectivity was established through the DSP ground communications network.[44]

The SLOW WALKER assignment would become a permanent part of the DSP mission in succeeding years. Beginning in 1985, Navy personnel at Nurrungar began manually extracting the "tactical events" of interest from the data being transmitted to the ground station and then manually entering each event into a communications processor for dissemination to the fleet. This selection and dissemination constituted the SLOW WALKER Reporting System (SWRS). Subsequently, a JOGGER reporting system was estab-

---

*The Navy had a similar intraservice issue arise when it was discovered that the Sound Surveillance System (SOSUS) underwater arrays, deployed to monitor Soviet submarine movements, could also provide significant intelligence on ocean surface activity. Concern about diverting SOSUS from its primary mission inhibited its full use in an ocean surveillance role.

lished to convey similar data obtained by the DSP-A/HERITAGE infrared sensor. According to one account, "The [SLOW WALKER] process was arduous and tedious, but [the Navy] personnel managed to reduce event handling time to virtually nil. The resulting product was much more tactically relevant to end users."[45]

By 1986 the Air Force Space Command expected that the SLOW WALKER program would evolve to include a real-time and detection capability by 1989. In 1989 the SWRS was integrated into the Tactical Event Reporting System (TERS) processing software at Woomera and the CONUS Ground Station at Buckley ANGB, Colorado, where a second Navy contingent, Detachment Buckley, had been established in 1987.[46]

By that time the Navy contingent in Australia had been designated Naval Space Surveillance Center (NAVSPASUR) Detachment Echo. The seven Navy personnel were integrated into all available positions in the space operations center. Similarly, Detachment Buckley personnel were fully integrated into CONUS Ground Station operations.[47]

An incident that occurred on May 5, 1984, in the early days of the SLOW WALKER mission, illustrated yet another of DSP's capabilities. According to a message posted on the Internet:

> At 1126Z, 5 May 1984, a DSP platform detected an object with heat in the 9,000 KW/SR range coming out of deep space and passing within 3KM of the DSP. Its star tracking telescope first detected the object. . . . The observation lasted 9 minutes. A detailed investigation failed to explain what caused the sensor reading, other than a real object of some type.[48]

The message had been written by Joseph Stefula, former state director of the Mutual UFO Network (MUFON). According to Stefula and others, the May 5 sighting by Flight 7, the Eastern Hemisphere satellite, was one example of a number of possible DSP detections of UFOs. Indeed, there was even the suggestion that there was a code name for such detections. According to an article in *Omni* magazine:

> UFOlogists knew, they would have to tap the most sophisticated information-gathering technology available: Department of Defense spy satellites, like the Defense Support Program (DSP) satellites. . . . In fact, rumor had it, heat, light, and infrared sensors at the heart of the satellites were routinely picking up moving targets clearly not missiles and tagged

"Valid IR Source." Some of these targets were given the mysterious codename of "Fast Walker."[49]

There were, in fact, two elements of truth to the claims of Stefula and *Omni*. There was a May 5, 1984, detection of a space object, and it was an example of a class of sightings designated FAST WALKER. According to Air Force Space Command Regulation 55-55, FAST WALKER denotes detection of a space object, which includes satellites and their debris, in a DSP satellite's field of view.[50] The designation was chosen because the dots associated with DSP observations of spacecraft would be more widely spaced than those associated with aircraft, given the spacecraft's much greater speed.

The May 5 event was alleged to have resulted in a 300-page internal report, which failed to explain the occurrence by natural or man-made means. In fact, the FAST WALKER of May 5 was explained. The object that came perilously close to Flight 7 was not a UFO but a signals intelligence spacecraft, probably the VORTEX satellite launched on January 31, 1984, from Cape Canaveral by the publicity-shy group of terrestrials who constituted the NRO. The spacecraft in question, had failed to enter its proper geostationary orbit. Whatever aggravation those at the NRO felt in the months since the launch over the uselessness of a key piece of intelligence hardware was probably surpassed by the short-lived fear that it might also obliterate a key warning satellite.[51]

Most FAST WALKERs have been routine observations of foreign spacecraft. The infrared readings obtained by DSP, resulting from the reflection of sunlight off the spacecraft, provided analysts at the CIA, DIA, and Air Force Foreign Technology Division (now the National Air Intelligence Center) with data on spacecraft signatures and movements. Such data allowed analysts to estimate where the satellite was going and its mission.[52]

In addition, DSP sensors would provide data concerning the reentry of satellites or other man-made space platforms. Two valuable detections occurred several years prior to the May 5 incident. In January 1978 DSP sensors detected the reentry of COSMOS 954, a Soviet ocean reconnaissance satellite with a live nuclear reactor, that the Soviets could not control. Rather than being able to boost it into an orbit that would keep it in space, they could only watch as the satellite's orbit decayed until reentry took place. At the Aerospace Corporation, the DSP track of the reentry was subjected to mathematical analysis and the impact point determined. A field crew was dispatched to the approximate location in Canada where the analysis concluded the satellite would have crashed to Earth and found a hunting party that had stumbled on the debris. In 1979 DSP sensors provided data on the reentry of

Skylab, the 130,000-pound U.S. space station whose reentry threatened various populated areas. According to one of those present at the Nurrungar site, the falling space station "sure lit up our screens."[53]

Among the satellite movements and debris that DSP detected from the early 1970s to the early 1980s were those associated with the Soviet antisatellite program. Between 1972 and 1982 the Soviets conducted sixteen antisatellite tests, using infrared and radar guidance systems. After being placed in orbit by an SL-11 booster, the intercept satellite would be maneuvered by ground controllers so that after either one or two orbits it passed sufficiently near the target satellite to permit its own guidance system to take over. When in range an explosive charge aboard the intercept vehicle was detonated, sending a cloud of shrapnel at high speed to destroy the target. DSP's monitoring of the launch, satellite movement, and aftermath contributed to U.S. intelligence analysis of the Soviet program.[54]

An event that occurred very shortly after the near collision of May 5, 1984, illustrates DSP's potential ability to serve as a source of infrared intelligence about terrestrial events. In mid-May an explosion rocked the Soviet naval weapons depot at Severomorsk, located about 900 miles north of Moscow on the eastern shore of a long bay leading to Murmansk. The explosion was so massive that it was first believed that a nuclear explosion had decimated the principal ammunition depot and home port for the Soviet Northern Fleet. The explosion was sufficient to destroy about 580 SA-N-1 and SA-N-3 naval surface-to-air missiles, about 320 SS-N-3 and SS-N-12 ship-to-ship weapons and about 80 nuclear-capable SS-N-22 missiles.[55]

Possibly the first indication of the blast was its detection by the DSP-E satellite, along with communications intelligence and seismic detection stations. In the absence of cloud cover, DSP's infrared detectors would have picked up the dramatic infrared signal instantaneously. Once it was known that an event of interest had occurred a KH-11 imagery satellite was instructed to send back imagery on its next pass over the region.[56]

A number of other explosions during the 1980s may also have been detected by the DSP-E's infrared sensor, with only cloud cover being able to inhibit DSP's detection of the events. During the Soviet occupation of Afghanistan, a number of mujahideen operations resulted in tremendous detonations. On one occasion the main Soviet ammunition depot at Bagram was attacked, resulting "in a spectacular series of bangs," according to Pakistani intelligence official Mohammad Yousaf. Reportedly, over 30,000 tons of ammunition were destroyed. Similarly, on August 27, 1985, the ammunition stockpile at Kharga garrison, which contained all types of ammunition, including the bulk of surface-to-air missiles, "went up in a spectacular fireball that rose a thousand feet in the air."[57]

On May 12, 1988, several buildings at a Soviet propellant plant in Pavlograd, located about 500 miles southwest of Moscow, were destroyed as a result of a major explosion. Among them was the only facility used to manufacture the main rocket motors for the then new SS-24 ICBM. Although the blast did not involve any nuclear warhead, "it sure tore than plant up," according to a *Washington Post* source.[58]

On August 17, 1989, a massive explosion took place at a defense industrial complex in Iraq. The Iraqis claimed that a fire had broken out at a petroleum products depot and factory south of Baghdad and that, "by pure coincidence, a truck loaded with explosives for use in engineering projects" was passing by.[59]

Even less-than-massive explosions could be detected by DSP's infrared detectors. Thus, a midair collision over the Grand Canyon was among the events that would be identified in DSP returns during the 1980s, just as the explosion of TWA Flight 800 was detected in July 1996 and the September 1997 collision of U.S. and German military planes off the Atlantic coast of Africa.[60]

DSP can also provide intelligence on missile development prior to test launchings. Thus, in late 1997 it was reported that "satellite reconnaissance" had picked up the heat signature of an engine test for a new generation of Iranian ballistic missile capable of carrying a 2,200-pound warhead more than 800 miles. While the report did not specify that DSP was involved, it is the most likely candidate given the nature of its sensors and orbit.[61] According to Brigadier General William King, former head of the Air Force Office of Special Projects and subsequently Aerojet's vice president for space surveillance operations, DSP serves as a worldwide infrared intelligence system—one that provides "credible, unavoidable" ground truth.[62]

Model of a Missile Defense Alarm System (MIDAS) satellite. Before the end of 1963, the MIDAS program had demonstrated the feasibility of using infrared sensors to detect missile launches. (U.S. Air Force)

An early 1960s launch of a MIDAS satellite from Cape Canaveral AFS, Florida. (U.S. Air Force)

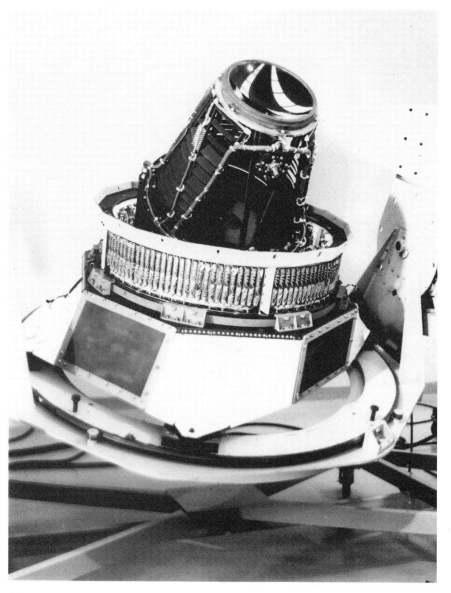

The sensor carried on the Program 461 satellites launched in 1966. The data produced by those launches contributed to U.S. knowledge of Soviet missiles and was a key element in the design of the DSP satellites. (Aerojet)

A DSP satellite in the cargo bay of the space shuttle *Atlantis*. Launched in November 1991, DSP Flight 16 has been the only DSP satellite placed in orbit using a space shuttle rather than an expendable launch vehicle. (U.S. Air Force)

Buckley Air National Guard Base, Colorado, 1986. Buckley has served as the CONUS Ground Station since the beginning of the DSP program. (U.S. Air Force, via Desmond Ball)

The Joint Defense Facility–Nurrungar. Nurrungar served as the main overseas ground station for the DSP program since 1970. In 1999 it was closed and a relay station established at the Pine Gap facility. (Desmond Ball)

A Soviet Tu-22 Backfire bomber in flight. The discovery that DSP satellites could detect Backfires in flight led to the establishment of the SLOW WALKER program. (Department of Defense)

A cutaway illustration of the Sensor Evolutionary Development sensor. The cones at the left are for the Advanced RADEC I nuclear detection sensors. The bar running across the center of the telescope contains the 6,000 infrared detectors. The lens mounted halfway up the telescope is meant to protect against blinding by lasers. The cone at the bottom is a star sensor, used to determine satellite altitude. (Aerojet)

Two Soviet-built Iraq-modified Scud-B missiles and their launchers. DSP was crucial to providing first warning of Scud launches during the Persian Gulf War. (Department of Defense)

The Patriot air defense system deployed in Saudi Arabia during Operation Desert Shield. During Desert Storm DSP satellites would prove capable of providing quicker warning of incoming Scuds than the Patriot radar system. (Department of Defense)

Artist's conception of the SBIRS-Low system being proposed by Lockheed Martin and Boeing. The Space-Based Infrared System is to consist of satellites in geosynchronous and highly elliptical (SBIRS-High) and low-earth orbit. (Lockheed Martin/ Boeing)

Artist's conception of SBIRS-Low system being proposed by TRW and Hughes. (TRW/Hughes)

Evolutionary Developments
and Suicidal Lasers

During 1983, while the U.S. Navy contingent at Nurrungar had been seeking to determine DSP's value for aircraft detection, it had relied on the infrared detections of Flight 7, the Eastern Hemipshere satellite. By July 1983 Flight 7 had been in orbit over six years, substantially longer than its expected design life. But, as in the past, the Air Force was reluctant to give up on a spacecraft already in orbit. Not surprisingly, given its age, problems emerged that led to serious concern over the satellite's continued ability to function.[1]

Problems had also developed with Flight 10. Erratic variations in the speed of its data downlink transmissions had, on at least one occasion, caused an outage at the CONUS Ground Station. Continuing concern over F-7 and F-10 and the potential for further problems prompted a decision to launch a replacement satellite in February 1984.[2]

It would not be until April that the new satellite would actually be placed in orbit. In the interim, on March 27, 1984, Flight 10 was shifted in an easterly direction, to 37W, toward Europe, thus extending its coverage—a position that became standard for future DSP-Atlantic satellites. Like a number of other developments in the DSP program, the shift resulted from a change in the Soviet "threat profile." This time the concern was not with new ICBMs or advanced SLBMs but with an intermediate-range missile. The move, which improved DSP coverage of Europe, followed a meeting of the JCS earlier in the month, during which the threat from 3,100-mile-range, three-warhead, SS-20 Soviet missiles was considered. The SS-20, whose Soviet code name was PIONEER, had been a key element in the intermediate nuclear forces arms control talks between the United States and Soviet Union that had collapsed in late November 1983, when the Soviets walked out of the negotiations. Eliminating, or significantly reducing, the 243 SS-20s targeted on Western Europe had been the primary U.S. objective in the negotiations. With the breakdown it was undoubtedly feared that the missiles would be targeted on America's European allies and U.S. forces in Europe for a significant time to come.[3]

The decision to move the satellite was a joint one involving the JCS; the Commander-in-Chief, SAC; the commander in chief, Atlantic Command; and the commander in chief, NORAD. Arriving on station on April 5, the satellite remained in that orbit for the rest of the year. There were no resultant gaps in Flight 10's previous surveillance zone because of its overlap with the coverage of the Pacific satellite, Flight 9.[4]

On April 14, Flight 11 was launched into geostationary orbit, becoming the first of two 1984 DSP launches that signaled the upgrading of DSP capabilties. The satellite had been retrofitted with the Advanced RADEC I package, to improve its ability to detect nuclear detonations in the atmosphere, and with extensively modified star sensors. F-11 was handed over to the Air Force Space Command on April 26 and placed in operation over the Pacific, replacing Flight 9, which became the primary Eastern Hemisphere satellite.[5]

But a number of problems—including the misidentification of several missile launches by Flights 6 and 7 on April 13—led General James V. Hartinger, the commander of NORAD, to request another satellite launch. Of even greater concern were problems that occurred after F-9 became the primary Eastern Hemisphere satellite. As a result, F-7 was moved from its limited reserve slot to take over for Flight 9 as the primary satellite. On November 9, after a team from the Air Force Space Command had succeeded in restoring Flight 9 to full operational capability, it returned to service as the primary DSP-E satellite.[6]

In the interim, General Hartinger's launch call for Flight 12 had been seconded by the Air Force Space Command. After JCS approval, the launch was scheduled for December 15, 1984. Problems with the booster caused the launch to be postponed until December 22, 1984 when the satellite was sent into its near-geostationary orbit (22,267 by 22,437 miles with a 3.4-degree inclination). The launch heralded another enhancement of DSP capabilities motivated by the change in the American vision of a possible nuclear war.[7]

The spacecraft that became Flight 12 was actually the sixth produced, but, along with satellite 5, it had been selected for extensive retrofitting. The remodeled spacecraft would represent the first attempt to alter the DSP design in response to the concerns over survivability that had been raised in December 1979 as well as the shift in U.S. nuclear strategy first specified in PD-59 and then confirmed in NSDD-13. The retrofitted satellites were to serve as the baseline for a completely new DSP generation. Satellites 14 through 17, which would be designated DSP-1 satellites, were to incorporate these improvements as well as others. Again, TRW would be responsible for the spacecraft, while Aerojet-General handled the payload and infrared sensor, and Sandia Labs the nuclear detonation detection sensor.[8]

The most significant alteration to the satellites, designated 5R and 6R, was the Sensor Evolutionary Development (SED) sensor, with over 6,000 infrared detectors, an almost threefold increase over previous versions of DSP. The increase in detectors was linked to U.S. counterforce strategy. With 6,000 detectors it was expected that the individual silos that missiles were fired from, rather than just the missile fields, could be identified. Thus, concentrating U.S. retaliatory strikes on silos still containing Soviet missiles would be, at least theoretically, feasible.[9]

Other changes to the sensor systems included a second wave-band capability, at 4.3 microns, known as a "second color" demonstration, also bolted to the focal plane, and the addition of an above-the-horizon capability. The second wave band would increase the capability to detect infrared emissions and reduce DSP's vulnerability to laser attack, a subject that had been investigated in tests from Kwajalein Island between 1980 and 1983. The approximately 750 over-the-horizon infrared detectors were designed to give F-12 the ability to "detect the second-stage exhaust signatures of missiles launched beyond earth horizon as viewed by the sensor" to permit global/polar coverage. Such a capability would allow DSP to promptly detect Soviet SLBMs, such as the SS-N-8, that might be fired from Arctic waters—although the lack of first-stage data (consisting of several data points) would preclude determination of the launch point. Another part of the new satellite was the Advanced RADEC I package that first flew on Flight 11.[10]

An additional improvement to the sensor was a more efficient thermal control system for the focal plane, which would do a better job of cooling the detectors in the infrared sensor, allowing them to maintain their sensitivity longer. Along with the sensor modifications came modifications to the spacecraft, many of which were necessitated by the changes in the sensor—greater pointing accuracy, increased power output, larger solar panels, and increased momentum in the reaction wheels. The spacecraft was also modified for compatibility with two new launch vehicles, the Titan 34D/IUS and the space shuttle.[11]

As a result of these modifications, Spacecraft 6R was longer than its predecessors (278 as opposed to 259 inches) and weighed considerably more (3,500 compared to 1,950 pounds). It also had a longer design life (three years rather than 21 months), although the increased design life was still far less than the lifetime actually exhibited by the satellites once in orbit.[12]

New software, developed by IBM, was to be installed in October and November 1984 to allow both ground stations to accommodate the data from both the below-the-horizon and above-the-horizon sensors. It would also provide for command and control of the upgraded sensor. However, Buckley did not receive the software to handle above-the-horizon data until June 1986.[13]

As a result of the extended testing, Flight 12 did not take its place in the operational constellation until May 29, 1985, when it became the DSP West (Pacific) satellite. During that period two of the older satellites, F-6 and F-7, had lost their earth-pointing capability and were dropped out of the constellation. In contrast to the fiery end awaiting low-earth-orbiting imagery satellites, which are sent back into the atmosphere to burn up, geosynchronous satellites such as DSP were simply boosted into a supersynchronous orbit, about 180 miles above the operational sensors' geosynchronous orbit.[14]

The successful launch of Flight 12 also set in motion other changes in the primary DSP constellation. Flight 9 continued in service as the Eastern Hemisphere primary satelite until April 4, when it became the Eastern Hemisphere LRS, with Flight 11 replacing it as the primary eastern satellite. Flight 10, despite some problems of its own, remained the DSP-West (Atlantic) satellite, while Flight 8, which had also experienced some problems, remained the Western Hemisphere LRS. The DSP constellation as it existed on June 1, 1985, is summarized in Table 8.1.[15]

The Air Force intended that the June constellation would be altered by the addition of a new satellite only months later. That month the Space Division began planning to launch Flight 13, Satellite 5R, on September 25.[16] But

TABLE 8.1. *DSP Constellation, June 1, 1985*

	Longitude	Satellite
Atlantic	35W	F-10
Pacific	135W	F-12
Eastern	65E	F-11
WHLRS	NA	F-8
EHLRS	NA	F-9

it would be over two years beyond the intended launch date before another DSP reached orbit. In expectation of a launch, 5R was shipped to Cape Canaveral on August 17 and mated to its booster, a Titan 34D/Transtage. However, on August 28 a Titan 34D carrying a KH-11 imagery satellite failed in a launch out of Vandenberg. Any DSP launch would have to be delayed until confidence in the booster was restored or the urgency of putting the satellite into orbit increased sufficiently to make the additional risk worthwhile. Launch was rescheduled for the spring of 1986, but on April 18, 1986, a massive explosion destroyed a second Titan 34D booster, this time carrying a KH-9 imagery spacecraft. Once again the DSP launch was postponed until confidence in the Titan could be restored. In anticipation of a long wait, the spacecraft was demated from its booster at the end of June and sent back to the satellite contractor, TRW.[17]

The long wait for a new satellite finally ended in November 1987. Despite the fact that both backup satellites (Flights 8 and 9) and two primary satellites (Flights 10 and 12) had experienced problems during the almost three-year gap between launches, the Air Force Space Command had been concerned that there be no doubt that booster problems had been solved prior to the next DSP launch. In a Febuary 1986 message to the Air Staff, it argued that somebody else's satellite should be the guinea pig used to test whether launch problems had actually been solved. In the message the Space Command pointed out that "failure of Flight 13 would result in severe consequences, because Flight 14 (scheduled for launch in May 1987) can only be flown on the shuttle. If the shuttle failure is not resolved by that time, the only other available option for the launch of Flight 14 is to await the [Titan IV's] availability in the fall of 1988."[18]

As DSP's managers had desired, Satellite 5R did not serve as the guinea pig to determine if the corrective measures taken with respect to the Titan 34D were effective. That distinction went to an NRO/CIA KH-11 that was successfully launched from Vandenberg Air Force Base on October 26, 1987, into low-earth polar orbit. Flight 13 followed just a month later, on November 28. The launch from Cape Canaveral's Pad 40 went smoothly, and the spacecraft was boosted into a 22,018-by-22,046-mile orbit, with a 3-degree inclination.[19]

At the beginning of 1988 on-orbit testing of Flight 13 continued. During what had been planned as the final days of testing, the Space Division determined that handover would have to be delayed. That left Flight 11, whose infrared sensor was expected to last through the year, as the primary satellite (at 65E) covering the Eurasian landmass. Meanwhile, Flights 10 and 12 continued as the primary Atlantic and Pacific satellites, at 35W and 155W, respectively.[20]

The delay in turning Flight 13 into an operational asset was not the only problem with the DSP constellation in early 1988. Flight 12 proved to be somewhat of a disappointment to its operators. Processing of above-the-horizon returns from Flight 12 became possible in June 1986 when the necessary software was installed, but the CONUS Ground Station experienced difficulty with airglow and auroral interference phenomena.[21]

The two limited reserve satellites were also experiencing signs of aging. Flight 9, with subsystem control problems, had a life expectancy of no more than three months. Flight 7 had similar problems but was expected to provide longer service if called back into operation.[22]

The problems with Flight 13 were sufficiently resolved for it to become operational in March 1988, and it became the primary DSP-Atlantic satellite at 35W. Flight 10 moved to 165W to cover the threat of SLBM launches from the Pacific, while Flight 12 became the primary eastern satellite at 65E. Meanwhile, responsibility for Flight 11 was assumed by the SPS located at Kapaun AS, Germany.[23]

In late April Flight 9 was retired from its job as Eastern Hemisphere reserve satellite and boosted into supersynchronous orbit. In the meantime, Flight 7, which had apparently replaced Flight 8 as the Western Hemisphere backup satellite, remained on station.[24] By August 10 the deployment scheme directed by the U.S. Space Command had been achieved, as shown in Table 8.2.

By late November, analysis of the constellation led to the conclusion that there was a high probability that Flight 10 or Flight 11 or both would fail during 1989. That made launch of Flight 14 "at the earliest high confidence opportunity" a matter of "operational necessity," given that it could require

TABLE 8.2. *DSP Constellation, August 10, 1988*

	Satellite
Atlantic	F-13
Pacific	F-10
Eastern Hemisphere	F-12
WHLRS	F-7
EHLRS	F-11

up to 120 days after launch to complete operational checkout of F-14 and redeployment of the constellation, as well as the standing requirement that at least one spare sensor be available.[25]

The standards that the DSP constellation would have to meet in 1989 had been in place for a considerable time. ICBM launches were to be reported to users within two minutes of launch. Launches of SLBMs occurring within 1,700 nautical miles of the continental United States were to be reported within sixty-five seconds of the time the missile broke water.[26]

The new satellite had originally been scheduled for launch in December 1988. Although the spacecraft had been designed for launch using the space shuttle and an inertial upper stage, it was instead scheduled to ride on the first Titan IV—a 204-foot-tall, 10-foot-diameter rocket capable of placing a payload of up to 10,000 pounds in geostationary orbit. The switch was a result of the *Challenger* disaster of January 1986 and the long-standing Air Force aversion to relying on manned space vehicles to place satellites in orbit. In 1987 the Titan IV had been designated as the primary booster for DSP spacecraft.[27]

However, delays caused by booster problems, weather, and scheduling conflicts with other programs threatened to postpone the mission until 1990. The spacecraft first arrived at the Cape on January 4, 1989. At midmonth a March launch was planned, but in February it was rescheduled for June. If launch did not occur by the end of June, it would not take place in 1989, because the facility was scheduled to undergo six months of modernization. After additional delays and cancellations pushed the launch perilously close to the deadline, Flight 14 was successfully placed into geostationary orbit on June 14.[28]

As with all previous DSP launches, the identity of the payload was officially secret. Hence, Air Force officials offered few comments after the launch beyond praise. "We had a beautiful launch," said Colonel Ron Rand, public affairs director for Air Force launchings in Florida. Donald Rice, Secretary of the Air Force, echoed that sentiment, stating that "this Titan 4 launch is a tremendous milestone for the nation" and that the launch was "vital to our national security space program."[29]

After successful completion of its on-orbit checkout, which took over sixty days, Flight 14 was assigned to the Pacific station, where it replaced Flight 10. F-12 remained as the primary Eastern Hemisphere satellite. In the past, Flight 10 would have become an LRS, backing up one of the newer satellites. However, its movement marked the beginning of a new DSP constellation, consisting of four rather than three primary satellites, and would provide increased and improved coverage of the Eurasian continent, where

Russia and China were testing their intercontinental missiles and Israel, Iraq, and India were busy testing intermediate-range missiles. Flight 10 became the Atlantic satellite, while Flight 13 was moved to 5E, where it became the first DSP European satellite, controlled by the SPS (which subsequently became the European Ground Station) at Kapaun.[30] With the Atlantic satellite at approximately 37W and an additional satellite at 5E, the DSP system would have an enhanced capability to cover not only Soviet and Chinese missile activity but also that of other nations such as India, Israel, Iraq, and South Africa.

The successful launch of Flight 14 was significant not only because it represented new blood and an expansion of the DSP constellation, as well as a successful debut for the Titan IV, but also because Flight 14 was the first of the nine satellites in the DSP-1 series, which added to the improvements first flown on Flights 12 and 13. Like those two spacecraft, Flight 14 carried the SED sensor, with its 6,000 detectors and increased resolution, with a portion of those detectors devoted to the above-the-horizon mission. It also was equipped with a second color/second wave-band capability sensitive to the 4.3-micron wavelength, which addressed not only long-standing concerns about Soviet laser jamming but also more recent ones about detecting future missiles with low exhaust flare glow intensity. Flight 14's sensor was expected to be capable of viewing a greater area, provide better discrimination of Soviet and Chinese missile launches, and have improved accuracy and an improved nuclear detonation detection capability (due to advanced radiation detectors).[31]

Other modifications were intended to improve survivability. The DSP star sensor had been redesigned to improve reliability and increase its resistance to nuclear radiation. The infrared sensor's focal plane was designed with two sections to protect its electronics from any Soviet ground-based lasers that might be directed against the satellite. On the outer surface of the satellite were impact detectors. Inside were computer and software improvements that would allow the spacecraft to manage its systems and determine its orbital position without support from the ground stations. The satellite would be able to return missile-warning data even though the ground stations might be neutralized by battle damage or other action.[32] The additions and modifications produced a much larger spacecraft—33 feet long, with a 13.7-foot span, and weighing in at 5,200 pounds. Part of the weight was the result of an increased fuel supply, which gave the satellite an expected lifetime of five years.[33]

In addition to the infrared and nuclear detonation sensors, Flight 14 carried three experiments. The Three-Color Experiment (TCE) was bolted onto the focal plane, and collected additional missile signature data—more for the Strategic Defense Initiative Organization than for anyone else.[34]

The Visible Ultraviolet Experiment (VUE) was also onboard, presumably to collect data on the ultraviolet emissions created by the shock heating of air immediately in front of an ICBM. Such data could be used to assess the possibility of tracking ICBMs using their ultraviolet emissions. In addition, observations must be made against Earth's bright infrared background. Since Earth's ultraviolet emissions are absorbed by the ozone layer, the planet appears as a relatively dark body when viewed in the ultraviolet spectrum from space. However, because of a mechanical problem, the experiment apparently proved unsuccessful.[35]

More mysterious was the third experiment, EP3, which possibly stands for Experimental Package 3. While EP3 is, according to an Air Force Space Command memo, not an SDI experiment, it was necessary to obtain permission from the Australian government, which had a policy of not allowing the DSP East satellite to be used for SDI experiments, before it could be turned on, suggesting that the uncharitable might interpret EP3 as such an experiment.[36]

Flight 14 carried room for eight communications links. Some of those links transmitted data from the infrared and nuclear detonation sensors as well as data on the health of the spacecraft. Others received commands from the ground station, including commands to reposition the satellite.[37]

Two links were missing from Flight 14: Links 5 and 6. Link 5 had been reserved for the transmission of sensor and telemetry from the Eastern Hemisphere satellite directly to the Pacific satellite, without relying on the Overseas Ground Station at Nurrungar. Link 6 had been reserved for commands for the Eastern Hemisphere satellite that would be transmitted through the Pacific satellite.[38]

The absence of such links reflected the inability of the McDonnell-Douglas Corporation to produce the promised Laser Crosslink System (LCS), also known as the Advanced Space Communications System, which had been planned as a key element of the DSP-1 satellites. The LCS, a canister-like system, was to weigh about 300 pounds and protrude from the side of the satellite. It was to be a satellite-to-satellite communications package that would allow the Eastern Hemisphere satellite to transmit its infrared mission data to Western Hemisphere satellites, which could then downlink

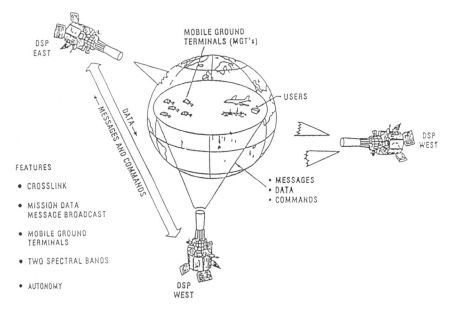

8.1. Survivable DSP-1 concept

the data to primary, backup, or mobile ground stations in the United States.[39] The cross-link concept is illustrated in Figure 8.1.

A cross-link had two potential benefits for the DSP program. First, it would increase the probability of DSP mission data getting back to the United States in event of war. If the Nurrungar station was destroyed by the Soviet Union, and the U.S. mobile ground system worked as planned, DSP-E data could be relayed to the DSP Pacific satellite and then on to a mobile ground terminal. Thus, the United States' ability to conduct a prolonged nuclear exchange, as mandated by the Carter and Reagan directives, would be improved. Second, the cross-link would eliminate the need for overseas ground stations, particularly the Australian station. Aside from reducing costs, Air Force and DSP program officials would not have to worry about the consequences of antibases group protests or concern themselves with the possible impact of developments in Australian politics on the DSP program.

According to an Air Force history, the decision to add a *laser* cross-link to the DSP satellites was based on the belief that laser transmissions would be hard to jam or intercept; a laser communications system was considered ideal for deployment on a satellite because of its small size, light weight, and low power requirements. McDonnell-Douglas had been selected to build it because of a successful history in developing laser technology. However, the problems the company encountered in trying to apply that technology to space systems had produced increased costs and a stretched-out production schedule.[40]

Alternatives, including a radio cross-link option as well as mobile terminals, had been suggested in a December 16, 1983, briefing to Edward C. Aldridge, Under Secretary of the Air Force and NRO director. However, according to an official history, the mobile terminals did not offer as much survivability, while the radio cross-links had higher costs and greater risk, and would cause a production delay of at least eighteen months. Aldridge concluded that the laser cross-link should be put into production as quickly as possible. On January 19, 1984, he notified the Defense Department of his decision.[41]

The next month, on February 8, 1984, Donald C. Latham, Deputy Under Secretary of Defense, sent the Air Force a memorandum stating that the cost of the laser cross-link was excessive. He suggested that the Air Force drop plans to produce the cross-link and instead accelerate development of a follow-on DSP system that would be cheaper and better.[42]

An impasse developed between the Air Force and the Defense Department, but in mid-February it was broken because the planned follow-on system to DSP, the Advanced Warning System, was folded into the SDI program and, consequently, its development was delayed indefinitely. As a result, an acceleration of the follow-on was no longer a possibility, and the Defense Department acquiesced in production of the laser cross-link.[43]

In 1985 hearings before the Senate Appropriations Committee, the Air Force gave the distinct impression that all was well with the LCS program, telling the committee in a written report that the "DSP laser cross-link program has cleared all technological difficulties," and noting that "the cross-links have been in production since April 1984." Despite this optimistic report, the Air Force also noted that the first cross-link system might not be ready to fly on Flight 14.[44]

And it wasn't. One problem the developers were able to successfully resolve was aiming the narrow laser beam at other DSP satellites, operating approximately 50,000 miles apart in space. But another problem proved more difficult: making the laser reliable enough to last for the five-year design life of the satellite. Despite their efforts, an eighteen-month lifetime for the lasers was the most that could be achieved.[45]

The first LCS canister was to have been delivered on July 1, 1987, for incorporation on Flight 14. But, as of July 1989 it had yet to arrive. Not only did the Space Division's DSP program office not believe it would arrive in time to make it on Flight 14; they also believed it was likely to miss Flight 15. The extent of the problems was illustrated by the titles of two Air Force memos ("List of 16 Biggest LCS Program Problems" and "LCS Top 10 Concerns") as well as the comments of a former TRW senior manager who said the system "has been fraught with troubles, troubles all the time."[46]

In August 1989 the *St. Louis Post-Dispatch* reported that the LCS was being reevaluated "due to problems encountered in the development of the laser cross-link technology and costs associated with those problems."[47]

That review, by the DSP program office at the Space Systems Division, concluded that there were five options. The program could continue and an attempt could be made to carry the cross-link on Flight 16, which would entail high technical risks. Or the schedule could be altered so the first cross-link would be not be scheduled to fly until Flight 18. A third option was to terminate the program altogether and develop a radio cross-link system. According to a Space Systems Division history, this would have delayed the first launch until Flight 21 and added "risk and uncertainty," which presumed that it was only a relatively short matter of time before the laser link was perfected. The fourth option was to forgo any cross-link system and rely on the combination of mobile ground stations and communications satellites to get data back to the United States—which would still leave the United States dependent on foreign sites. Finally, the delivery schedule for the laser link could be altered and work begun on a 60-GHz radio cross-link as a backup.[48]

Colonel John R. Kidd, who had recently become the Space Systems Division DSP program director and would at one point, in exasperation, pronounce himself "the proud owner of the world's only suicidal laser," briefed several officials, including General Bernard F. Randolph, the commander of the Air Force Systems Command (SSD's parent organization). Randolph favored the second option, and Kidd briefed John J. Welch, Assistant Secretary of the Air Force for Acquisition, on October 12, giving Randolph's position. Welch endorsed that position and sent a message officially approving the new schedule on December 11, 1989.[49]

Once this decision was made, the next step was to incorporate the new schedule into TRW's spacecraft production contracts. It was to include not only a relaxed schedule but also improvements to the laser cross-link that would make it more robust and enhance its performance. In addition, the first cross-link would be mechanically and electrically integrated with Satellite 17 and put through functional tests to make sure it could acquire and communicate with the cross-link on another satellite. The DSP program office believed that any problems could be identified and eliminated before the cross-link was integrated and tested with Satellite 18. Once the testing was concluded, the LCS would be removed from the spacecraft.[50]

The DSP program office and TRW completed negotiations on the restructured program on January 5, 1990, with the two parties agreeing on June 1, 1990, as the delivery date for the LCS. On January 17, however, TRW notified the program office that McDonnell-Douglas would not accept the terms that TRW had negotiated with the Air Force. McDonnell-Douglas wanted February 1, 1991, as the delivery date for the first cross-link, $18 million to redesign it, and a special provision that would recognize the technical risk of producing the cross-link. The program office interpreted the request for such a provision to mean that McDonnell-Douglas would not accept any arrangement that might result in a termination for default.[51]

When negotiations resumed following this setback, they were impeded by TRW's need to obtain McDonnell-Douglas's concurrence with the terms of any agreement with the government. The main issue was the delivery date for the first laser cross-link. A definitive date would establish the government's right to terminate the effort if the unit were not delivered on time, and McDonnell-Douglas was extremely reluctant to take the risk of termination. As a result, the negotiations proved to be long and difficult, and they did not conclude until July 20, 1990.[52]

The agreement specified that the second through eighth LCS canisters would be delivered between October 15, 1991, and February 15, 1994. A compromise with respect to the first LCS specified that it would not have to be delivered in flight-ready form. Rather, some specifications were waived so it could be delivered in a form that, while not flight-ready, allowed it to be tested on Satellite 17. After testing it would be brought up to flight configuration.[53]

There was also progress on the technical front. The first cross-link unit was shipped to TRW on August 29, five months ahead of the current schedule. Just days later, on September 1, the lasers for the first flight model cross-link passed their tests. Writing in 1991, Air Force Space Systems Division historians observed that "as the fiscal year ended, it appeared that a problem that had been around for eight years and cost $400 million was on its way to solution."[54]

Not quite. A second system was delivered to TRW in December 1991, but it failed the test process. In mid-1992 the electronic parts on a third unit, made by a subcontractor to McDonnell-Douglas, failed during ground tests.

Over a year later, in October 1993, it was reported, and acknowledged by the Air Force, that cancellation of the entire laser cross-link program was being considered. Cancellation soon followed.[55]

A similar fate awaited another system associated with the DSP-1 series and intended to help implement the Carter-Reagan revision to U.S. nuclear strategy. System One was the designation for the software that would replace the System Eight software at the ground stations. It was specifically designed to allow full exploitation of data from the DSP-1 sensor. It would have permitted greater discrimination between closely spaced events, such as massive missile attacks, and increased threat coverage during periods of intense infrared activity. But in May 1990, almost a year after the launch of Flight 14, it was still undelivered. Even worse, it was not expected to achieve initial operational capability at the CONUS Ground Station until September 1992. According to one former industry executive involved in the DSP program, it was "a real mess." In 1993 the Air Force accepted the recommendation of the Aerospace Corporation and canceled the program.[56]

NINE

Australia, Germany, and New Mexico

By the late 1980s the DSP overseas ground station at Nurrungar had been in operation for over fifteen years. For much of that time, the United States had been concerned about two threats to the station's viability. One was the possible actions of the Soviet Union in the event of war or acute crisis. Threats that would be evaluated in periodic threat assessments included Soviet use of ICBMs, SLBMs, bombers, air-to-surface missiles, or cruise missiles against the facility. In addition, intelligence analysts would examine possible threats from Soviet unconventional warfare, electronic warfare, and chemical and biological warfare operations.[1] The other threat came from domestic political opposition in Australia. The closest the United States had come to being dispossessed from Nurrungar (and the other facilities in Australia) was during the Labor administration of Gough Whitlam (1973–75). Whitlam had been among the critics of secrecy attached to the original agreement, but not necessarily the bases.

Shortly after his election, Whitlam informed the United States that discussion of the future of the bases would be part of the process of working out a new relationship between the two countries. But, on April 4, 1974, he told Parliament, "The Australian government takes the attitude that there should not be foreign military bases, stations, installations in Australia. We honour agreements covering existing stations . . . but there will not be extensions or proliferations."[2]

Such statements led Henry Kissinger, U.S. national security adviser and Secretary of State, to issue, on July 1, 1974, National Security Study Memorandum 204 on "U.S. Policy Toward Australia." It called on the NSC's Interdepartmental Group for East Asia to conduct a "review of U.S. policy toward Australia in light of recent changes in the Labor Government." Among the subjects the group was directed to consider were the "prospects for keeping U.S. defense installations in Australia, and the policy options for trying to prolong their existence there."[3]

It is not yet known what recommendations were contained in the interdepartmental group's report, or what the United States did about them. In any case, before Whitlam could carry out his pledge to remove the U.S. bases he was removed from office by the Queen's representative, Governor-General Sir John Kerr. On November 11, 1975, Kerr sacked Whitlam, ostensibly over a budget impasse with the opposition, and made Malcolm Fraser, leader of the opposition, caretaker prime minister. The opposition-controlled Senate had refused to pass appropriations bills until Whitlam agreed to early elections.[4]

Many left-wing Australians considered the events, or at least Kerr's action, to be a coup orchestrated by the United States and Britain to avoid the damage those nations' leaders believed Whitlam would do to the security relationship. Aside from Whitlam's statements, an incident early in his administration, when the attorney general had demanded access to the files of the Australian Security Intelligence Organization (ASIO), had set off alarms at the CIA, leading to concern over the viability of the U.S.–Australian intelligence relationship. Kerr's background helped fuel speculation that he was acting on behalf of the Americans and the CIA. He had been associated with two organizations, Law Asia and the Australian Association for Cultural Freedom, which had been characterized as "CIA fronts." He was also reputed to have had contact with CIA officers serving at the U.S. embassy in Australia.[5]

Had Whitlam stayed in power and carried through on his promise to oust the U.S. facilities, it undoubtedly would have had a significant impact not only on the intelligence and defense relationship but also on the overall U.S.–Australian relationship. The United States had been dispossessed from or forced by circumstances to abandon other valuable intelligence facilities before and would be again. But exiting the central Australian bases in 1974 or 1975, given the importance of the programs controlled from Pine Gap and Nurrungar, and the unique protection those facilities offered from Soviet eavesdropping and electronic warfare activities, would have been viewed as a significant blow.

American use of Nurrungar, Pine Gap, and other bases in Australia would not be fundamentally challenged by subsequent Australian governments, including Labor governments. However, despite the secrecy attached to the bases, investigative research by academics and journalists, disclosures by former intelligence officials, and the 1977 espionage trial of TRW employee Christopher Boyce ensured that the salience of the U.S. bases issue in Australian domestic politics had remained high. (Boyce, along with childhood friend Andrew Daulton Lee, were charged with selling the Soviets material on a CIA agent communications satellite code-named PYRAMIDER.

In addition, as the United States ascertained but did not charge publicly, the two had also sold the KGB information on the RHYOLITE program. Part of Boyce's explanation for his activities, which along with details of the RHYOLITE program ultimately appeared in the press, was that the United States was interfering in Australian politics and deceiving the Australian government about the planned follow-on for RHYOLITE, code-named ARGUS.)[6]

In 1980 Desmond Ball, of the Australian National University's Strategic and Defence Studies Centre, had published *A Suitable Piece of Real Estate: American Installations in Australia,* which included a chapter dealing with the Pine Gap and Nurrungar facilities. In November 1982 the *Sydney Morning Herald* carried a three-part series on space warfare, with one article focusing on "Australia's nuclear targets"—the U.S. bases. A 1983 *National Times* article, "Putting Mobile American Targets on Australian Highways," raised the prospect of Mobile Ground Terminals roaming Australian roads in a nuclear crisis. The *National Times* of April 27–May 3, 1984, devoted more than a page to an article entitled "Strike One!" Directly underneath the headline was the following statement: "The satellite ground station at Nurrungar in South Australia has been transferred to the newly created Space Command where it plays a key role in nuclear war-fighting plans." At the beginning of the next month the same paper ran another Nurrungar story—"Nurrungar's Nuclear Attack Role Admitted."[7]

Two themes running through such articles and books were Australian government secrecy and duplicity regarding the exact nature of the facilities and the possibility that their presence would lead to Australia's becoming embroiled in a nuclear war. Thus, the *National Times* "Nurrungar's Nuclear Attack Role Admitted" asserted: In documents obtained by *The National Times* this week, the Pentagon . . . states that the Nurrungar ground station is expected to feed data on Soviet missile launches and nuclear explosions to the US President's flying command post after a nuclear war has begun. These facts . . . contradict official statements by previous Australian governments."[8]

Shortly after the article's publication, in early June 1984, Prime Minister Bob Hawke and Minister for Foreign Affairs Bill Hayden acknowledged, as Hayden had in the fall of 1983, that Nurrungar and Pine Gap did place Australia at some risk of nuclear attack. Hawke informed the public that "the Government believes that hosting the facilities does bring with it some degree of added risk of nuclear attack." Meanwhile, former critic Hayden told a Sunday morning television audience that the facilities were "nuclear targets in certain circumstances, I would think high-priority ones in an all-out exchange." However, he had come to believe that *"the risk is worthwhile because they contribute to nuclear stability and make the likelihood of such an exchange much, much less."* (emphasis added).[9]

But the support of former critics such as Hayden and Defence Minister Kim Beazley did not persuade those who had long opposed the presence of the bases. Indeed, in the view of one Australian commentator, because the bases were a fundamental, nonnegotiable element of the Australian–U.S. alliance, they were, for the Australian antinuclear movement, the key target "in its bid to break the alliance."[10]

The continued salience of the issue was of particular concern to U.S. officials in the spring of 1987, with the approach of negotiations over continued American use of the bases. The subject of the bases had almost certainly come up during Prime Minister Hawke's April 1986 visit to the United States.* In the early spring of 1987, Beazley and the Department of Defence issued statements regarding the evolution of the U.S.–Australian relationship in the joint management of the Nurrungar, Pine Gap, and Northwest Cape sites.[11]

The Australian government position was challenged that August in a new book by Desmond Ball, *A Base for Debate: The US Satellite Station at Nurrungar.* Described by the Air Force Space Command as "an articulate critic of the American military presence in Australia," Ball recommended that the United States should be given notice that the Nurrungar facility must be closed in 1989. He argued that DSP did not require an Australian ground station for its operation and that the system's capabilities were "increasingly extended further from the essentially unobjectionable mission of early warning to the support and enhancement of U.S. nuclear war-fighting capabilities."[12]

Ball also charged that the Australian government was being less than fully candid with its citizens concerning the implications of the Nurrungar operations—stressing its positive contributions without discussing its drawbacks. As an example, he cited a 1984 statement by Hayden that Nurrungar's operations contributed to the "deterrence of nuclear war" by providing "early warning information received from space satellites about missile launches." But, Ball argued, DSP's ability, albeit limited, to provide launch attack characterization and assessment contributed to the Reagan administration's "de-

---

*A four-and-a-half-page National Security Decision Directive, classified SECRET, "U.S./Australian Relations: Policy for the Official Working Visit to the United States of Australian Prime Minister Bob Hawke, April 17, 1986," noted that "Australia hosts joint defense and intelligence facilities that: . . . enhance strategic early warning in the event of Soviet ballistic missile attack." Approximately three pages of the directive have been deleted under FOIA exemptions, so it is not possible to determine what further discussion took place concerning Nurrungar and Pine Gap.

stabilizing" nuclear war-fighting strategies, including launch-under-attack and launch-on-warning options.[13]

In addition, Ball charged that "the nuclear detonation (NUDET) detection sensors aboard the DSP satellites are designed more to contribute to nuclear war-fighting than to the verification of arms-control agreements." The sensors, he alleged, would allow not only verification of test-ban agreements but assessment of the impact of nuclear strikes against the Soviet Union—an essential requirement for a war-fighting strategy that required retargeting.[14]

Third, Ball challenged Australian government assurances, such as the February 1985 statement by Prime Minister Hawke that "we have already been given the unqualified assurance by the Government of the United States that none of the joint facilities has any role in the research undertaken under the Strategic Defense Initiative." Skepticism was justified, he argued, for a variety of reasons—including the fact that the SDI program was in its early stages, while the DSP system was continually evolving.[15]

Furthermore, according to Ball, such assurances ignored the vast database accumulated during sixteen years of DSP operations from Nurrungar. He observed:

> Nurrungar has detected and recorded data on more than 6,000 Soviet missile and satellite launches. . . . This data is stored on magnetic tape and sent to Aerojet for further processing and analysis. This data includes information on the burn-times, burn-rates and spectral intensities of the various liquid and solid propellants used in Soviet missile and satellite launch vehicles at various altitudes and atmospheric densities under all reasonable meteorological conditions.[16]

That data, complemented by extensive simulation and analysis programs and limited observations in the medium-wavelength infrared band, were given to the Strategic Defense Initiative Organization (SDIO) in 1983. Such data, Ball observed, would, according to a study produced in reponse to President Reagan's National Security Study Directive 6-83, "Study on Eliminating the Threat Posed by Ballistic Missiles," "provide high confidence that a space-based infrared sensor system can be developed to provide the sensitivity, clutter rejection, resolution, and booster trajectory accuracy to support boost-phase intercept requirements."[17]

Finally, Ball suggested that Nurrungar made an attractive nuclear target, even in the absence of all-out war, as a means of degrading the American early warning and attack assessment system without attacking U.S. territory. He noted that the Pentagon had produced a *Satellite Mission Survivability Study*, which considered a Soviet attack on Nurrungar as a plausible means of demonstrating Soviet resolve as well as degrading the U.S. C³I system, while avoiding the consequences of an attack on the CONUS Ground Station at Buckley.[18]

The data and arguments in *A Base for Debate* produced significant press coverage in Australia. Newspapers published lengthy articles with headlines such as "Close US Base at Nurrungar, Says Professor," "Shut US Base at Nurrungar, Says Defence Expert," and "Australia Joins the Star Wars Program—in Secret."[19] The book also produced a response from Beazley in a television interview conducted on the same day as the book's public release, in which he defended the legitimacy of the Nurrungar site and the Australian role.[20] An August 25, 1987, message from the U.S. Defense Attache Office in Canberra reported on Beazley's talk show appearance. Beazley's major points, according to the attaché's report to the DIA, were as follows:

> "Like many of Des's books, it is a very good piece of detective work on the public record. The problem is that the public record is not all there is on these matters. He has taken as existing what may have been foreshadowed developments and has attributed to the DSP functions that belong to other programs. His assumptions that DSP contains war-fighting and SDI functions are simply incorrect."*

Responding to Ball's allegations that data collected at Nurrungar have been passed to SDI planners: "The point is that under the joint facilities

---

*Beazley's assertion may be accurate with respect to Ball's characterization of the DSP NUDET sensors as part of the Nuclear Detonation (NUDET) Detection System (NDS). The ability of that system to locate nuclear explosions to within 100 meters stems, in part, from their placement on the Global Positioning System constellation, and its ability to provide precise locational information through use of several satellites simultaneously to determine location. It should be noted that the Air Force Space Command's August 31, 1983, *Security Classification Guide (SCG) for Defense Support Program* did, on page 7, did specify that "any association of DSP with IONDS" was Secret. IONDS was an earlier designation for NDS. The first GPS satellite to actually carry a NDS package was launched in July 1983. However, the SCG did not specify the nature of the association and at that time the guide also specified that *any* reference to DSP's nuclear detection capability was classified.

On the other hand, DSP could only considered to have no war-fighting functions if it was assumed that the system would be neutralized in a Soviet first strike. Aside from the fact that the improvements in the SED sensor were specifically intended to enhance the ability to use DSP data for war fighting, by allowing identification of empty silos and thus of silos still containing missiles, any system that provided launch detection information during the course of a nuclear war would be used in conducting that war.

agreements, both Australia and the U.S. have access to the information collected and they can use it for whatever purpose they see fit. There is a distinction between that and the DSP program having been tasked for SDI experimental purposes. The U.S. has been sensitive to our position that we will not be directly involved in the SDI program."

As for the charge that Nurrungar would soon become redundant, Beazley said it would take longer than Ball's projection of five years to develop the necessary technology.[21]

Public interest was heightened when the *Canberra Times* published excerpts from the book from August 22 to 24. Beazley turned to the U.S. Defense Department, requesting assistance in formulating a detailed reply. The Deputy Chiefs of Staff for Plans and Operations at Air Force Space Command headquarters took part in efforts to frame an American contribution to the document, which was planned for release to the Australian Parliament in September.[22] But the document was never introduced because, according to an Air Force Space Command internal history, "by that time public interest in the issue had begun to wane, and the Australian press was devoting more coverage to the possible linkage of trade negotiations and the joint defense facilities in negotiations with the United States and the continuing decline of the domestic 'Peace' movement than to any sustained discussion of the Ball book."[23]

The Space Command, however, considered that the situation could become untenable. Its 1987 history observed, that "Despite the failure of the book to generate any groundswell of public opinion against the presence of the OGS, the command continued to weigh its options in retaining or discarding the installation as the DSP system evolved."[24]

Among those who certainly hoped the situation would become untenable, and were trying to make it so, were the members of the Anti-Bases Campaign S.A., who wished to see all U.S. bases removed from Australia. The campaign, in addition to holding rallies, published the *Nurrungar News* every other month. In October 1987 it staged a rally outside the Nurrungar facility. Over 120 unionists, peace activists, and their supporters traveled from Adelaide to Nurrungar, where they were stopped by a barrier about six miles from the base. Margaret Colmer, the campaign's spokeswoman, said that the protesters "totally objected to the Nurrungar base because it is involved in Star Wars." She and the other protesters proceeded to issue an "eviction notice" against the Americans and warned that failure to act would "recklessly endanger the peace and security" of Australians.[25]

While the Anti-Bases protesters focused on closing down Nurrungar, the Woomera Community Board expressed its support of the U.S. presence by granting "Freedom of Entry" to the unit. This medieval custom has, which a long history in Australia, permits a military unit the privilege of an armed march through a city—although only Australian troops bore arms in the parade, which coincided with the fortieth anniversary of Woomera.[26]

On November 16, 1988, a little over a year after the Anti-Bases Campaign served its eviction notice on the U.S. personnel at Nurrungar, an exchange of diplomatic notes between Gareth Evans, the Australian Minister for Foreign Affairs and Trade, and Laurence W. Lane Jr., the U.S. ambassador to Australia, set the stage not for eviction but for long-term renewal of the lease at Nurrungar. In his note Evans conveyed the Australian government's proposal that in the future the station would be known as the Joint Defence Facility, Nurrungar. As a result, the new agreement would be titled "Agreement between the Government of Australia and the Government of the United States of America Relating to the Establishment of a Joint Defence Facility at Nurrungar."[27]

Evans also suggested the removal of the vague description of Nurrungar's activities in the preamble—"to support defence activities"—and substitution of "to provide ballistic missile early warning and other information related to missile launches, surveillance and the detonation of nuclear weapons." In addition, Evans proposed that a sentence be added to Article I that would prohibit use of the facility for other purposes "except with the express consent of the Australian government."[28]

With respect to the length of the agreement, Evans suggested ten years, although after seven years each government could give three years' notice of intent to end the arrangement. That same day, Ambassador Lane informed Evans that he had "the honor to confirm that the Government of the United States accepts the proposals."[29]

Thus, on November 22 Prime Minister Hawke informed Parliament that Australia and the United States had signed a new ten-year agreement covering the operation of Nurrungar and the other U.S. bases in Australia. He also told the legislators that "the DSP, through Nurrungar, would give the earliest warning of an ICBM attack on the U.S., or its allies [and] because the DSP gives longer warning of an attack than any other system it reduces the chances that U.S. forces could be destroyed in a surprise attack. And that makes it extremely unlikely that anyone would ever try such an attack."[30] Hawke also announced that, in addition to Australians constituting some "40 per cent of the staff in the key operational areas," a RAAF Wing Commander would henceforth fill the deputy commander position and would share with the U.S. commander responsibility for "the management of the station and its physical security."[31]

But many specifics would have to be negotiated before the actual implementing agreements would be ready for signature. A November 28, 1988, letter from Secretary of the Air Force Edward C. Aldridge Jr. to Defence Minister Kim Beazley expressed optimism that such issues would be easily settled. In his letter, Aldridge thanked Beazley for making his recent trip to Australia "such a pleasurable experience." He went on to inform the Australian that "the new arrangements outlined in the proposed diplomatic note attached to your letter of 13 October 1988 meet with my complete agreement." The letter also noted that "Air Force Space Command is preparing a proposal for further negotiations on details to amend the classified Implementing Agreement of 4 December 1969 and to outline responsibilities of the Deputy Commander position."[32]

When completed, the agreement would consist of the main text of the implementing agreement, five pages in length, and Annexes A through H: Administrative Purposes, Defence Support Centre Woomera Cost-Sharing, Security, Manning, Transport and Communications, Real Property Management, a classified annex, and Mobile Ground Station Deployment. The process of producing the agreement began in November 1989 when U.S. Air Force headquarters assigned to the Air Force Space Command responsibility for negotiating the new agreement with the Australian Defence Department. Two months later, Space Command representatives met with Jim Nockels, the Counsellor, Defence Policy, in the Australian embassy. All the major issues were quickly resolved with one exception—the role of the Australian deputy commander.[33]

The Australians, in the person of Nockels, insisted that the United States make an exception to a long-standing policy regarding control of strategic systems by permitting control of DSP by the Australian deputy commander when the U.S. commander was not available. The Australians maintained that such an exception was inherent in the September 1988 agreement between Aldridge and Beazley.[34] Air Force Space Command indicated a willingness to discuss language on the issue that would be acceptable to both sides. However, both sides acknowledged that the underlying dispute regarding the actual intent of the 1988 agreement would have to be resolved before further progress could be achieved, and that Space Command did not have authority to alter standing U.S. policy.[35]

The Space Command apprised Air Force headquarters and the Defense

Department of the situation and requested policy guidance. During the first week of April, the Australians were advised that U.S. (Department of Defense and Air Staff) policy remained unchanged. On April 17 and 18 Nockels informed the Directorate of International Affairs that no progress could be made on other command initiatives until the deputy commander issue had been resolved.[36] The Space Command was advised by Air Force headquarters that the Department of Defense might negotiate a compromise if the Space Command could not. Policy, not language, was the issue, and exceptions had been made over the years for allies.[37]

The Air Force Space Command believed that a similar relationship should be recognized with Australia—that the operation of mixed U.S.–Australian crews at Woomera had, in fact, given Australian crew members full access to station operations for many years and the deputy commander had had de facto control of station operations in absence of the U.S. commander ever since the deputy commander's position had been established in 1988.[38]

Negotiations between the Air Force Space Command and Australia concluded on June 18 and 19 in Colorado Springs, with the U.S. Defense Department and Air Staff agreeing to an exception, acknowledging that the deputy commander had operational control in the absence of communications with the commander, although they refused to sanction a written agreement.[39]

The senior Australian representative at the negotiations, Barry Davies, estimated that Australia would be ready to sign in two to three months and expressed his government's desire to hold a ceremonial signing in Australia. But those plans ran afoul of a failure to resolve some minor points, which still had not been settled at the end of 1990.[40]

While those concerns eventually would be resolved, the new agreement did not end the United States' concern with the Australian opposition to Nurrungar. In 1989 the Air Force Office of Special Investigations (OSI) had prepared a study, classified CONFIDENTIAL NOFORN (Not Releasable to Foreign Nationals), entitled "Australian Anti-base Groups." A 1991 Air Force Space Command threat assessment noted that "in peacetime, the primary unconventional warfare threat would include terrorism, vandals, dissidents, and protest groups. . . . Activity at this level would most likely be intended to harass or cause inconvenience, as opposed to destroying the facility or halting operations."[41]

The Space Command study, however, did express confidence in the Australian government's willingness to neutralize threats to the base from protesters. It noted that "the Australian government has shown the resolve to meet the security needs of the site. When it became apparent during previous demonstrations that security personnel were undermanned for the size and aggressiveness of the demonstration, increased security measures were taken to provide more personnel and equipment."[42]

Neither the conclusion of a new agreement nor the prospective end of the Cold War, and the clearly reduced likelihood that the station would ever be employed in nuclear war fighting, eliminated Nurrungar as a source of controversy. Indeed, in 1991, the vastly improved relations between the United States and Mikhail Gorbachev's Soviet Union caused another controversy. In a televised speech on September 27, 1991, President George Bush announced what he described as a historic set of initiatives "affecting every aspect of our nuclear forces." Bush ordered termination of the MX and SRAM nuclear weapons programs and informed his listeners that "to foster cooperation, the United States soon will propose additional initiatives in the area of ballistic missile early warning."[43]

Elaborating on Bush's statement, a Department of Defense statement of October 1 noted that the additional initiatives could include the following:

- Confidence-building measures, like mutual visits to early warning facilities and possible officer exchanges at those facilities;
- Possible measures to enhance early warning capabilities, including actual sharing of early warning information; and
- Discussions regarding the concerns and limitations of both sides regarding early warning.[44]

While Bush's initiatives were intended to further reduce tensions with the Soviets, they increased tensions with some Australians. In November 1991 Desmond Ball wrote that the Bush proposal concerning sharing of early warning information with the Soviet Union was "simply another instance in a litany of U.S. disdain for Australia with respect to the operation of Nurrungar and the use of early warning intelligence collected and processed at the station."[45]

At approximately the same time, an Australian defense official, Brigadier Adrian D'Hage, told an Australian newspaper that his government "was well aware" that Soviet military personnel might inspect the joint facility at Nurrungar. "The Australian and U.S. governments are aware that the statement might have implications for the joint facilities," he said. D'Hage refused to say whether the United States had invited the Soviets to inspect Nurrungar without consulting the Australian government.[46] Ball's article also brought a response from the foreign affairs minister, Senator Gareth Evans, who stated that the question of access to early warning information from Nurrungar arose after the Bush statement. But Evans noted: "In that particular context, at that particular time, there was a communication from the U.S. government to our Government saying that the U.S. was thinking about discussing sharing that warning information with the Soviet Union but that the U.S. would of course consult with Australia before pursuing the matter further."[47]

Evans also denied that there was any truth to reports that the Soviets had been invited to visit Nurrungar. He stated, "So far as the question of physical access goes . . . there is no truth whatsoever in the report . . . that access has been offered to the Soviet Union" and emphasized that no Soviet visit could take place without Australian government permission. Meanwhile, an American embassy spokesman said that no invitation had been issued or would be issued without consultation.[48]

A November 4 cable from the American embassy in Canberra noted that the Australian Department of Defence "hoped to keep this press flurry isolated from plans for an impending public announcement on the role of the joint facilities with respect to the Gulf War." It also noted that "for the present, we are taking the line that, as a matter of course, the U.S. and Australia consult on issues related to the joint facilities" and requested public affairs guidance "ASAP." In addition, it reminded the State Department that the Australian Department of Foreign Affairs and Trade had raised a question at the time of Bush's speech about the implications of the speech for sharing of early warning information with the Soviets and mentioned that the department had made a follow-up call in response to the Australian press articles.[49]

In a November 21 cable the embassy informed Washington that an Assistant Secretary of the Department of Foreign Affairs and Trade requested any details on an upcoming meeting between U.S. and Soviet officials related to the Bush intiative. It also noted that the Assistant Secretary "took the opportunity to recall the U.S. commitment to prior consultations concerning the early-warning part of the President's nuclear initiative."[50]

A November 23 response from Secretary of State James Baker indicated that the United States would not be presenting concrete proposals on early warning at the meetings and that it was possible that establishment of a U.S.–Soviet early warning working group would be discussed. But it asserted that "prior to proposing any early warning measures to the Soviets which would involve the Joint Defense Facilities in Australia we would, of course, undertake consultations with the Government of Australia."[51]

Australia was not the only part of the world where there was concern in the late 1980s and early 1990s over the impact political turmoil might have on a DSP ground station. Until early 1981 the Simplified Processing Station remained in Cornhusker, Nebraska, as a result of insufficient funds being available to transport it. Beginning in early 1979, the ADCOM spent a great deal of time considering where the station should be moved. By March 2, 1979, ADCOM's first choice was "no longer viable."[52]

It was clear to ADCOM that a European location was preferable and had four operational advantages, including being able to support satellites in a greater number of positions. Such flexibility would allow "worldwide satellite deployment and could be important for such changes in the DSP system as increased force structure, threat changes, or a change in deployment requiring a satellite to cover the NATO countries more efficiently."[53] Of course, such conclusions were academic without the funds to get the station from Nebraska to Europe. Finally, on March 26, 1981, after the intervention of Senator Ted Stevens (R-Alaska), the Air Force could inform Congress that the SPS was to be moved to a location in Europe. It stated that the "purpose of moving it overseas is to improve the availability of [data from the DSP-E satellite] and to support computer enhancement at the Overseas Ground Station that is scheduled for February–June 1983."[54]

But relocation did not come immediately. Finally, in September 1982 the commander in chief of the Air Force Space Command approved proceeding with the move to Europe—specifically to Kapaun Air Station in Vogelweh, near Kaiserslautern, about thirty-one miles west of Mannheim.[55]

At the time of the move, of course, no one expected that the Soviet Union and the governments of its East European allies would collapse within a decade. The fall of those regimes, including the East German regime, changed the political landscape of Europe. In adjusting to the new reality, the United States began closing down a number of signals intelligence installations in Germany whose value had dropped—particularly those that had focused on Warsaw Pact forces stationed in Eastern Europe. But the changes in Eastern Europe did not lessen the value of the DSP facility at Kapaun, Germany, which could continue to play a valuable role in backing up Nurrungar, as it had in 1986.[56]

Since that time, the facilities status and capabilities had been enhanced. During the 1988 fiscal year the Air Force Space Command had decided to scrap a plan, designated SPS Ugrade, and replace the SPS with a new ground station called the SPS Replacement (SPS/R). SPS/R was turned over to the Air Force on November 28, 1989, and on January 10, 1990, it was redesignated the European Ground Station (EGS) and took control of the new European satellite in the DSP constellation.[57]

But the Air Force Space Command feared that a leftward swing in Germany might threaten continued operation of the EGS. In February 1990 Colonel Fred H. Hirsch, Air Force Space Command's director of missile warning, wrote another Space Command official that "political uncertainties caused by recent events and recommendation for possible relocation will depend heavily on your assessment of political stability and estimated time to complete host country negotiations. Two possible locations in mind."[58]

A March 1990 Air Force Space Command memo noted:

> Further complicating the situation is that both East and West Germany are holding national elections this year. . . . In both Germanys, the left (Social Democratic Parties) could well become dominant. Of course, in the east this is more probable than in the west. Should the left gain ascendancy there will be a de-emphasis on defense with a concomitant increase on restructuring the economies of both states and expanding social programs. Even if Mr. Kohl retains power (rated 50-50 at best) he will still need to be more sensitive to the left.[59]

A related concern that arose as the movement for unification picked up speed was that unification might come at the expense of continued German participation in NATO. A German withdrawal would logically be coupled with the departure of all foreign forces from the country.[60]

Given the uncertainty felt by the Air Force Space Command, and the fears of its commander that changes in U.S. troop allocations might result in directions to quickly relocate the EGS, contingency plans were developed to move the station. A preliminary site evaluation was conducted in July 1990. Preparations were also started for a site survey team to visit the EGS and several possible bases that could serve as a new home for the ground station; however, the Iraqi invasion of Kuwait placed such plans on hold. Subsequently, the focus shifted to building a German commitment to retain the EGS at its Kapaun location, where it remains today.[61]

In contrast to Nurrungar, the presence of a DSP ground station at Kapaun from the early 1980s never became an issue of contention between the U.S. and German governments or a domestic political issue in Germany. Whereas Australia was far away from the European battlefield on which any NATO–Warsaw Pact conflict would be fought, Germany was where the two armies would first clash. The multitude of U.S. bases—for military, intelligence, and warning purposes—were an essential element of the U.S. commitment to the defense of Germany. The sheer number of bases made it unlikely that one or two would take on any particular significance. And Germany was a nuclear target, with or without a DSP ground station.

★ ★ ★

While the SPS wound up, despite ADCOM's original plans, as a fixed ground station, the Mobile Ground Terminals did become mobile assets. In 1984 the MGTs arrived at their main operating base: Holloman Air Force Base, located on the edge of Alamogordo, New Mexico, near the White Sands Missile Test Range. The base, located between the Sacramento and San Andreas Mountains, was home to a number of fighter squadrons and would eventually become the main operating base for the F-117A stealth fighters.[62] It was expected that in the event of a sufficiently threatening crisis the MGTs would leave Holloman—not for New Mexico's highways and main roads but for secondary roads. The key was not to be high speed but their obscure location.[63]

Operating the MGTs was the responsibility of the 1025th Satellite Communications Squadron (Mobile), subsequently the 4th Satellite Communications Squadron, and the 4th Space Warning Squadron. Whatever its formal designation, the unit was charged with operating

a Mobile Ground System (MGS) to produce and report worldwide missile warning and nuclear detonation data to the Defense Support Program (DSP) users and their survivable counterparts. The MGS is tasked to provide missile warning whenever conditions arise that preclude the use of non-survivable DSP elements. These conditions could result from a variety of threats ranging from natural disasters to hostile enemy attacks. The MGS was intended to survive an initial attack and be the primary source of DSP data during the [trans-] and post-attack periods.[64]

The Mobile Ground System consists of the MGTs, the Mobile Communications Terminal (MCT), and the Crew Support Vehicles (CSV). The MGT, an eighteen-wheeled tractor/tractor trailer, receives data from DSP satellites through an antenna on the side of the van, while the hardware inside the truck processes the data into usable form. In addition to the antenna system, known as the Phased Array Subsystem (PASS), the MGT consists of a Data Processing Subsystem (DPSS), a Transportation Subsystem (TSS), and a Communications Subsystem (CSS), the last consisting of a UHF radio transmitter and an AFSATCOM terminal.[65]

The MGT is complemented by the MCT, which would broadcast processed data to users via the Defense Satellite Communications System. During the Cold War, the communications systems would have permitted MGTs to pass data to various airborne command posts.[66]

The most prominent of the airborne command posts was the National Emergency Airborne Command Post (NEACP), or "Kneecap." The NEACP, stationed at Andrews Air Force Base, Maryland, would take the

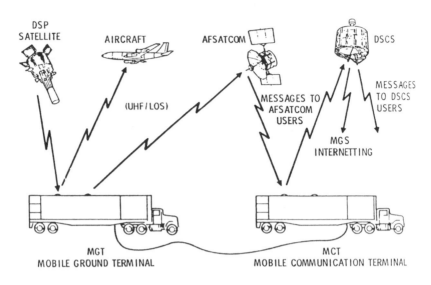

9.1. MGT, MCT operations

president and/or the Secretary of Defense and other key officials away from Washington if nuclear war was or seemed imminent. From there the president or Secretary of Defense would, in theory, direct some, although not all, aspects of U.S. nuclear operations.[67]

Other command posts that might have received data from the MGTs included the SAC's EC-135 LOOKING GLASS, which had a variety of methods of launching U.S. ICBMs. With no direct means of accessing the data from DSP satellites, the MGTs meant, at least in theory, a higher probability that launch detection data would be received by the SAC commander in chief. Also among the command posts that could receive data via the MGT was the Pacific Command's BLUE EAGLE aircraft.[68] Figure 9.1 illustrates the operations of the MGT and MCT.

At the end of March 1985 a two-day Operations Security Workshop addressed the threat from the Soviet intelligence services that the squadron would face, for their mobility would be of diminished value if the Soviet Union were aware of where they were and what they were doing at all times. Given that the Soviets had made checking for any signs of attack the top priority for their intelligence services, it would be unlikely that the 1025th's operations would be ignored. The workshop included briefings from visiting speakers who addressed a variety of topics, including "Intelligence Threats

to the 1025th SCS Mission." Those in attendance were informed of the proliferation of Soviet agents, Soviet intelligence-gathering equipment, and Soviet operating methods.[69]

That the Soviets had a number of deep-cover illegal agents, posing as
Americans and distributed through the Southwest, was a reasonable assumption—indeed, more than an assumption. National Security Agency intercepts
of communications from the Soviet STRELA satellite constellation, which
transmitted information to and received information from Soviet illegals, indicated the presence of such agents.[70]

The methods discussed, in addition to human sources, certainly included
the Soviet Union's photographic satellites, which undoubtedly took frequent
images of Holloman and surrounding areas, as well as Soviet space and
ground signals intelligence resources. By 1985 the Soviet Union was operating its fifth-generation imagery satellites, which have near-real-time electro-
optical imagery capability. They could thus provide GRU analysts of the Soviet General Staff's Chief Intelligence Directorate (GRU) with virtually
instantaneous imagery of activities at Holloman and the surrounding areas
when they were overhead. Among the ground signals intelligence resources
that the Soviets could devote to monitoring Holloman were self-contained
mobile homes belonging to personnel from the Soviet embassy in Mexico
City. Such vans were a common sight at various places along the U.S.–
Mexico border, including Ciudad Juarez, which is only about forty miles
south of Holloman.[71]

Beyond the threat of hostile intelligence collection, there was also concern with the threat from Soviet special operations forces—particularly the
GRU's SPETsNAZ troops.[72] While the purpose of the MGTs was to ensure
the flow of DSP data by being untargetable by Soviet strategic weapons, the
SPETsNAZ could provide a means of destroying the stations prior to both
Soviet employment of nuclear weapons and the MGTs' leaving Holloman in
the midst of a serious crisis. It is also conceivable that the SPETsNAZ could
have planned to track the MGTs after their departure and attempt to destroy
them when they were outside the protection of Holloman's security forces.

Some SPETsNAZ might already have been living in the area. According
to a study of the SPETsNAZ sabotage threat "the use of sleeper, deep cover,
or in-place SPETsNAZ sabotage agents in the enemy country would facilitate
SPETsNAZ activity. . . . They assimilate into the communities where they
live and undertake no activity which would bring attention. In essence, they
become, and are, members of the community."[73]

In the event of a prolonged crisis, SPETsNAZ might also have been
infiltrated into the United States to perform sabotage missions. A study

conducted by R&D Associates for the Defense Advanced Research Projects Agency noted that "the possibility of moving paramilitary units through Mexico, into the U.S., and to their targets, without detection, seems serious when one considers the limited effectiveness of the U.S. internal security forces in stopping illegal traffic in aliens and narcotics."[74]

On October 1, 1986, the MGTs were turned over to the Air Force Space Command for operational use. Before the year was out, one was employed in the NORAD Cheyenne Mountain Complex (NCMC) POLO HAT exercises. Deployed in the vicinity of Holloman, it operated continuously for periods of a week, receiving and processing data from the DSP-Atlantic sensor and relaying them to users.[75]

During 1987 the squadron, which consisted of about 250 personnel and was by then known as the 4th Satellite Communications Squadron, deployed an MGT and associated personnel overseas to allow the Nurrungar ground station to conduct computer upgrades. Operations began by the end of March and were expected to continue for eight to ten months. The following January at least one MGT was employed, in an exercise designated PET SEMETARY, to support the power upgrade at the Buckley ground station.[76]

While the MGTs were employed in 1987 and 1988 to support the peacetime modifications to the Buckley and Nurrungar stations, the Air Force Space Command considered the question of where Mobile Ground Support Units (MGSUs)—consisting of an MGT, MCT/LCV, a crew support vehicle, and a crew quarters vehicle—might be predeployed for use in wartime. And just as the fixed ground station to support the DSP-East satellite had to be located overseas, in an area with direct line of sight to the satellite, the same restrictions applied to any MGSU that would support that satellite.[77]

These requirements imposed significant restrictions on possible MGSU locations, unless the DSP constellation were adjusted. It was believed that the best location was the northeast coast of Australia. Other possibilities included Venezuela, the Antilles, and New Guinea. However, as an Air Force Space Command history observed: "Negotiating the rights to locate a MGSU in a foreign country promised to be a long and delicate process. Even a valued ally such as Australia had been reluctant in the past to accept any system that might be connected to the" (remainder of sentence deleted, but probably "the conduct of nuclear operations" or a similar phrase).[78]

One place the United States could send the MGTs without negotiation was Puerto Rico. The ability of the mobile terminals to function in the midst of war would be tested there, to the extent that such things could be tested, in the summer of 1990 during MGS Deployment Exercise ANCHOR READY.[79]

Finding a foreign government willing to permit the presence of MGSUs for war-fighting purposes was not the only problem facing the Air Force with respect to the mobile systems in the late 1980s and early 1990s. As with the large processing stations, it was necessary to upgrade the MGT software to allow the full exploitation of the data obtained by the DSP-1 satellite. The original schedule had called for the contractor, IBM, to refurbish the two vehicles and return them to the Air Force by May 27, 1990. That deadline passed, and by the end of 1990 the Air Force anticipated it would be March 1991 before all the problems exhibited by MGT-14 Vehicles 1 and 2 would be resolved.[80] In August, Colonel James Knapp, AFSPACECOM's assistant deputy chief of staff/requirements, described the main deficiencies as "system crashes, loss of mission data, high false event rate, radiation hazards." As a result, the vehicles were returned to IBM.[81]

Apparently, IBM was able to make the necessary corrections. On March 29, 1991, the Air Force certified the first two MGT vehicles as ready for initial operational testing and evaluation, even as Air Force Space Command leaders expressed unhappiness with the system reliability demonstrated during in-plant testing, and logistic experts cautioned against turnover in the absence of technical data. On April 2, after having sat unused for approximately one year, the two trucks left IBM Boulder for Holloman. However, one diesel tractor was unable to average more than six miles per hour or, at any time, to exceed twenty-five miles per hour. As a result, both vehicles stopped at Peterson Air Force Base in Colorado Springs for assistance.[82]

Because of the delays, the testing phase did not begin until April 22, 1991. The evaluation then lasted sixty days, with initial unsuccessful results, caused by inadequate operator training and software deficiencies, that later were superseded by more successful results. The vehicles were able to transmit both high- and low-speed messages to users via the Ground Communications Network (GCN) on May 1. On August 22 the Air Force Space Command accepted Vehicles 1 and 2, and began preparing Vehicles 3 and 4 for shipment to Boulder, with the expectation that IBM would complete refurbishing them around September 1, 1992. The final two vehicles were to be modified by late in the 1993 fiscal year.[83]

One vehicle was tested in Australia during the summer of 1992. It arrived on a C-5 Galaxy transport carrying semitrailers containing mobile satellite communications equipment, including terminals, antennas, data-processing and support equipment, and other supplies. The plane's cargo was to be used in a joint U.S.–Australian exercise, designated ANCHOR READY 92-1, announced by Defence Minister Robert Ray on July 17, 1992, and designed to test the ability of the MGTs to take over Nurrungar's mission "in the event of unforeseen circumstances such as natural disaster." The exercise did not involve dispersal of the MGT and its consequent employment outside of the Nurrungar area—possibly due to the terms of the relevant annex of the implementing agreement and almost certainly because of feared political repercussions. Rather, the vehicles operated within the confines of the Nurrungar facility. An additional test using the MGT was performed by the Australian Department of Defence. Designated HAIRPIN-3, it focused on determining the value to Australia of using DSP's SLOW WALKER aircraft detection capability.[84]

During the September 28–29 Third Nurrungar Review Conference, held in Colorado Springs, Colorado, an American representative reported that the exercise had produced lessons concerning overseas deployment and gaining visual access to the satellite. According to minutes of the meeting, one Australian representative, Martin Gascoigne, a longtime analyst for Australia's defense intelligence organization, expressed particular pleasure that the deployment had "attracted minimal public attention, thereby reducing host nation security concerns."[85]

Of course, a year and a half earlier, the overseas ground station in Australia had been "exercised" in actual wartime conditions.

TEN

Desert Storm

On the morning of August 2, 1990, the mechanized infantry, armor, and tank units of the Iraqi Republican Guard invaded Kuwait and seized control of that country. The invasion triggered an American response, Operation Desert Shield, to deter any invasion of Kuwait's oil-rich neighbor, Saudi Arabia. On August 7 deployment of U.S. forces began. United Nations Security Council Resolutions 660 and 662, which condemned Iraq's invasion and annexation, called for the immediate and unconditional withdrawal of Iraqi forces. On August 20 President Bush signed National Security Directive 45, "U.S. Policy in Response to the Iraqi Invasion of Kuwait," outlining American objectives, which included the "immediate, complete, and unconditional withdrawal of all Iraqi forces from Kuwait" and the "restoration of Kuwait's legitimate government to replace the puppet regime installed by Iraq."[1]

A UN ultimatum, Security Council Resolution 678, followed on November 29. It stipulated that if Iraqi dictator Saddam Hussein did not remove his troops from Kuwait by January 15, a United States–led coalition was authorized to drive them out.[2]

As U.S. and allied forces poured into Saudi Arabia, allied intelligence organizations began to update their database on Iraq, with a particular view to identifying potential targets for a bombing campaign. At the same time, the Iraqi dictator and his spokesmen threatened to unleash retaliation of their own.[3]

One possible means of retaliation was terrorism, which theoretically could reach any country—including the United States. Officials of numerous governments feared that any UN response would unleash a wave of terrorism. As a result, in December the U.S. Joint Special Operations Command, which controlled the U.S. military's counterterrorist units—the Army Special Forces Operational Detachment–Delta and the Navy's SEAL Team 6—formed three task forces to combat the possible terrorist onslaught. One task force focused on southwest Asia, a second on Europe, and the third on the United States.[4]

Another possible means of Iraqi retaliation, which had a shorter range but could still reach Israel, Saudi Arabia, and several other countries, with possibly devastating effect, was the Iraqi arsenal of Scud IRBMs. It was feared that in addition to carrying conventional warheads the Iraqis might also be able to equip the missiles with chemical and biological warheads.

The Iraqi Scuds came in several varieties: the Soviet Scud 1C/Scud-B and two modified versions, the Al-Hussein and the Al-Abbas/Al-Hijarah. The Scud-B, which was exported by the Soviet Union to Iraq and a number of other countries under the Soviet designation R-17E, is approximately thirty-seven feet long and almost three feet in diameter. The range of the missile is estimated at 175 to 185 miles; its circular error probable (CEP), the radius of the circle around an intended target within or on which half of the warheads fired at the target would be expected to land, at between 1,475 and 2,950 feet; and its warhead weight at 2,200 pounds. It could carry over 1,000 pounds of chemical agent.[5] The basic Soviet Scud was described as a "clumsy, obsolete Soviet missile which had originally been designed to lob a half-ton warhead 190 miles . . . to . . . within a half-mile of its target—close enough for Soviet purposes because their Scuds would carry nuclear warheads."[6]

The Al-Hussein, named for the grandson of the prophet Muhammad, had been produced by reducing warhead weight, increasing the length of the missiles by adding midsection pieces from cannibalized Scuds, and increasing the size of the fuel tanks. As a result, the Al-Hussein was longer (at forty-one feet) heavier (16,148 to 13,860 pounds), had at least twice the range (372 miles), and traveled at Mach 5 (3,500 mph). At the same time it was considerably less accurate, with a CEP of almost 10,000 feet, and could carry only a 1,000-pound conventional warhead. The Al-Hijarah ("Stones"), originally known as the Al-Habbas, had the greatest range (465 miles), could carry a conventional warhead of only 650 pounds, and also suffered from the same high degree of inaccuracy as the Al-Hussein.[7]

Whereas the Scud missiles came in three varieties, launch sites came in two basic varieties: fixed and mobile. Fixed sites included the areas designated H-2 and H-3 in western Iraq; mobile launchers included transporter-erector-launchers (TELs) and indigenous mobile erector-launchers (MELs). A TEL consists of a tractor trailer with an erector-launcher assembly, while the simpler MEL consists of simple launch rail, elevated hydraulically. Scud launches using mobile launchers often involved the launchers moving to presurveyed locations prior to beginning operations, since knowledge of the precise coordinates the missile is being fired from increases its expected accuracy.[8]

*   *   *

Based on recent experience, Iraqi use of Scuds, as well as their detection by the DSP-E satellite, was to be expected. On September 22, 1980, Iraq invaded Iran, setting off an eight-year war. During the first weeks of the war, Iraq initiated the use of surface-to-surface missiles when it fired mobile FROG-7s, whose maximum range is approximately forty-four miles, at Iranian positions. While the FROGs proved ineffective, Iraq made better use of its Scud-Bs beginning in October 1980. The inaccurate Scuds were of little use against military targets, but Iranian population centers were a different story. Initial target cities included Dizful, Ahwaz, Khurramabad, and Burujird (118 miles from the border). Tehran, about 120 miles beyond the range of the Scud-B, was initially safe from attack, which led to a crash program to extend the missile's range.[9]

In August 1987, with the introduction of the Al-Hussein, Iraqi IRBMs fired from south of Baghdad could reach Tehran and Qom. Iraq soon began firing an average of three Al-Husseins a day, launching as many as 160 missiles at Tehran between February 29 and April 18, 1988, when the war ended.[10]

Iran made much less use of IRBMs, employing them most prominently in March through June 1985. On March 5, 1985, Iraqi missiles hit a steel factory in Ahwaz, as well as an unfinished nuclear power plant in Bushahr. Charging that Iraq had breached a June 1984 UN-sponsored agreement to abstain from attacking civilian targets, the Iranians shelled Basra. The Iraqis responded with air attacks on a number of Iranian cities and towns, including Isfahan, about 250 miles from the border. Iran retaliated by bombing a Baghdad suburb on March 11, the same day it launched a major land offensive. The next day Iraq attacked sixteen Iranian cities and towns, and gave one week's notice that Iranian airspace would be treated as a war zone—implying that civilian aircraft might be shot down. Iran then hit Kirkuk on March 14 with a surface-to-surface missile, probably a Scud-B obtained from Libya. In response to an Iraqi March 15 raid on Tehran, Iran fired four Scud-B missiles at Baghdad. Altogether twelve surface-to-surface missiles would be fired by Iran and would land in Baghdad during the period from March to June 1985 that became known as the "war of the cities."[11]

The Iraqi-Iranian missile exchanges demonstrated DSP's ability to monitor a regional missile war. During the course of the war, the Nurrungar ground station received data from the DSP-E satellite indicating an Iranian or Iraqi launch of an IRBM a total of 191 times. Of the 191 launches, 153 were Iraqi and 38 Iranian.[12]

The Iran-Iraq war was by no means the only conflict that featured frequent Scud launches that were seen by DSP's infrared sensors. The war in Afghanistan, which followed the Soviet invasion in December 1979, featured the frequent use of intermediate-range missiles, both during the Soviet presence and in the aftermath of the Soviets' withdrawal. Indeed, several hundred Soviets had remained as advisers to service and fire the missiles. They were particularly busy during the battle for Jelalabad in mid-1989, when over 400 Scuds were fired into the vicinity and helped stop the mujahideen from seizing the city during four months of fighting. Altogether over 1,000 Scuds were detected by the DSP-E satellites during the war.[13]

As the Iran-Iraq and Afghanistan conflicts wound down, another set of intermediate-range missile launchings were also observed by DSP's infrared eyes. The infrared signatures that would register with DSP's sensors were signs of the impending proliferation of the missiles.

In September 1988 Israel conducted a secret test of its Jericho-II missile, whose range is at least 500 miles. The first test had been conducted in May 1987, with the warhead splashing down in the Mediterranean Sea south of Crete. In late May 1989 the Indian Minister of State for Defence and other Defence military officials watched as a missile they had named after the Hindi word for fire, AGNI, roared out over the Bay of Bengal from its launch site, 750 miles southeast of New Delhi. The two-stage, fourteen-ton missile was estimated to have a range of 1,000 to 1,500 miles and the capacity to carry nuclear, conventional, or chemical warheads. Prime Minister Rajiv Gandhi observed, "We must remember that technological backwardness also leads to subjugation. Never again will we allow our freedom to be so subjugated."[14]

The Indian test was followed, in short order, by South African and Israeli missile tests. On July 5, 1989, a DIA "Special Assessment" noted that a "probable SRBM was launched from South Africa's Arniston Missile Test Range on 5 July." Other reports indicated that the missile would have a 900-mile range, enabling the South Africans to strike frontline states, including Angola and Tanzania.[15] Then, on September 15, Israel launched a Jericho-II from the Israeli air force's research and development facility at Palmikhim, near the town of Yavne. The missile traveled approximately 800 miles, landing in the Mediterranean approximately 250 miles north of Benghazi, Libya.[16]

It had been widely believed and reported that Israel was providing assistance to South Africa—one of many components of a covert military relationship. Evidence supporting that belief was obtained from the satellite imagery of the South African and Israeli launch sites and DSP data obtained during the launches. Equipment at the South African test site resembled that used for Israeli Jericho tests. Further, analysis of the infrared returns from the rocket plume of the July 5 test indicated that the Israeli and South African plumes bore a striking resemblance to each other.[17]

The importance of employing DSP to monitor such tactical ballistic missile launches was acknowledged in a 1990 U.S. Space Command order, which stated:

> Continuous DSP coverage of the Soviet fixed ICBM complexes remains the number one DSP pre-attack surveillance priority. Continuous coverage of Soviet mobile ICBM and SLBM deployment is an extremely high second priority due to increased numbers, accuracy, and MIRV'ed capability of these missiles. DSP NUDET coverage over North America, the Soviet Union, China, and potential proliferating countries is a third DSP pre-attack surveillance priority. In addition, the rapidly increasing tactical threat and third world threat—including tactical ballistic missiles . . . requires DSP to add a fourth pre-attack priority to its mission operations.[18]

That same year the U.S. Space Command reported that with respect to medium-range and short-range ballistic missiles "DSP has demonstrated a good probability of detection [deleted] SS-1C (Scud B), CSS-2 and CSS-1 missiles. Launch of these missiles contains launch time, launch location, missile class (SRBM or I/MRBM), missile azimuth, and operator validity assessment."[19]

The year 1990 brought not only recognition of the value of DSP spacecraft for providing data on such launches but also a fundamental change in policy concerning their use. As result of a decision by the JCS in the first half of the year, DSP data on tactical missile launches could be employed not only to provide data to national security decision makers and scientific and technical intelligence analysts in the United States but also to support tactical commanders during combat.[20]

A few months after that decision there was a real possibility that U.S. military commanders would require such data. As the UN deadline approached, Saddam's adamant refusal to move his troops out of Kuwait brought war closer. As January 15 neared, each side continued to strengthen its military position, while hoping that the other would back down. With respect to the United States' ability to detect Iraqi missile launches and other relevant infrared events, two events of importance occurred.

At the beginning of November 1990, DSP Flights 10, 12, 13, and 14 were all in operation.[21] Flight 14 served as the primary Pacific satellite, Flight

TABLE 10.1. *DSP Primary Satellite Positions, November 1, 1990*

	Station	Position
F-12	Eastern Hemisphere	65E
F-13	European	5E
F-10	Atlantic	35W
F-14	Pacific	165W

10 as the primary Atlantic satellite, and Flight 12 as the primary Eastern Hemisphere satellite. F-13's position reflected the previously noted 1989 establishment of a four-satellite constellation, with a European station at 5E. The stations and positions are given in Table 10.1.

These satellites would soon be joined by Flight 15, the second DSP-1 spacecraft. The launch vehicle, the Titan IV, had arrived at Cape Canaveral on June 25 and was mated with the inertial upper stage (IUS) on July 6. On July 26 the IUS was mated with the spacecraft. By that time the launch, which had originally been planned for the summer, had been rescheduled for mid-September. By September 18 everything was ready for the final countdown to begin, and the launch was scheduled for September 20. At the last minute, booster problems forced a postponement until November 12.[22] At 7:37 P.M. on November 12, the Titan IV/IUS combination carrying the DSP spacecraft was launched from the Cape and successfully placed the spacecraft in geostationary orbit.[23]

Saddam himself also may have inadvertently contributed to the U.S. capability. On December 2 Iraq launched three missiles from a military base near Basrak, in southeast Iraq. The first missile was launched at 4:00 A.M., and the subsequent two missiles were fired at 4:30 and 5:00 A.M. American intelligence assets detected indications of the first missile about six minutes into the seven-minute flight toward a remote area in western Iraq; while its infrared signal was detected by DSP satellites, it was not recognized as such when the signal was being detected. The three missiles all landed about sixty miles from an Iraqi oil-pumping station and military airfield known as H-3. The tests, according to several media accounts, "allowed U.S. Space Command to tweak its detection systems and improve warning time to the theater."[24]

On January 15, 1991, President George Bush signed National Security Directive 54, "Responding to Iraqi Aggression in the Gulf." In the directive's first paragraph, Bush observed, "Economic sanctions . . . have had a measurable impact upon Iraq's economy but have not accomplished the intended objective of ending Iraq's occupation of Kuwait. There is no persuasive evidence that they will do so in a timely manner." Subsequently, after noting a number of U.S. and international authorities, he stated, "I hereby authorize military actions designed to bring about Iraq's withdrawal from Kuwait."[25]

The first step of the campaign began on January 17, with predawn bombing raids on targets in Baghdad and other key locations in Iraq. The targets included command and control facilities, airfields, communication centers, early warning radars, chemical and biological weapons bunkers, and Baath Party headquarters. Also on the target list were Scud storage sites and the fixed sites from which Iraq would launch the missiles.[26]

Iraqi retaliation was not long in coming. While the fixed sites may have been put out of commission, the allies would learn just how difficult it could be to locate and destroy mobile missiles. Late in the afternoon of January 17, Iraqi rocket troops launched two Al-Hussein Scuds (the Scud used in all instances by the Iraqis during the war). The missiles, however, plunged into the ocean. Just before 3:00 A.M. Baghdad time on January 18, the first successful Scud launch against Israel took place. Another seven launches at Israel were conducted in the next half hour. Four more missiles followed on January 19. The next day Iraq commenced firing Scuds at Saudia Arabia, launching eight. The first attacks on Israel injured forty-seven people and did extensive damage to civilian property.[27] Table 10.2 provides data concerning the launch dates, times, locations, and targets.

Throughout the war the United States and its allies would have two goals with respect to the Scuds. The first was to put them out of commission, by whatever means necessary—whether by air strikes or special forces operations. The second, in the event that the Iraqi capability could not be completely eliminated, was to limit the damage they caused. The greatest danger was that repeated Scud attacks on Israel would result in Israeli intervention in the war, possibly producing the very rupture in the allied coalition that Saddam was trying to engineer. Thus, President Bush "directed that unprecedented steps be taken to persuade Israel not to exercise its unquestioned right to respond to Iraqi attacks."[28]

DSP would be "the primary Scud detection system during Operation Desert Storm," according to the Pentagon's report to Congress, and would contribute to both efforts. By providing tactical warning of launches (for which there was often no intelligence warning) and by providing an estimate of launch location (which when combined with knowledge of the trajectory allows an estimation of impact point), DSP satellites, along with detection and tracking radars in Turkey, helped provide warning to the targets of Scud

TABLE 10.2. *Scud Launches During Desert Storm*

	Date	Z Time	Launch Position	Target
1	Jan. 17	2359	330148N401708E	Israel
2	Jan. 18	0005	331055N411841E	Israel
3	Jan. 18	0007	341537N410106E	Israel
4	Jan. 18	0007	341537N410106E	Israel
5	Jan. 18	0007	341414N405960E	Israel
6	Jan. 18	0018	330112N405726E	Israel
7	Jan. 18	0019	340103N403528E	Israel
8	Jan. 18	0027	340124N403609E	Israel
9	Jan. 19	0514	330551N410020E	Israel
10	Jan. 19	0516	340153N403615E	Israel
11	Jan. 19	0516	340149N403545E	Israel
12	Jan. 19	0516	330117N405710E	Israel
13	Jan. 20	1843	305811N472714E	Saudi Arabia
14	Jan. 20	1843	305833N472849E	Saudi Arabia
15	Jan. 20	2139	310400N472429E	Saudi Arabia
16	Jan. 20	2139	310335N472507E	Saudi Arabia
17	Jan. 20	2142	300229N472329E	Saudi Arabia
18	Jan. 20	2148	300154N471450E	Saudi Arabia
19	Jan. 20	2149	300253N471222E	Saudi Arabia
20	Jan. 20	2149	300330N471322E	Saudi Arabia
21	Jan. 21	1918	314509N471023E	Saudi Arabia
22	Jan. 22	0041	300350N474208E	Saudi Arabia
23	Jan. 22	0041	300345N474160E	Saudi Arabia
24	Jan. 22	0041	300413N474223E	Saudi Arabia
25	Jan. 22	0049	311111N471401E	Saudi Arabia
26	Jan. 22	0410	310708N471450E	Saudi Arabia
27	Jan. 22	0410	304941N472724E	Saudi Arabia
28	Jan. 22	1827	340658N404342E	Israel
29	Jan. 23	1954	310517N472416E	Saudi Arabia
30	Jan. 23	1954	300351N474747E	Saudi Arabia
31	Jan. 23	1954	310549N472351E	Saudi Arabia
32	Jan. 23	1954	300528N474836E	Saudi Arabia
33	Jan. 23	2000	341622N410018E	Israel
34	Jan. 25	1555	341817N410122E	Israel
35	Jan. 25	1555	341817N410122E	Israel
36	Jan. 25	1555	330503N410703E	Israel
37	Jan. 25	1555	330503N410703E	Israel
38	Jan. 25	1555	330502N410620E	Israel
39	Jan. 25	1555	330860N410719E	Israel
40	Jan. 25	1559	340224N403236E	Israel
41	Jan. 25	1604	340209N403332E	Israel
42	Jan. 25	1922	300324N474534E	Saudi Arabia
43	Jan. 25	1922	300205N474736E	Saudi Arabia
44	Jan. 26	0028	310829N472430E	Saudi Arabia
45	Jan. 26	1946	300606N475034E	Saudi Arabia
46	Jan. 26	1958	341343N405304E	Israel
47	Jan. 26	1958	341451N405225E	Israel
48	Jan. 26	1959	341754N410405E	Israel
49	Jan. 26	2017	330844N411559E	Israel
50	Jan. 28	1755	300342N475109E	Saudi Arabia
51	Jan. 28	1904	341313N405049E	Israel
52	Jan. 31	1555	341548N405659E	Israel
53	Feb. 2	1824	341649N405806E	Israel

TABLE 10.2. *Continued*

	Date	Z Time	Launch Position	Target
54	Feb. 2	2141	301251N474360E	Saudi Arabia
55	Feb. 2	2333	341509N405760E	Israel
56	Feb. 7	2254	301414N474320E	Saudi Arabia
57	Feb. 9	0036	341536N405908E	Israel
58	Feb. 11	1654	341625N405721E	Israel
59	Feb. 11	1920	301526N474242E	Saudi Arabia
60	Feb. 11	2324	341729N410257E	Israel
61	Feb. 14	0842	334333N444354E	Saudi Arabia
62	Feb. 14	0842	334107N443717E	Saudi Arabia
63	Feb. 14	0842	334253N443758E	Saudi Arabia
64	Feb. 14	0842	334028N443652E	Saudi Arabia
65	Feb. 14	0842	334021N444612E	Saudi Arabia
66	Feb. 15	2300	315037N470815E	Saudi Arabia
67	Feb. 16	1812	342323N411335E	Israel
68	Feb. 16	1812	341422N405416E	Israel
69	Feb. 16	1813	341403N405121E	Israel
70	Feb. 16	1813	341503N405142E	Israel
71	Feb. 19	1751	341609N405922E	Israel
72	Feb. 21	1406	332644N442709E	Saudi Arabia
73	Feb. 21	1406	332228N442948E	Saudi Arabia
74	Feb. 21	1800	332335N442356E	Saudi Arabia
75	Feb. 21	2331	304807N473310E	Bahrain
76	Feb. 21	2331	304548N473538E	Bahrain
77	Feb. 21	2331	304712N473531E	Bahrain
78	Feb. 23	0159	313810N471608E	Saudi Arabia
79	Feb. 23	0159	313810N471608E	Saudi Arabia
80	Feb. 23	1647	341616N410047E	Israel
81	Feb. 24	0132	301520N474221E	Saudi Arabia
82	Feb. 24	0916	332036N443136E	Saudi Arabia
83	Feb. 24	1823	301529N474325E	Saudi Arabia
84	Feb. 25	0126	341516N405718E	Israel
85	Feb. 25	0326	341455N405752E	Israel
86	Feb. 25	0326	341516N405757E	Israel
87	Feb. 25	1731	310720N472634E	Saudi Arabia
88	Feb. 25	2226	302834N480024E	Saudi Arabia

attacks. The same data also could be used in directing the aircraft used to try to put the missiles out of commission.[29]

In preparation for Scud attacks, the United States had deployed sixty Patriot missile launchers to Riyadh and Dhahran. The Israelis also received two batteries of Patriots in December 1990, as a result of Saddam's repeated threats to attack Israel in the event of war. After the initial attack on Israeli targets, the United States redeployed more Patriot batteries from Germany to supplement the launchers already in place. In the end, seven Patriot batteries (four American, one Dutch, two Israeli) were deployed in Israel, although one (the Dutch) did not arrive until very late in the war and took part in no Scud engagements.[30]

Each Patriot battalion includes six batteries, with each battery consisting of one ground-based radar unit for surveillance and target tracking and engagement; an Engagement Control Station (ECS) for command and control of the interceptors; eight missile launchers with four of the seventeen-foot long, 2,000-pound MIM-104 missiles on each; and a Communications Relay Group for communications support. An Information Coordination Center controls the batteries and coordinates their operation with other battalions and higher-level U.S. command authorities.[31]

The heart of the Patriot system is the weapons control computer, which performs the system's main tracking and control functions. The system's radar detects an airborne object that has the characteristics of the Scud. The range gate—an electronic detection device within the radar system—calculates an area in the air space where the system should next look for the missile, based on the Scud's known velocity and time of last radar detection. The range gate filters out information about airborne objects outside its calculated area and processes only the information needed for tracking, targeting, and intercepting Scuds. Finding an object within the calculated range-gate area would confirm that it was a Scud.[32]

Originally, the Patriot had been designed for European operations against Soviet medium to high-altitude aircraft and cruise missiles traveling at speeds of no greater than Mach 2 (1,500 mph). To avoid detection, it was designed to be mobile and to shift locations after only a few hours. But an increasing Soviet tactical ballistic missile threat in the 1970s and early 1980s had led to its modification for use against those missiles.[33]

Use of the missiles to neutralize incoming ballistic missiles in a Persian Gulf war had been part of a strategy developed in April 1990. The strategy did not, however, envision using DSP data to cue the Patriots.[34] Among those involved in developing that strategy was Charles Horner, Central Command Air Force chief. Horner recalled after the war: "I was already aware of the danger from [Iraqi] Scuds before we went to the Gulf but it never occurred to me to use DSP to provide warning of Scud attacks, because I was ignorant of space. The space guys figured they could help, and they sent a team to CENTCOM and set up a Scud warning system."[35]

Normally, the Patriot's radar would detect an incoming Scud 4.5 minutes after it had been launched—and about 2.5 minutes before the Scud would impact. The ability of the DSP constellation to provide advanced warning depended, then, on its being able to report Scud launches within less than 4.5 minutes.[36] By January 17 there were three DSP spacecraft in place to attempt to accomplish that mission. Flight 12 was still the primary eastern satellite, with Flight 15 still in the checkout phase. At the same time, the data from Flight 13 were processed by both the European Ground Station and Buckley. To facilitate the transmission of data from Flight 13 to Buckley, a "bent-pipe" arrangement was set up on Ascension Island in the Atlantic. Mission data from Flight 13 were received at the station and retransmitted to a DSCS satellite and then on to the CONUS Ground Station.[37]*

Providing warning to Patriot crews, and other interested officials, involved more than simply detecting an infrared event. The process began with detection. With the DSP telescope sweeping the earth once every ten seconds, it would take no more than that long to first detect a launch, provided the atmosphere was cloudless and completely dry. Given cloud cover, detection would be delayed until the missile broke through the clouds.[38]

The raw data obtained by the satellites would then be processed. As noted earlier, processing took place at all three ground stations—Buckley, Nurrungar, and Kapaun. According to General Horner, the requirement that the data be routed through the ground stations and then the NORAD Cheyenne Mountain Complex (NCMC) was due, in part, to the fact that the system "was designed to support strategic nuclear warfare—we wanted to have very tight constraints on the [warning] data so we wouldn't respond to a false alarm." In order to separate the SCUD-alerting mission from that of providing warning of any ICBM/SLBM attack on the United States, SCUD alerts were issued through the Space Command Center (SPACC). The mission of issuing any warning of an attack on the United States remained with the Missile Warning Center.[39] In addition, because the infrared signatures of Scud-type missiles are of lower intensity than ICBMs and SLBMs (other than the SS-N-6), which have larger engines and burn longer, processing of the data required special techniques in order to optimize the probability of detection and reporting.[40]

Once the data were processed at one of the ground sites, warning was provided in one of two ways to Central Command wartime headquarters in Riyadh. One means was a simple phone call, routed through the appropriate DSCS satellite. The second, and more commonly used, link was the Tactical Event Reporting System (TERS). Data were transmitted from Skaggs Island, California, to a DSCS satellite and downlinked to Tactical Related Applications (TRAP) terminals, which retransmitted the data. A set of second elements are the Navy's Tactical Receive Equipment (TRE), Air Force's Con-

---

*In addition, in case Iraq unleashed terrorist attacks on one or more of the fixed ground stations, the Air Force Space Command dispersed the six MGTs across the country. (Dwayne A. Day, "The Air Force in Space: Past, Present and Future," *Space Times,* March–April 1996, pp. 15–21.)

stant Source, and Army's Success Radio receivers. Emanating from the receivers are several output lines, each tailored to focus on data of concern to a specific warfare specialty.[41]

After the information arrived at Central Command headquarters, it would be transmitted through the Air Force Satellite Communications System (AFSATCOM)—a series of transponders on a variety of U.S. satellites, including DSCS satellites—to the Patriot's battalion-level Information Coordination Central and then by UHF links to the individual Patriot batteries. Upon receipt of the warning, the Patriot engagement control station would send a message to the radar set to initiate the various systems. The ECS then designated the targets to be tracked, issued missile firing alerts, and assigned target-via-missile (TVM) guidance priorities.[42]

During the first four days of the war, January 17 through 20, when twenty missiles were fired, the average time between detection and reporting to the theater was 5.4 minutes. Hence, during this period the Patriot radar would have already picked up the incoming Scud by the time warning via TRAP arrived. After the initial four days, when the remaining sixty-eight missiles were fired, DSP timeliness improved considerably, to a mean of 3.3 minutes. Hence, Patriot batteries would receive, on average, an extra 1.2 minutes of notice concerning incoming missiles, including where to point their radars.[43]

After the war, in assessing the performance of the Patriots and the utility of DSP in supporting them, Henry Cooper, director of the SDIO, observed that DSP "was designed for different systems, obviously. . . . We were not able to hook DSP into Patriot . . . but [we could] tell the people where to look and start to track." DSP's advance information about the direction from which the missiles were coming allowed the ground crews to point their radars in that direction; that, "in effect, helped them to intercept the Scuds farther out," Cooper told the Electronic Industries Association.[44]

Cooper's view, expressed shortly after the war, has been challenged in one way: postwar analyses of Patriot performance have questioned the system's wartime effectiveness. A study by the General Accounting Office, apparently based on classified Army data, concluded that claims for warhead kills higher than 9 percent could not be supported by reliable evidence.[45]

Israeli officials, who might have a motive for downplaying the Patriot's value, claimed the Patriot batteries in Israel intercepted next to none of the Scuds. General Dan Shomron, chief of staff of the Israeli Defense Forces during the war, described accounts of the Patriot's success as a "myth." Haim Asa, a member of an Israeli technical team that worked with the Patriots during the war called them "a joke."[46]

Defense Minister Moshe Arens, when asked in an Israeli documentary how many Scuds were intercepted, said that "the number is miniscule and is in fact meaningless." Shomron, Asa, and Arens all agreed with a 1991 Israeli Air Force report, which concluded that "there is no evidence of even a single intercept," although there is "circumstantial evidence for one possible intercept."[47]

Irrespective of the controversy over performance of the Patriot, it is clear that the DSP provided earlier warning of incoming Scuds than was available from other sources, and that the ability to relay that data to the relevant military units in sufficient time to be of use was established. In addition, DSP data provided to CENTCOM were used to trigger air-raid alerts, warning military personnel and civilians alike of incoming Scuds. If Iraq had used chemical warheads with the Scuds, the additional warning would have allowed greater numbers to don gas masks.

American warships, equipped with TRE, also found the DSP data useful. According to the leadership of the USS *Worden,* "Scud launches were displayed almost instantaneously on the JOTS II screen. From launch locations, destinations could be estimated, and using this information it was occasionally possible for *Worden*'s organic sensors to identify and track the Scud missiles to their targets."[48]

In addition, in accordance with the Bush mandate to persuade Israel not to take military action, information on Scuds targeted at Israel was passed to Israeli officials through a special communications link designated HAMMER RICK. They, in turn, could sound air-raid sirens, five minutes prior to a missile's arrival, to alert civilians in the cities that were targeted by Iraq—Tel Aviv and Haifa.[49]

The data were also undoubtedly used to warn the Israeli government of the three apparent launches, two on January 25 and one on February 16, in which the missiles were not aimed at civilian centers but at Israel's Dimona nuclear facility. Among the measures Israel has deployed to protect the facility is a missile battery linked to radar screens. Whether those missiles, deployed to prevent aircraft from intruding into the surrounding airspace, would have had any success against an incoming missile is not clear, but certainly Israel would have employed every missile in the battery in an attempt to prevent a Scud from hitting the facility.[50]

At 8:23 P.M. on February 26, a Scud missile fragmented in the atmosphere above Dhahran and tumbled downward. Its warhead tore an eight-foot-wide crater into a transformed warehouse that was housing American soldiers. That attack, two days into the ground war, inflicted more casualties—28 killed and about 100 wounded—than the Iraqi army did during the first two days of ground combat. Whether or not a Patriot would have intercepted or diverted the Scud will never be known with certainty. At the time the Patriot battery in the area was out of commission for a computer repair.[51]

If the battery had been operational, it would not have had a full warning from the DSP network. A DSP satellite did detect the Scud launch, as did the radar system at Pirinclik, Turkey, and provided an alert warning within fifty seconds. However, a full missile warning report was not forwarded from the DSP network because of a "worst-case combination of events."[52]

An Air Force investigation determined that all three DSP satellites with a view of Iraq observed the Al-Hussein version of the Scud, which had been fired from a mobile TEL vehicle. Flight 12, whose data were being processed at Nurrungar, had three infrared returns of the missile during the satellite's scan of the region. However, Flight 12 had "a low aspect view geometry"—a nose-on view of the missile—and there was insufficient target motion to produce an "event report" warning.[53]

Returns from Flight 13 were being processed at the European Ground Station and the Buckley facility. According to an Air Force Space Systems Division study of DSP performance in Desert Storm, "high clouds over the launch site obscured early scans for Flight 13." Infrared returns were detected during two scans of the spacecraft, which was sufficient to produce a "quick alert" message of a probable missile launch. The message was transmitted, resulting in the sounding of air-raid sirens in Dhahran five minutes before the missile's impact.[54]

Meanwhile, Flight 15 had "good viewing geometry and collected three scans of data," according to the SSD report. However, the satellite's data were being processed at Nurrungar in an off-line mode known as Hot Shadow, even though it had the best view of the area after it had been moved into position at 105E on January 31. Although Flight 15 produced infrared returns during three scans, it was not being monitored by ground operators because of its being in hot shadow standby mode. The Air Force concluded:

> The overall result was that due to [a] worst case combination of events, this launch was not reported [as a final missile warning report] by the composite [DSP] system in spite of the fact that sufficient data existed for the detection and reporting. If data from all satellites had been fused at a common processing center, this launch would have been easily detected.[55]

* * *

On the first night of the war, U.S. F-15Es struck at the fixed Scud installations in western Iraq, designated H-2 and H-3, with the objective of removing the Scud threat, particularly to Israel. Describing the aftermath of the raids, General Norman Schwarzkopf, commander in chief of the U.S. Central Command and commander of the coalition forces, wrote that "our bombers had obliterated every known Scud site in western Iraq, destroying thirty-six fixed launchers and ten mobile ones."[56]

But "the initial hope of the air planners in Riyadh that heavy attacks on the fixed Scud sites during the opening hours of the air campaign would largely eliminate Iraq's capability to launch ballistic missiles against Israel or regional members of the U.S.–led Coalition proved to be illusory." The United States would discover that it did not have to worry about the fixed sites but about the Iraqi mobile Scud force, which U.S. intelligence agencies would ultimately conclude might have been as large as fifteen battalions, each with fifteen launchers.[57]

The Scud hunt that would last throughout the war involved 2,493 sorties, the majority of which took place during the first three weeks of the war. Operations took place around the clock. The targets were the missiles, their TELs, and the Scud support infrastructure. Aerial assets used in the campaign included F-15Es, F-16s, A-6Es, A-10s, B-52Gs, F-117As, and RAF Tornadoes.[58]

A crucial part of the attempt to locate and destroy the Scuds or their infrastructure was determining where to look. A variety of intelligence assets, from electro-optical and radar imagery satellites to aerial imagery and signals intelligence platforms, were employed to locate the missiles. DSP was best suited to provide data on the location of the launch points, which could be combined with intelligence data in planning prelaunch Scud hunting missions or employed to guide aircraft toward the Scud units in the immediate aftermath of a strike.

Thus, on the first day of the war, DSP satellites pinpointed the initial eight launch points at four distinct locations: one missile launched from Rutbah; three from a sixty-mile stretch of the Baghdad-Amman highway running east of Rutbah and south of Qasr Amij; another two from virtually identical locations about ten miles from Syria, at Wadi Rutgah on the Al-Qaim Shab Al Hiri road; and two from Wadi Al Jabariyah south of Al Qaim, approximately eight miles from Syria. The data were used to direct F-15Es and F-16s back to the launch sites.[59]

Throughout the course of the war, DSP data would continue to provide estimated locations for Scud launches. Such data were passed, via AFSAT-COM to JSTARS aircraft, to be used in directing aircraft in their hunt for

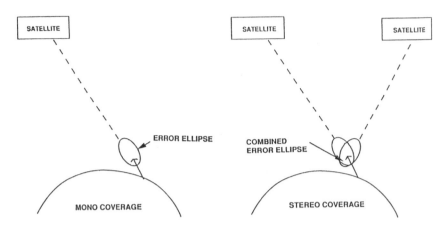

10.1. Mono versus stereo coverage

Scud launchers. Based on DSP data, CENTCOM would also direct aircraft to attack launchers.[60] An innovation that was essential in providing the required data—the ability to employ stereo processing for the DSP data—was significantly refined as a result of the work of an Aerojet DSP system analyst, Nina Zubiel, who would subsequently receive the company's R. B. Young Technical Innovation Award. In June 1990 a U.S. Space Command internal memorandum noted that "impact location and range to impact cannot be provided with TERS." Stereo processing (Figure 10.1) allows analysts to better pinpoint launch sites by correlating the reading from two DSP satellites. Employing the observations from two different angles produces more precise estimates of the missile's trajectory, which translates into more accurate estimates of the missile's launch site and expected impact point.[61]

In the end, the Scud-hunting job proved more difficult than anticipated. The signs of forthcoming launches, which had been expected on the basis of years of monitoring Soviet Scud launch procedures, were largely missing, the result of Iraqi actions to suppress the signatures that might be detected by imagery or signals intelligence.*

---

*Due to a lack of imagery confirming destruction of mobile launchers, neither the CIA nor the DIA credited any kills to allied aircraft. Some argue that a drop in the average daily launch rate from 4.7 per day during the war's first week to 1.5 per day for its remainder shows that the campaign took its toll. Some also suggest that the campaign had a qualitative effect, that the Scuds were less accurate than they would have been otherwise because the Scud units, in trying to keep a low profile, avoided using radar or weather balloons, which would have improved Scud accuracy. Others

Destroying Scud launchers subsequent to launch was also a serious problem despite the fact that, according to a postwar DIA assessment, "by the end of the first week of hostilities, it became apparent that the Iraqis were operating from seven discrete 'launch baskets.' (The launch baskets fell into two Scud boxes: a western set near Ar Rutbah, Al Qa'im, and H-2 aimed at Israel and southwestern Saudi Arabia; and an eastern set near Qal at Saih aimed south at Saudi Arabia and the other coalition states.)[62]

The difficulty was the result of a number of factors: the time it took for coalition aircraft to arrive at the estimated launch point (approximately twenty minutes when already flying as part of a Combat Air Patrol and forty minutes if on ten-minute alert status); the ability of the Iraqi Scud teams to "fire a missile, drive away, and hide in a culvert all within five minutes"; the difficulty of detecting or distinguishing launcher infrared and radar signatures from those of other vehicles; and the Iraqi use of decoys.[63]

In addition to providing data on Iraqi missile launches, the DSP detected, weather permitting, other infrared events of interest to the U.S. military and intelligence community. During operation Desert Shield, an accident at an Iraqi arms depot at As Shuaybah, which the Iraqis were using to supply their troops in southern Kuwait, demolished nine revetted storage bunkers and eight storage buildings. American intelligence experts estimated that at least 10,000 metric tons of ammunition were blown up, sufficient to supply six Iraqi heavy divisions for three days of medium- to high-intensity fighting. That explosion generated more than enough infrared energy to be detected by a DSP satellite. And, as would be expected, Iraqi fighters operating on afterburner were detected by DSP satellites.[64] On January 28, a coalition air attack on another Iraqi ammunition depot triggered an even larger explosion—so massive that both Israel and the Soviet Union queried the United States regarding whether she had employed tactical nuclear weapons. Once again the explosion registered with DSP infrared sensors.[65]

---

argue that the increased tempo of Scud attacks during the final weeks of the war—focused on Hafar, al-Batin, King Khalid Military City, and Dimona rather than Tel Aviv, Haifa, Riyadh, and Dhahran—and that the Iraqis never launched as many Scuds in a single day at the beginning of the war as they were capable of is possible evidence that the campaign had little effect. (U.S. Air Force, *Gulf War Air Power Survey, Draft Summary*, pp. 3–30; Michael R. Gordon and Bernard E. Trainor, *The General's War: The Inside Story of the Conflict in the Gulf* (Boston: Little, Brown, 1995), p. 230; Norman Schwarzkopf, *It Doesn't Take a Hero* (New York: Bantam Books, 1994, p. 488.)

★ ★ ★

In the aftermath of the war, DSP received extensive praise from a number of sources. Lieutenant General Thomas S. Moorman, commander of the Air Force Space Command, observed that "the political, psychological, and military value of providing early warning was incalculable. These systems provided warning to threatened countries and served to contain the conflict. On the military side, these systems clearly enhanced the effectiveness of our missile defense."[66]

The defense secretary's assessment was also full of praise. In his 1992 report to Congress and the president, Cheney wrote that DSP "proved to be of immeasurable political, psychological, and military value in detecting tactical missile launches in the Persian Gulf region. During Operation DESERT STORM, the DSP system detected the launch of Iraqi Scud missiles and provided timely warning to civilian populations in Israel and Coalition forces in Saudi Arabia, helping them to protect lives and property."[67]

The enthusiasm of Australia's Minister for Defence Robert Ray matched that of his American counterpart. On November 5, 1991, he told Parliament:

> During Desert Storm, DSP detected the launch of Iraqi Scud missiles, provided timely warning to the civilian populations and coalition forces in Israel and Saudi Arabia, and helped safeguard property. The coalition forces who were protected can be justifiably proud of the system's outstanding defensive performance.
>
> The DSP's superb operational performance conclusively proved that the system provides significant early warning of enemy tactical missile attack. This recent experience confirms the continued relevance and flexibility of DSP. . . . the deterrent value of DSP provides a continuing contribution to regional peace and stability, the avoidance of nuclear conflict, and international arms control efforts.[68]

Praise also came from former Aerospace Corporation employee Newton Mas, who had worked on DSP and had moved to Israel after his retirement. His assessment was: "If it hadn't been for the stellar performance of DSP during the SCUD affair, half the people in central Israel would have become completely paranoid. The five minutes of warning provided by DSP meant that we didn't have to stay sealed up for the whole period."[69]

DSP's performance in the Gulf War represented a significant step in the program's evolution. The system demonstrated its ability to provide improved missile warning to military theater commanders, as well as intelligence data, during combat. The need to improve the accuracy of launch-point determination served as the catalyst for improvements in stereo processing during the war. In addition, the Gulf War stimulated postwar research and actions to improve the quality and dissemination of DSP data in a future theater conflict. One key element of the history of DSP for the last decade is the focus on adapting the program to the new international environment, in which it is far more likely that U.S. troops and allies will be threatened by intermediate-range missiles than that the United States will be threatened by ICBMs and SLBMs. The second key element is how DSP performance, despite the universal praise it received immediately after the war, became a target for criticism as part of the debate over the need for a new system—one that would provide additional capabilties and correct some of the limitations, real and imagined, of DSP.

ELEVEN

False Starts

While the DSP constellation was providing crucial data during Desert Storm, plans also were being made to orbit another satellite. The launch was to represent two firsts: the first DSP on a space shuttle, as well as the first time a DSP launch was acknowledged in advance. Original plans called for a launch in March 1991, but the date continued to slip—first to July, then to November. After being taken in and out of storage with changes in the launch date, the satellite was finally shipped to Cape Canaveral on August 1, where, eight days later, it was mated to the IUS that would send it into geostationary orbit.[1]

On October 21 the satellite was taken to the launch pad. The DSP constellation that existed at that time was slightly different from the one at the end of the Persian Gulf War. Flights 10, 13, and 14 remained as the Atlantic, European, and Pacific satellites. However, Flight 15 had taken over as the Eastern Hemisphere satellite, replacing December 1984's Flight 12. Flight 10 outlasted its three immediate successors as part of the operational constellation, but it would soon join them in retirement or semiretirement as a result of the impending failure of its reaction wheel, which was instrumental in keeping the satellite's infrared sensor pointed toward earth.[2]

At 6:44 P.M. on November 24, 1991, the space shuttle *Atlantis*, commanded by Fred Gregory and carrying a crew of six, lifted off from Cape Canaveral. Shortly after reaching orbit, the *Atlantis* crew opened the payload bay doors and began checkout procedures for launching DSP satellites. After receiving a "go" from mission control, they fired the explosive bolts, which separated the umbilical cables from the IUS carrying the third DSP-1 spacecraft. The crew then moved a tilt table in the payload bay to an angle of 58 degrees for deployment.[3]

After deployment the crew used the engines to maneuver the shuttle away from the DSP/IUS payload. Using onboard computers, the IUS began to boost the DSP into orbit. At 1:03 A.M. the IUS second stage ignited for about 108 seconds. Once at proper altitude, the IUS stabilized the DSP satellite, separated from the spacecraft, and performed final procedures to prevent collision or contamination before deactivating.[4]

Deploying F-16 was not the only surveillance-related aspect of the mission. Several others concerned the potential utility of manned observation from space. A Navy experiment, Terra Scout, involved onboard observation and analysis of various sites on earth by crew member Tom Hennen, who had studied those sites. Hennen was able to obtain reconnaissance assessment data on twenty-two targets. The Air Force M88-1 Battleview and Maritime observation experiment involved identification by Navy Lieutenant Commander Mario Runco Jr. of ships and some aircraft, including C-5s, C-141s, and 747s sitting on their airfields. The experiment was also designed to evaluate the ability of a space-borne observer to gather information about ground troops, equipment, and facilities. Another experiment was the Air Force Maui Optical Site (AMOS) Calibration Test, in which the AMOS used the shuttle orbiter as a calibration target for its ground-based electro-optical sensors.[5]

After completing its missions, *Atlantis* and its crew landed at Edwards Air Force Base at 2:35 P.M. Pacific Standard Time on December 1. Originally, they had been scheduled to land on December 4 at Kennedy Space Center, but failure of the shuttle navigational unit led to an early landing at Edwards, whose runway is surrounded by desert, allowing a greater margin for error.[6]

On December 25 the satellite they had helped place in orbit completed early-orbit testing five days ahead of schedule and was turned over to the Air Force Space Command. With F-16 ready for operations, the process of revising the constellation began. F-16 was moved to the Eastern Hemisphere station at 65E, replacing F-15. F-15, in turn, drifted over to 5E to become the European satellite, displacing Flight 13, which was maneuvered westward to assume the Atlantic station at 35W that had been Flight 10's responsibility since late 1989. In the midst of all the movement, Flight 14 stayed put as the Pacific satellite.[7]

The satellite movements reflected the changes in the international environment that had taken place over the last fifteen years and represented another aspect of the evolution of the DSP system. In that earlier period the newest and most sensitive satellite was most likely to be sent to occupy the Atlantic station. There it could detect the launches of Soviet SLBMs, which burned less intensely than Soviet ICBMs and might be fired first in an attempt to quickly wipe out U.S. command and control facilities and key personnel in the Washington area. The concern was no longer a Soviet sneak attack. Indeed, the Soviet Union's existence came to an end on the very day that Flight 16 became operational. Now the primary concern was with tactical missiles fired in anger or as part of a test program in the Middle East, Asia, and southwest Asia—areas that would be viewed by F-15 and F-16.

Thus, on August 19, 1992, concern that enforcement of the UN resolutions against Iraq would lead to conflict led the commander in chief, CENTCOM to ask the U.S. Space Command to "provide optimum DSP coverage of potential Iraqi Scud launch areas." After consultation between the U.S. and Air Force space commands, the Joint Staff, and the commander in chief of the Pacific Command, a DSP satellite was relocated to improve coverage of the Scud launch areas. (In addition, the DSP MGT that had been deployed to the Overseas Ground Station at Nurrungar for exercise purposes was kept in place until mid-November.)[8]

In addition to carrying the standard DSP-1 launch and nuclear detection sensors, F-16 carried two scientific instruments, which had also been carried on F-14 and F-15. With the launch of F-16, three primary satellites were capable of returning the desired scientific data. One of the instruments on the satellites is the Synchronous Orbit Particle Analyzer (SOPA), which registers the presence of electrons and electrically charged atoms in space near a DSP satellite, thus permitting determination of when levels of potentially dangerous electrical charges or radiation near the spacecraft are on the rise.[9] A second instrument, the 9.5-pound Magnetospheric Plasma Analyzer (MPA), provides scientists at Los Alamos National Laboratory, who designed the instruments, with an unprecedented method for monitoring changes in Earth's magnetic field. David McComas, a Los Alamos scientist involved in the project, who led the development of the MPA, described it as a "new and unique capability."[10]

The MPA, which will be carried on all DSP flights until the program concludes, produces extremely accurate measurements of the extent of highly heated gases known as plasmas at geostationary orbit. Such plasmas are caused by the interaction between fast-moving particles from the sun and Earth's magnetic field. Flights 14 through 16 could simultaneously produce a highly detailed map of the behavior of plasmas around Earth, from which researchers can deduce the activity of Earth's magnetic field.[11]

Researchers had thought that the plasma would sport a regular teardrop shape, but the measurements by the MPA indicate that the plasmapause— the edge of the plasma surrounding earth—is roiled with waves and has a much more irregular shape than predicted. "It's an awful lot more complicated than the simple picture people have been drawing," McComas told an interviewer.[12]

The refinements in scientific understanding of the behavior of the earth's magnetic field are important because geostationary orbits are used by a wide variety of defense and commercial satellites. Fluctuations in the earth's magentic field, caused by disruptions from the sun, can interfere with or even damage those satellites.[13] However, data from the satellites generally are not provided directly to space weather forecasters, who seek to predict when magnetic disturbances might hit specific satellites. Rather, the data are used to refine the computer algorithms that, in turn, are used to produce such forecasts.[14]

Sometime early in the twenty-first century the first of the DSP successor satellites will be launched. The projected characteristics of those satellites have varied substantially over the last fifteen years. At the beginning of 1983, the projected successor to DSP was to be the Advanced Warning System (AWS). AWS development would focus on further increases in sensor performance, developing onboard data processing that provided warning messages directly to users, and increased survivability. Also to be explored was the ability of an AWS to perform several other tasks that had become subsidiary DSP missions: technical intelligence, tactical theater warning, and aircraft detection and tracking.[15]

However, the requirements for a DSP successor were to change drastically after the evening of March 23, 1983. On that night President Ronald Reagan addressed the American public from the Oval Office. In his televised speech Reagan announced that he intended to "embark on a program to counter the awesome Soviet missile threat with measures that are defensive." He was dissuaded from showing satellite photographs of installations in the Soviet Union to illustrate the Soviet threat, although he did show aerial photos of Soviet installations in Cuba and Nicaragua. Two days later the president signed National Security Decision Directive 85, "Eliminating the Threat from Ballistic Missiles." The directive announced the beginning of "an intensive effort to define a long-term research and development program aimed at an ultimate goal of eliminating the threat posed by nuclear ballistic missiles." The effort, dubbed Star Wars by the media, would be officially known as the Strategic Defense Initiative (SDI).[16]

Less than a month later, on April 18, President Reagan signed National Security Study Directive 6-83, "Study on Eliminating the Threat Posed by Ballistic Missiles." The directive specified that two separate efforts be conducted: a Future Security Strategy Study and a Defense Technology Plan

Study. The former would focus on the roles that defense against ballistic missiles would play in future U.S. and allied strategy, while the latter would focus on producing a long-term research and development program that would provide the basis for selection and deployment of defensive systems if they proved feasible.[17]

With the creation of the SDIO in 1984, headed by Lieutenant General James Abrahamson, all existing service research and development programs of any relevance were swept into the new organization. In the case of AWS, its formal transfer came when the Deputy Secretary of Defense directed that ballistic missile detection and warning requirements of the Strategic Defense System and standard tactical warning/attack assessment functions be satisfied by a single system.[18] The proposed detection system was to be named the Boost Surveillance and Tracking System (BSTS). Whether the spacecraft would operate in geosynchronous or elliptical orbit, employ short-wave or medium-wave infrared frequencies, or have a mosaic or scanning sensor was not initially specified.[19]

A second issue concerning BSTS was its mission requirements. It could be used to simply cue another system to continue surveillance in support of a weapon system that would seek to destroy incoming missiles, or it could cue a combined weapons and surveillance system. The most stringent, yet feasible, requirement for a sensor would be to detect ICBMs to the end of their final boost phase and provide sufficiently accurate information on the missile's trajectory and velocity to allow handoff to post-boost-phase/midcourse interceptors. The latter became the requirement for the new system. While DSP could operate as part of a missile defense system in which it served as "bell ringer," with the more stringent monitoring tasks being performed by lower-altitude surveillance satellites or weapons sensors, it could not satisfy the more stringent requirement that temporarily, and apparently somewhat arbitrarily, became the requirement for the BSTS.[20]

In November 1985 the SDI was still a going concern, but Abrahamson had changed his view concerning BSTS requirements. He observed: "As we've looked at our systems analysis and our architecture studies, we're finding that some of [our] original assumptions . . . in terms of requirements for these systems may have been more stringent than we really do need. . . . Maybe all we need is what's called a 'bell-ringer' type of system at geosynchronous [orbit]."[21]

Easing the requirement for BSTS had brought together the Air Force

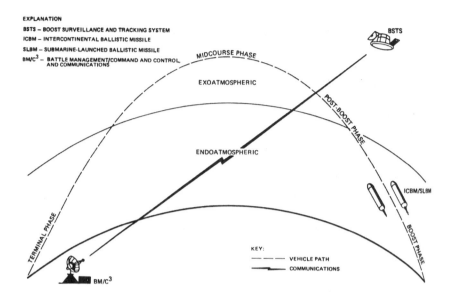

11.1. BSTS operations concept

and SDIO, Abrahamson believed, although subsequent events would demonstrate that the Air Force was not particularly interested in significantly reducing requirements for a follow-on system to DSP. In any case, at the time, questions remained about when his organization would turn over responsibility for the program to the Air Force. It would be several years before that turnover would occur, and in 1987 SDIO was still in charge of the program. The system's functional concept, as indicated in an August 1987 SDIO publication, is depicted in Figure 11.1.[22]

In 1988 two contractors were pursuing the BSTS contract, using different approaches. If one attended a 1988 press briefing by Mel Brashears, the Lockheed Missile and Space Division program manager for BSTS, it would appear that requirements for BSTS had expanded. At the briefing Brashears explained Lockheed's vision for BSTS: it would be a twelve-satellite constellation, including on-orbit spares, with a twenty-year, $9 billion life-cycle cost. Deployment would begin in the mid-1990s, going from initial to final operational capability over a two- to three-year period.[23]

The satellites would be three-axis-stabilized in a highly inclined elliptical orbit with an apogee of 24,000 miles and a perigee of 16,000 miles. They would be launched on Titan IV boosters and would weigh 10,250 pounds—

almost twice the weight of the DSP-1 satellites. The main bus for the satellite would measure 10 by 10 feet, while its solar array wingspan would be 100 feet. The BSTS ground station would be located in the southwestern United States, but the satellites would be able to downlink to mobile systems operating in the same area. It would also be able to communicate by laser cross-link via other BSTS satellites.[24]

The most important job for the BSTS would be to discriminate between different boosters and launch vehicles, using information such as plume intensity and launch profile to categorize targets. According to Brashears, BSTS would identify targets within seconds, and more quickly than DSP, after the boosters cleared certain atmospheric effects.[25]

To perform its main detection and tracking mission BSTS would have a two-color capability, operating in two wave bands. Brashears also envisioned that the BSTS constellation would have no fewer than seventeen auxiliary missions, all classified, but including electro-optical and multispectral imaging. Its use of integrated circuit technology would allow it to scan four times faster than the DSP—once every 2.5 seconds. Seventy-five percent of the data processing was to be for the primary launch detection mission, with the remainder for auxiliary missions.[26]

The competing contractor was Grumman Aerospace, whose approach differed from Lockheed's in two crucial ways. Lockheed's system would rely on the traditional scanning sensor used in the DSP satellites. Grumman had selected a staring sensor, which would remain continually focused on its field of view. Such a sensor required more detector elements than the scanning sensor, with the elements arranged in a large mosaic, which made it potentially more difficult and expensive to produce.[27]

Whereas Grumman was more adventurous with respect to the sensor, it was more conservative in approaching the cross-link requirement. It was planning to employ a more traditional, radio-frequency cross-link system, while Lockheed was planning on using a laser cross-link it if was awarded the BSTS contract.[28]

By 1989 the Air Force had scaled back some of its ambitions for the BSTS. By that time the constellation was to consist of six to ten satellites, and the first of those satellites would not be able to hand off targeting data to antimissile weapons. It was expected that battle management and communications upgrades would be made on later satellites.[29]

BSTS would also have the Survivable Ground Terminal, the primary center for BSTS operations, a 55-foot van, weighing 68,143 pounds. A Crew Support Vehicle (CSV) would be employed for sleeping, eating, personal tasks, and storage for a crew of eight to thirteen. A Multi-Purpose Tanker would carry 4,410 gallons of potable water and 232 cubic feet of ready-to-eat meals, while the 50-foot fuel tanker would carry 6,500 gallons of diesel fuel. Finally, an 8,000-pound, 19-foot-long security/logistics vehicle would complete the contingent.[30]

In May 1990 it was reported that plans for the BSTS would be completely restructured; Defense Department officials intended to simplify the BSTS mission and transfer the program from the SDIO to the Air Force. That report reflected the results of an April 25 BSTS strategy meeting attended by Assistant Secretary of Defense C³I Duane Andrews, Air Force Assistant Secretary for Space Martin Faga, Lieutenant General George L. Monahan Jr., director of the SDIO, and three other senior officials. At that meeting the officials agreed to return BSTS to the Air Force virtually immediately, where it would be the successor to DSP. A classified report submitted to Congress by the Office of the Secretary of Defense outlined the new strategy, which envisioned full-scale development beginning in 1991. The report stated that the system would be focused on the traditional tactical warning and attack assessment mission, partially due to fears that tying it into antimissile targeting would violate the ABM treaty.[31] BSTS also was undercut by another system, Brilliant Pebbles, which was intended to track and attack warheads. Such a system would make BSTS no longer essential, SDIO director Monahan told the House Armed Services Research and Development Subcommittee.[32]

Thus, in the summer of 1990 the House and Senate voted to prohibit full-scale development of BSTS, denying a requested $265 million. The Senate Armed Services Committee recommended that the Pentagon be ordered to consider alternatives to BSTS, such as modernized versions of DSP and the Brilliant Eyes missile detection system, the latter being described as "attractive because it distributes sensors over a large number of spacecraft, which enhance survivability, and increase launch flexibility."[33]

With BSTS in jeopardy, one of the two companies vying for the BSTS contract openly released previously classified details of the schedule for the first BSTS satellites. Briefing materials distributed by Lockheed at a display in a Senate office building on July 24 and 25, 1990, indicated that the construction of the first BSTS satellite would be completed in 1995, with a second the following year. Two satellites per year would then be produced for the following three years, and the ninth in 2000.[34]

Lockheed claimed that BSTS would enable the United States to identify an accidental launch of an enemy missile and to identify attacks from countries other than the Soviet Union and thus "avoid [launching] massive retaliation for [a] small attack."* The satellite would allow U.S. officials to determine which U.S. missile fields were likely to be destroyed by incoming warheads, which, according to Lockheed, would permit "knowledgeable launch decisions"—presumably by enabling the U.S. president to order the launching of targeted missiles before their destruction in their silos.[35]

The Lockheed briefing materials also indicated that the satellite's design had been altered to ensure "added compliance" with the 1972 Anti-Ballistic Missile Treaty. The alterations included the elimination of an antenna for communicating with satellites in low-earth orbit, apparently to prevent any communication between BSTS spacecraft and space-based defensive weapons. Also dropped was strategic defense–related computer hardware and software.[36]

Exactly what the Air Force thought of Lockheed's public disclosure is not clear. However, a spokesman for the Space Systems Division said that official details concerning BSTS deployment were classified and that Lockheed's briefing materials did not "reflect Air Force plans for BSTS."[37]

Lockheed's going public did not prevent Congress, in the 1991 defense authorization bill, from requiring the Air Force to hold a new competition pitting BSTS against a number of other concepts, including one named Brilliant Eyes.[38] Brilliant Eyes was described in detail by the House Committee on Appropriations, which reported that the hypothetical satellites would

> carry a suite of passive sensors including short-, medium-, and long-wavelength infrared and visible light sensors. These sensors acquire and track strategic and longer range tactical ballistic missiles in the boost phase and continue to track and discriminate the reentry vehicles from debris and penetration aids throughout the midcourse flight of the missiles. The satellites are in low earth orbits which afford relatively short ranges to the missiles compared to early warning geosynchronous satellites allowing missile and warhead tracking after their engines stop burning necessary to provide accurate track data for interceptor support.[39]

Congressional reluctance to approve BSTS in the summer of 1990 was followed by an August 31, 1990, memo from Deputy Secretary of Defense Donald J. Atwood, which informed its readers that part, but not all, of the BSTS budget would be transferred to the Air Force, that the program would

---

*Such claims, while true, if it is assumed that a single launch would represent an accidental launch, did not represent any improvement over DSP, which could also detect the launches of a single missile and identify the nation from which the launch emanated.

again become a tactical warning/attack assessment program, and that its previous title—Advanced Warning System—would be restored. On September 5 David Chu, Assistant Secretary of Defense for Program Analysis and Evaluation, officially transferred management responsibility for BSTS to the Air Force and called for the Defense Acquisition Board to review "options for reducing the size and cost of BSTS satellites while preserving essential tactical warning/attack assessment capability" in November.[40]

By the end of the fiscal year the AWS program office at the SSD had worked out a new acquisition plan and schedule for the program. It called for awarding contracts on April 10, 1991, that would provide funding to permit contractors to attempt to demonstrate feasibility. A contract for full-scale development was to be awarded on January 15, 1992. In the meantime, the program office submitted its acquisition plan to the Air Force Systems Command on September 14, 1990, and briefed the acquisition strategy to the Air Force Secretariat's Acquisition Executive, John J. Welch Jr., on September 24. The program office planned to release a request for proposal to potential bidders around the middle of November 1990.[41]

On April 11, 1991, Deputy Defense Secretary Atwood addressed the National Space Symposium, informing his audience that work on a DSP follow-on would begin in 1992 and predicting deployment within a decade of a system that would provide "increased accuracy and more timely warning" to American military forces. "It will also enhance the nation's ability to deal with the increased short-range missile threat of the future," Atwood said, noting the importance of satellites in the Gulf War and technical developments made in SDI missile defense research.[42]

Atwood declined to give details on how many satellites or what types of sensors would be involved, but he said the new system "will replace our Defense Support Program around the turn of the century" and would take advantage of the experience gained in earlier work on SDI missile tracking. He also said that the system was intended to provide more accurate and timely warning of missile attacks to American military forces.[43]

But that system was not to be called the Advanced Warning System because funding for that system was deleted from the Air Force's fiscal year 1992 budget. Nor would it be a version of Brilliant Pebbles or Brilliant Eyes. Duane Andrews explained to the Senate Armed Services Committee in May 1991, "We concluded that the technology that was available in the Brilliant Pebbles–Brilliant Eyes concept was not mature enough for us to commit to a decision on one of our most critical assets—our ability to provide missile warning." He further explained that Brilliant Eyes was intended "as a mid-course sensor system and needs cueing from a boost-phase sensor which would provide impact point prediction and target handoff data to ground-based interceptors, which would aid theater defense." Thus, the new program was the DSP Tactical Warning/Attack Assessment Follow-On. However, it would not survive as such long enough to become "DSP-TWAAFO."[44]

TWELVE

The Unwanted Option

In early 1991 the projected DSP follow-on acquired yet another designation, one more susceptible to being turned into a acronym—the Follow-On Early Warning Satellite (FEWS). FEWS was to be a somewhat scaled-down, less costly version of the AWS, with some capabilities being deferred to the later versions of the system.

While various Air Force and Defense Department officials focused on a completely new launch detection satellite, others in the Defense Department, armed services, and contractor organizations explored ways in which the DSP system could be improved—particularly with regard to detecting Scud-type missiles and providing timely information to military commanders. In addition to certain parts of the Air Force, the Army and Navy focused more on improving DSP than on procuring a new system. The Tactical Surveillance Demonstration (TSD), a project of the Army's Strategic Defense Command, commenced at White Sands Missile Range, New Mexico, in the late 1980s. The TSD, a prototype that directly receives data from DSP satellites and employs stereo processing, is able to process the signals in real time and generate launch point, impact point, and trajectory information to warn of missile launches and cue missiles such as the Patriot.[1]

The Navy's original concept was the Tactical Ground System (TaGS), which envisioned a transportable ground station, similar to the Army's TSD, that would be capable of directly receiving missile launch data and processing it in near real time. The ground stations would travel with naval expeditionary strike forces and joint forces in a theater of operations.[2]

In 1992 the Army and Navy decided to merge some of their efforts, resulting in the signing of a memorandum of agreement between Rear Admiral Roger Ailes, commander of the Navy's Space and Naval Warfare Systems Command, and Lieutenant General Robert Hammond, commander of the Army Strategic Defense Command, in April. The agreement committed the two services to jointly develop a portable ground station, the Joint Tactical Ground Station (JTaGS)—"an operational, testable, militarized, transportable prototype system for surveillance and reporting of tactical IR events to operational commanders."[3]*

Beyond being involved in building a mobile ground terminal the Navy also had initiated two programs, still in development, that sought to enhance the utility of DSP and other overhead systems for war-fighting purposes. RADIANT IVORY focused on developing a shipboard system for combining DSP data with data from other intelligence sources to improve defense against aircraft and missile threats. In August 1992, in the last step of a three-phase test process for the year, a RADIANT IVORY exercise involved the correlation, in near real time, of data from three DSP satellites. The data were then transmitted, via an encrypted commercial communications satellite link, from the European Ground Station to the Army's TSD processor at White Sands and then relayed to the USS *Saratoga*.[4]

Initially, the Air Force was not enthusiastic about modifications to DSP, or about the Army and Navy efforts. Once again, the ability of DSP to provide data beyond its primary mission of warning of strategic attack, which other services considered valuable and wished to exploit more fully, had produced a counterreaction from the Air Force. The same disinterest that had greeted Aerojet representatives when they attempted to convince the Air Force that DSP could be employed in support of missions such as aircraft detection and the same animosity that resulted from Aerojet's attempt to sell the Army and Navy on DSP applications related to their service needs were still present. Aside from any bureaucratic motivations, the Air Force was concerned that DSP's primary warning mission would in some way be compromised by its widespread use for other purposes. And the Air Force was not reluctant to throw up roadblocks to Army and Navy use of the system.

Indeed, a February letter from the Chief of Naval Operations to the head of the U.S. Space Command, accusing the Air Force Space Command of obstructing the Navy from using DSP data in Operation Desert Storm, resulted in a meeting, chaired by Brigadier General Robert Stewart, USSPACECOM director of plans, and hosted by Vice Admiral William A. Dougherty, the deputy CINC. It was a problem that also concerned the Army. Handwrit-

---

*To further enhance the accuracy of the JTaGS it is intended to employ the LLYNX-Eye, a real-time laser boresighting system that reduces the position errors of DSP satellites. More precise estimates of satellite location improves the accuracy of DSP launch and impact point estimation. The system was tested in May 1995, at Roving Sands, but, as of September "require[d] more work to make it a reliable tool." See "Army Expands Combat Support," *Aviation Week & Space Technology*, September 18, 1995, pp. 56–57.

ten comments on a February 22, 1991, USSPACECOM document, "Tactical Exploitation of DSP," refer to three specific cases of AFSPACECOM "noncooperation": in the summer of 1990 and in November and December of that year. The complaints included AFSPACECOM's failure to release necessary crypto keys to the Army in the summer of 1990. The meeting was intended to resolve issues concerning access to DSP data and technical parameters, mutual command support of tactical warning efforts, warning of strategic attack versus warning of theater missile attack, and the use of DSP to support specific service programs.[5]

The meeting opened with Vice Admiral Dougherty noting that, while "our utmost concern is the primary DSP mission. We cannot disturb it," "USCINCSPACE will provide and expand DSP tactical support to the extent that such support does not interfere with its primary (strategic) mission or TW/AA architecture." The admiral also noted that "the components must support each others' Tactical Support efforts," and that the "CINC has approved TSD. Impediments must be removed."[6]

As a result of the meeting it was agreed that AFSPACECOM would immediately release the crypto keys and other material necessary to support the TSD experiment. It was also agreed that the commander in chief of USSPACECOM had to approve all tactical exploitation experiments, demonstrations, or operations, and that such activities would be conducted in compliance with JCS policies. In addition, the Army and Navy representatives approved the concept of USSPACECOM, through AFSPACECOM, exploiting data at in-theater TSD and JTaGs facilities and having ultimate command responsibility for the facilities, although a USSPACECOM memo noted that "we do not believe that that was Navy's original intention."[7]

The same memo, reflecting the Air Force Space Command's aversion to the Army and Navy programs, observed that "AFSPACECOM will likely continue to argue that even if tactical data was processed independently [i.e., by Army and Navy personnel] that such processing should be done at a DSP ground station." It also noted that "AFSPACECOM was extended opportunity to take part in TSD (and other experiments), but representatives did not indicate their intention to do so." The Army TSD effort did spur the Air Force to begin the TALON SHIELD program, with the objective of fielding an intelligence center at Falcon Air Force Base that would combine DSP data with intelligence from overhead imagery and signals intelligence systems. According to Colonel John Hale, chief of the Air Force Space Command's Space Applications Center, "given its level of [computer] processing capability, it can handle a whole lot more than [just] DSP data."[8]

The key differences between the Army/Navy JTaGs and Air Force

TALON SHIELD programs were in the extent of centralization and the sources of data employed. TALON SHIELD envisioned a fixed centralized facility receiving data from all DSP satellites as well as from other sources, which would be geared toward redistributing the data promptly in support of theater commanders. The JTaGs were to receive data from three DSP satellites and would operate as stand-alone terminals that might be relocated anywhere in the world.[9]

A second Air Force program, TALON SWORD, focuses on providing DSP data, in real time, to tactical commanders. As noted in chapter 11, the data produced by DSP satellites during the Persian Gulf War were *relayed* to theater commanders, but even though air sorties would be launched within minutes to attack the Scud launchers, they were often gone by the time the aircraft arrived.[10]

In one sense, while JTaGS and all the RADIANTs and TALONs were intended to produce some very real enhancements in the value of DSP data, they were a sideshow to the Air Force. In 1991 and beyond, the real contest—the one that involved billions of dollars, careers, and reputations— was being fought within and between the Defense Department, the Air Force, contractors, and Congress.

Defense Department, Air Force, and industry officials told Congress and the press of the improvements that FEWS would bring and how the system was essential to satisfying future tactical warning and intelligence requirements, reducing cost, and supporting any theater missile defense system that might be deployed. In an April 15, 1991, report to Congress, Assistant Secretary of Defense for C³I Duane Andrews claimed that dual cross-links on FEWS would allow elimination of the overseas ground stations, which would reduce operating costs and manpower.[11]

But beyond saving money, FEWS would, according to Andrews and others, significantly improve launch detection and reporting capabilities. In May 1991 Andrews told the Senate Armed Services Committee that "FEWS [is] intended to provide worldwide surveillance for tracking missiles in the boost phase. If a space-based midcourse interceptor is deployed, FEWS would provide handoff and state vector data to Brilliant Eyes, which could do midcourse tracking and battle manage any mid-course interceptors."[12] That same month Gordon Soper, the Pentagon's director of strategic and theater nuclear command, control, and communications systems, announced that FEWS would enter full-scale development in 1994, with the launch of the first satellite scheduled for the end of the decade.[13]

Meanwhile, an industry official told the trade publication *Defense News* that FEWS was intended to correct a variety of DSP "deficiencies," including those revealed during Desert Storm. "Although they did a tremendous job, we might not be so lucky next time," the source said, claiming that the success was a function of a set of unique circumstances: the isolation of the missile launchers, the cooling of the desert at night, and the focus on limited areas. Other circumstances could have required more capability in discrimination of the rocket plume or in the processing of information, the source said.[14]

However, the year did not go by without some questions being raised about the extent of FEWS's value. A report of the Defense Department C³I Systems Committee noted that a cost and operational effectiveness analysis study by the Air Force Space Command "had difficulty in demonstrating clearly significant consequences of the performance differences in terms of *decision options available to the NCA and force commanders* in the three scenarios it examined: large nuclear, limited nuclear, and theater conventional."[15] The report also observed that "FEWS' greatest potential benefits appear to be in *theater conflicts*. Whether or not these differences are significant will depend on the threat, our capabilities and tactics, and other theater-specific conditions. FEWS' most important contribution could be *assisting theater missile defense.*"[16]

The committee met on October 21, 1991, to discuss FEWS, three days after the JCS Joint Requirements Oversight Council (JROC) "confirmed the FEWS performance objectives and thresholds will provide the operational capability necessary to satisfy the need." However, with the concerns discussed in its report in mind, it did not immediately pass the FEWS program on to the Defense Acquisition Board (DAB). But, after a second meeting on November 19, it was permitted to reach the DAB.[17]

However, that same month the General Accounting Office released a report concluding that funding for a follow-on system would be premature. Further, it observed "there are indications that an enhanced DSP could be nearly as effective [as FEWS] and would cost billions of dollars less than a fully capable FEWS. Five separate studies provide a basis for these indications."[18]

The following month the DAB, chaired by Under Secretary Donald Yockey, ordered the Air Force to clarify the thresholds for worldwide coverage and the probability of detecting missile launches. It also ordered the Air Force to assess the trade-offs between performance, cost, and risk in attempting to satisfying criteria with respect to timeliness, locating launch areas, and identifying impact areas.[19]

★　★　★

But the ardor of senior Air Force officials for FEWS did not di-minish. Another voice trumpeting the virtues of FEWS was that of Major General Garry Schnelzer, the program executive officer for space programs at Air Force headquarters. In late April 1992 Schnelzer told a meeting of the American Institute of Astronautics and Aeronautics in Washington that "you're going to have to be able to look at much dimmer targets" and that "you're going to have to get quicker updates," about three times faster than with DSP. In addition, the Air Force wanted FEWS to track missiles "right through burnout" so that the system will provide "an accurate azimuth of that threat" and its data fed into the SDIO's Brilliant Eyes and Brilliant Pebbles system.[20]

Echoing Schnelzer was Lieutenant General Thomas S. Moorman Jr., now Air Force Space Command vice commander.* Just a few weeks after Schnelzer's address, Moorman told the Senate Armed Services Committee that "FEWS will have greater sensitivity in picking up missile attacks on our theater forces as well as the United States." He concluded, "It will also have a literally global capability and be able to handle launches anywhere in the world."[21]

The value of FEWS's planned expanded onboard processing capability was explained by Duane Andrews in testimony to the House Appropriations Committee in spring 1992 hearings:

> If everything works perfectly, and you get to detection, the communica-tions delay is [deleted] but everything, because you have people in the loop, everything doesn't often work perfectly and you end up with [de-leted] delay in getting the notification forward. The thing the direct down-link buys you is you don't have to depend on that extra communications link so something else can't go wrong. The commander will have the ter-minal with him. That takes on-board processing to be able to do that.[22]

A key document stating the Air Force's case for FEWS was Schnelzer's *Air Force Space Sensor Study,* issued on April 12, 1993. The FEWS satellite system discussed in the study would consist of five satellites operating in el-liptical orbit and carrying visible short-, medium-, and long-wave infrared

---

*Moorman's "demotion" from Air Force Space Command commander to its vice commander was the result of a 1992 Defense Department decision to again have one individual serve as both commander of the U.S. Space Command and the Air Force Space Command. When General Donald Kutyna assumed the role of U.S. Space Command commander in chief in March 1992, and thus Air Force Space Command commander, Moorman became vice commander.

sensors.[23] Among the advantages of FEWS over DSP, according to the report, was "early detection and longer (and more accurate) tracking of dimmer strategic and tactical missiles." FEWS would also provide coverage of the north polar region, which, the report claimed, was not provided by DSP. In addition, Schnelzer wrote that "sensitivity, update rate, enhanced connectivity (through cross-links, on-board data processing, direct message to user) and affordability are the three key factors differentiating FEWS from the current DSP." Sensitivity and update rate, he then explained, "address our military capability to detect, classify, and respond against fast-burn theater through strategic ballistic missiles."[24]

In addition FEWS's greater short- and medium-wave infrared sensitivity would permit more reliable detection of short-range theater ballistic missiles than DSP and would permit the tracking of missiles until booster burnout. FEWS would also have a sample rate and a dim target tracking capability an "order of magnitude better than DSP," according to Schnelzer. Among the consequences would be "smaller launch and impact point prediction error than DSP."[25]

Schnelzer's paper went on to claim that during the Persian Gulf War DSP could locate a missile launch site only to an area the size of Washington, D.C. In contrast, he claimed, FEWS's ability to track dim upper stages to burnout and to track them more accurately would permit narrowing the location of the launch site to an area the size of Robert F. Kennedy Stadium. Another limitation he attributed to DSP was the inability to "see some tactical threats due to high clouds and/or short burn times," which would make it impossible to reliably locate launch locations.[26]

Once a mobile launcher moves away from the launch site, its location becomes ever more uncertain. Thus, the dramatic reduction in uncertainty about launch location, when combined with FEWS's faster report time and direct downlink capability, would help minimize the search area for a mobile launcher and result in a "much increased probability of launcher destruction," while simultaneously reducing the number of aircraft required to perform the mission, according to Schnelzer.[27]

Schnelzer's report also stated: "FEWS increases the warning time by faster revisit time and by stereo-processing the impact point on-orbit, with warning messages sent directly to all in-theater users. The greater sensor sensitivity, faster revisit time (more track data per missile) and dual processing results in 4-fold increase in heading accuracy over operation DSP performance."[28]

To make the case that FEWS would be superior even to an upgraded DSP, Schnelzer considered an upgrade designated DSP++. The DSP++ described by Schnelzer would have utilized FEWS sensor technology to upgrade the DSP sensor, would have increased average lifetime from 5 to 8.5 years, and would operate satellites in an inclined orbit to improve polar coverage. However, DSP++ would be limited by sensor revisit rate, sensitivity, and resolution, a lack of onboard processing, and the effect of solar impacts. The result, according to the study, would be less accurate impact point prediction, less timely reporting to tactical commanders (because the data would have to be routed from the United States), and an inability to satisfy technical intelligence requirements.[29]

Among Schnelzer's final observations were that "DSP++ provides minimal technical/operational merit. It does not have the ability to detect the new missile threats such as dim targets—targets which will comprise a significant portion of the 2005 threat. DSP++ offers no threat growth capability nor direct and timely messages to theater users." In addition, he argued that DSP++ does not produce data of high enough quality to be used in conjunction with the Brilliant Eyes missile defense sensor. He thus recommended that work should continue on TALON SHIELD for its short-term benefits, that no further work should be done on DSP++, and that the FEWS and Brilliant Eyes programs should be continued.[30]

That same spring the House Appropriations Committee heard supporting testimony from General Donald Hard, director of space and SDI programs in the Office of the Assistant Secretary of the Air Force for Acquisition. Hard told the committee that the DSP system had a number of limitations, the result of its being designed for "the 1970s threat of very large and bright ICBMs," and a result of the limited technology available when the system was designed. Hard stated:

The DSP satellite scans for targets by spinning the sensor. Since the sensor rotates at this fixed spin rate, the satellite can only see targets as often as the satellite spins around again. If the satellite would spin faster it could detect and construct missile tracks faster. However, this faster spin rate would result in less sensitivity, with the possibility of missing some of the targets which it is currently able to detect.

Once missile tracks are developed at the ground stations, warning messages can then be transmitted to National Authorities or the Theater Commanders. This process did not have to be as fast when the only threats were ICBMs targeted against the U.S. from halfway around the earth. For today's threats of shorter range missiles and theater interests, the resulting warning time is inadequate. Lives are at risk based on timely and accurate warning.

The basic sensor on the DSP satellites has limited sensitivity against many of today's targets. The detectors [are] just not capable of seeing the booster burn of some of today's threat missiles. The system was designed to operate primarily in the short-wave infrared frequency band. The medium-wave infrared frequency band offers the ability to track dimmer targets, but is not fully exploited in the DSP design.

The DSP system is susceptible to solar outages due to seasonal variations in the position of the sun relative to the satellites. This is a design limitation of the basic telescope which collects the infrared energy and can be corrected with better telescope design such as those planned for FEWS.[31]

Hard stated that there were other shortfalls that could only be discussed in closed session, which, along with the ones he could discuss openly, combined "to make the DSP system inadequate to meet today's threat and without the potential growth to meet tomorrow's threat of proliferated, dim, short-burning missiles." And, as did Schnelzer, he argued that DSP++ did not "have the growth potential to meet the future threat."[32]

General Hard's remarks concerning DSP's success in the Persian Gulf War were similar to those of General Schnelzer: "We found the Scuds in the Persian Gulf because the conditions were near optimum. We were able to tailor the constellations. The weather helped us. The Scud itself is fairly bright. There are dimmer targets than that to try to pick up." And FEWS, the general told his audience, would allow location of mobile launchers to "within an area about the size of RFK instead of an area the size of Washington, D.C."[33]

If the technical terminology used by Andrews, Hard, Schnelzer, and other FEWS advocates had been simplified, their message would have been that FEWS would see some targets (missiles and aircraft) better and more frequently than DSP, would see others that DSP would miss altogether, would provide more accurate information about the location of detected objects, and would get the data to the theater quicker. Such advanced capabilities, they argued, would be of significant value in a world in which Scud-type missiles and regional conflicts had become of greater concern than Soviet ICBMs and strategic nuclear war.

But while Air Force officials were offering such justifications and singing the praises of FEWS, two analysts at the Aerospace Corporation, a federally funded technical support organization that works for the Air Force, struck a very sour note. The analysts were Guido W. Aru and Carl T. Lunde.

At the time of the report, the thirty-year-old Aru, who had joined Aerospace in 1987, was project leader for the System Architecture and Integration Section of the Space-Based Surveillance Division. Before joining Aerospace he had served as lead engineer for DSP tactical applications at the Aerojet Corporation. In contrast, Lunde had been at Aerospace for over ten years.[34]

On April 23, 1993, just eleven days after Schnelzer's study was completed, Aru and Lunde finished a 500-page report entitled *DSP-II: "Preserving the Air Force's Options."*. The report, which was backed up by hundreds of additional pages of engineering studies, was not just an out-of-channels document produced without authorization or approval. Those signing off on its publication included Everitt V. Bersinger, Aerospace Corporation's principal director, Defense Support Program, and Colonel John R. Kidd, the system program director, Defense Support Program at the Air Force Space and Missile Systems Center, which was responsible for seeing launch detection systems through the research and development and procurement stages.[35]

Aru and Lunde wrote that they had two principal objectives: to examine the role of DSP as a safety net for FEWS in the event of delays in FEWS deployment or program cancellation, and to explore the benefit of potential near-term enhancements to DSP to provide improved interim capabilities prior to FEWS deployment.[36] They then reported on several alternatives to FEWS, all based on a DSP modification designated DSP-II, which they stated would satisfy 85 to 98 percent of the FEWS requirements at savings of from $6 to $8.8 billion. They argued:

> Because of the New World Order which changes the threat from protracted nuclear warfighting, which requires significant expense to achieve survivability (e.g., on-board processing, LASER crosslinks, etc.) to regional conflicts and limited nuclear war potential, changes to the satellite can be made to significantly reduce weight and power requirements. These changes would enable the use of a smaller launch vehicle, which could save approximately $150 million per launch. As a direct result of these investments, the life expectancy or Mean Mission Duration can be increased from 5 years to 8.5 years.[37]

The DSP-II spacecraft they proposed would use the same structure used by the retrofitted Sensor Evolutionary Development satellites (Spacecraft 5 and 6, Flights 12 and 13). That structure is significantly smaller and lighter than the DSP-1 spacecraft structure, which was built to accommodate the Laser Crosslink System (LCS), the power requirement of the Advanced RADEC I nuclear detonation detection sensor, and the mission data message rebroadcast capability. They argued that since "survivability is no longer paramount, the functions of these payloads can be accomplished via alternative means—bent pipes can be used to relay data to CONUS for consolidated processing at significantly less cost than crosslinks: the GPS endo-atmospheric NUDET detection capability can be used in place of the DSP based AR-I capability; and MILSTAR and other communications systems can be used in place of the Mission Data Message (MDM) subsystem."[38]

The DSP-II sensor would be a modification of the DSP-1 sensor. The overall system would consist of four or five geosynchronous satellites with coverage of the polar region provided by continuation of the DSP-AUGMENTATION program. Aru and Lunde asserted that with 78 percent of Earth's surface being water, "a properly positioned four-satellite constellation with [deleted] provides for stereo coverage of all current tactical threat regions and monocular coverage of all possible SLBM areas."[39]

In peacetime, data from all overseas DSP-II satellites would be relayed to the United States by redundant bent pipes, such as the one on Ascension Island, with the Australian and German ground stations being closed. The relay stations would be established at existing U.S. facilities overseas, including nodes of the Air Force Satellite Control Network. The existing station at Buckley would serve as the main DSP ground station, with its backup located at the National Test Facility at Colorado Springs, Colorado. The TALON SHIELD system would be employed to process DSP data for the theater mission. At the same time, the Mobile Ground System would be eliminated and replaced by a Mobile Theater System, consisting of five mobile systems, which would permit fusion of DSP data with other surveillance systems.[40]

Aru and Lunde offered the comparison of four DSP-II-based systems with FEWS shown in Table 12.1, comparing costs and the percentage of the FEWS requirements that would be satisfied by each system.

When the report came to the attention of General Charles Horner, now commander in chief of the U.S. Space Command, on May 20, he called Aerospace president Edward C. Aldridge that same day to complain. In a May 24, 1993, letter Horner asked Aldridge to "help me understand how such a document could be distributed," challenging "my top-priority program." He characterized the study as challenging the JROC requirements for FEWS; as flawed, technically, operationally, and politically; and as "grossly inconsistent with the warfighting requirements, underlying assumptions, and findings" of the *Space Sensor Study*.[41]

In a handwritten letter, also dated May 24, Horner stated: "My fellows are very hot about this—They see it as a short sighted effort of the DSP SPO to sell FEWS out. I, like you, will back any scheme that makes sense but I don't believe we can abandon the war fighters by taking a near term view. If

TABLE 12.1. *DSP-II Versus FEWS Cost Data*

	Cost to Year (in billions)	Requirements Satisifed (%)
5 DSP-II + other	12.4 to 2015	98
4 DSP-II + other	10.7 to 2015	95
5 DSP-II	11.3 to 2015	88
4 DSP-II	9.6 to 2015	85
FEWS	18.4 to 2010	99

I'm wrong educate me—if I'm right in support [of] FEWS help me get this thing back on track."[42]

Aldridge had already acted to recall the report three days prior to Horner's letter on the basis of the general's phone call. He wrote Horner on May 27 that "Aerospace clearly understands the operational needs of FEWS, has supported its development, and has established that the FEWS capability and design is the only concept that satisfies the needs of the operational commanders." He then wrote, without any hint of irony, that "too much program 'advocacy' had crept into the [DSP-II] report."[43]

On the same day that Horner sent his letter to Aldridge, Colonel Joe Bailey, the system program director of Space-Based Early Warning Systems at the Air Force Space and Missile Systems Center, an office formed by combining the DSP and FEWS program offices after the retirement of Colonel Kidd, weighed in with his first critique of the DSP-II study. His "Point Paper on DSP-II TOR" claimed that the DSP-II system did not meet FEWS requirements, that the unsatisfied requirements were "systems drivers" for FEWS, and that the savings realized under DSP-II were the result of satisfying different requirements. The specific requirements that Bailey claimed were unmet by DSP-II included the ability to track the upper stage of a missile until burnout, the required revisit rate, increased performance to support theater missile defense, full onboard processing, full constellation cross-links, nonmissile measurement and signature intelligence ("Mission E" in FEWS terminology), and area of interest capabilities. In addition, he claimed that the study's cost estimate and savings were "optimistic, flawed, and have not undergone the same level of scrutiny as other previous comparisons."[44] Supporters also told journalists that FEWS would allow operations with a much smaller constellation of Brilliant Eyes satellites—about twenty rather than the planned thirty-nine that would be required with DSP-II.[45]

Other claims for the advantages of FEWS were given in a July 13, 1993, briefing. The briefers claimed that "FEWS is 20 times more sensitive than DSP," "sees 99% of SCUD-class missiles" compared with the 50% to 80% seen by TALON SHIELD, and "sees most SLOW WALKERS," while "DSP/TALON SHIELD would see a few." The briefers also claimed that FEWS would be ten times more accurate than DSP and make "launch location targetable." In addition, FEWS would enhance counter-air operations by providing wide-area surveillance and would cue Airborne Warning and Control System and Combat Air Patrol to the presence of aircraft flying on afterburner.[46]

According to the briefers, staying with DSP, an option referred to as "DSP Forever," would cost only $1.8 billion less than a FEWS baseline program and $1.9 billion *more* than a low-cost FEWS program. The DSP Forever option, the briefers argued, would purchase DSP spacecraft with obsolescence upgrades, require indefinite continuation of the AUGMENTATION program, and require Titan IV launch vehicles. A FEWS baseline would also buy DSP Spacecraft 23 through 25 and involve the use of a Titan IV launch vehicle, but it would require continuation of the AUGMENTATION program only until 2005. The FEWS low-cost program would cancel DSP 23 through 25, restructure DSP-AUGMENTATION and use the money to offset the FEWS cost, and employ a lighter medium-launch vehicle. The cancellation of DSP 23 through 25 would not result in a coverage gap, according to the briefers, assuming a five- to seven-year average lifetime for the remaining DSP satellites.[47]

As for the DSP++ program, the briefers stated that it would cost about the same as FEWS, would be less sensitive and less accurate, did not meet the JROC requirements, and would require thirty-nine Brilliant Eyes satellites to achieve the same capability as twenty to twenty-four Brilliant Eyes associated with a FEWS program.[48]

While defending its plan to pursue FEWS rather than an upgraded DSP, the Air Force was also concerned with reducing the overall cost of a FEWS program to make it a more acceptable option for a Defense Department faced with budget cuts. The Defense Department was also concerned. In late June it ordered a review of tactical warning/attack assessment options, to be completed by mid-July. Options to be considered included a hybrid system of high-altitude and low-altitude warning satellites, including use of the DSP-II system proposed by Aru and Lunde.[49]

What impact, if any, the study had, is not clear. But in August 1993 it was reported that the Air Force was halving the size, weight (to about 5,000 pounds), and cost of the FEWS program. In addition to the direct cost savings of a smaller satellite, the Air Force would also be able to save billions of dollars by using medium-launch vehicles instead of the Titan IV.[50] The changes would cut into the capabilities of FEWS as well as its cost. At the time an Air Force official said that additional weight and cost savings would come from "eliminating [about six] ancilliary payloads that are covered by other [surveillance] systems."[51]

Measures planned to reduce weight and costs included using an inclined circular orbit rather an elliptical orbit, allowing the satellite to operate with smaller fuel tanks, eliminating the nuclear explosion detection payload weighing over 700 pounds, and reducing the size of a processor. While removal of the nuclear detection function on FEWS would still leave a superior capability on the Global Positioning System navigation satellites, the other two changes would decrease launch detection capability. The use of a circular rather than an inclined orbit would "open up coverage gaps [in the polar region] that are exploitable," according to an Air Force official, while the smaller processor would decrease the number of threats the system could simultaneously detect, track, and report.[52]

Despite the reductions, Air Force officials continued to extol the virtues of FEWS. According to General Horner, FEWS would "help work the cruise missile problem" and play a bigger role in supporting operations against manned aircraft. During preparation for bombing strikes and air-to-air attacks, aircraft would rely on extra speed, generated by their afterburners, to protect themselves from interception, but FEWS would be able to track the aircraft, Horner stated. If in order to avoid FEWS detection the aircraft go much slower, they would become victims of other systems.[53]

While the Defense Department and Air Force pushed ahead with FEWS, they decided to convert Brilliant Eyes into a research and development effort aimed at fulfilling future requirements for midcourse interception of ballistic missiles. The decision was motivated both by a concern over ABM Treaty compliance issues associated with any missile defense of which Brilliant Eyes would be a part and by budgetary pressure from Congress.[54]

Not everyone in the Defense Department and Air Force was convinced that FEWS was unequivocally the best option. A June 8 memorandum from William J. Lynn, director of program analysis and evaluation for the Office of the Secretary of Defense, to the Under Secretary of Defense (Acquisition and Technology) urged consideration of DSP-II/Brilliant Eyes by the department's Bottom-Up Review Steering Group. The memo noted that "although DSP-II would have some performance shortfalls relative to FEWS, it might be adequate to satisfy our essential post–Cold War global and theater needs if it were used in conjunction with Brilliant Eyes."[55]

There were also dissenters within the early warning program. One of the most prominent was Colonel Edward Dietz, Colonel Kidd's successor as head of the DSP program at the Air Force Space and Missile Systems Center. In a July 20 Defense Program Office interoffice memo, better known as a "Dietzgram," he challenged much of the analysis supporting FEWS over DSP. In his memo, addressed to Colonel Bailey, Dietz objected that "DSP performance assessments are pejorative, while FEWS is presumed to perform better than specifications"; that consideration of DSP-II is omitted, that costs of the AUGMENTATION program are added to all DSP options but not the FEWS option; that "it has been widely, and incorrectly, stated that FEWS is cheaper than DSP"; and that "the types of tactical targets that DSP can not see is widely and incorrectly quoted, as are the conditions under which DSP performance is degraded." Dietz also objected that "DSP mean mission duration [i.e., average lifetime] is generally overstated"—which would lead to an overly optimistic projection of how long it would be before the final DSP satellites lost their capability.

That complaint was followed by a September 7, 1993, "I am Concerned" memo from Dietz to Bailey, in which he wrote that "General Horner, his staff, and AFSPACECOM are clearly misinformed re: FEWS' and DSP's attributes. The errors are favorable to FEWS, giving the appearance of bias." He continued, observing that "General Horner erroneously believes FEWS is cheaper, ten times more accurate, capable of detecting cruise missiles and SS-21's etc., and DSP current life expectancy is 7–8 years." The responsibility for the misinformation, Dietz wrote, was that of the Space and Missile Systems Center, and, by implication, Bailey, since the center's "documentation is the orignal source of the errors and subsequent [center documentation] . . . reinforces the errors."[56]

Also questioning the analyses in support of FEWS was Major Roger L. Hall, chief of architecture and integration at the Space and Missile Systems Center. In an October 12 memo he wrote that

SPACECOM stated—advertised—publicized FEWS performance is becoming more and more overstated and incredulous (i.e. "Washington D.C./football stadium charts," statements made about FEWS cueing based upon single hits, and General Horner's statements related to SS-

21s, clouds, and low-altitude [cruise missiles], etc.) Some of the more recent claims are probably beyond the capability and capacity of any space-based asset and may damage SPACECOM's credibility.[57]

Hall's memo came during the same month that the Space-Based Infrared (SBIR) Sensors Technical Support Group, chaired by Robert Everett of the MITRE Corporation, produced the report that had been commissioned by Under Secretary John Deutch in early July. Also serving on the group were scientists from the Institute of Defense Analyses, Aerospace, MITRE, Johns Hopkins University Applied Physics Laboratory, and MIT Lincoln Laboratory.

Deutch had instructed the scientists to review and recommend options for future U.S. space-based infrared surveillance systems for the years 1993 to 2015, focusing on DSP, FEWS, Brilliant Eyes, or other appropriate acquisition options. They were to consider how the systems fulfilled requirements for tactical warning/attack assessment, theater missile defense, and global awareness and "identify cost-effective options for consideration by DoD executives."[58]

In the course of their study, the group was briefed on requirements by the U.S. Space Command, Air Force Space Command, Ballistic Missile Defense Office, Army, Navy, DIA Central Measurement and Signature Intelligence (MASINT) Office, JCS, and Office of the Secretary of Defense. Hughes, Grumman, Aerojet, Lockheed, TRW, Raytheon, and ITT provided briefings related to DSP upgrades, FEWS, Brilliant Eyes, ground-based radar capability, and theater missile defense. The DIA provided a threat briefing, while briefings on programs and initiatives already under way, such as JTaGs, RADIANT IVORY, and TALON SHIELD, were provided by a variety of Air Force and Navy organizations.[59]

The report followed eighteen meetings of the group, stretching from August 17 and October 7, during which time representatives from the Air Force Space Command and JCS received a "roasting." Two individuals who did not testify before the panel were Aru and Lunde. The panel had wanted to hear them, but the Air Force had vetoed the group's request.[60]

In both its overall and specific findings the group questioned the ability of DSP or modifications such as DSP-II to meet future requirements, as well as the need for FEWS. It noted that stereo DSP would provide an adequate near-term capability for missiles with ranges greater than 186 miles, but that a better system would be needed to permit launches on cheaper medium-launch vehicles rather than the expensive Titan IVs, to provide growth potential to guard against future missile developments, and to support other missions.[61]

However, the panel did not believe there was any urgency in replacing DSP, and it recommended that the original plan to purchase DSP Satellites 24 and 25 was a sound one. It did feel that since even a DSP-II system would require several hundred million dollars for research, development, testing, and evaluation, it was possible to use an equal amount of money to produce a better system.[62]

Such a system would improve the capability to detect short-burning missiles with ranges of 186 miles or above and to estimate launch and impact points by increasing the number of times a missile in flight would be detected—of particular concern to the panel in light of the short burn time of many theater missiles. Included in this group are the Scud-B, with its sixty-second burn time, as well as the Chinese M-9 and M-11, which have ranges of 372 and 186 miles, respectively, and burn times between sixty-two and sixty-eight seconds. Had it been built, the Condor II/Vector, with a 620-mile range, also would have fallen within this group. The ability to detect such missiles could be further limited by cloud cover. A cloud deck at six miles is not uncommon. It would take forty seconds for a Scud-B to break through, leaving time for only two DSP detections. If the cloud deck was located about four miles above earth, Scud-B burnout would take place thirty seconds after piercing cloud cover.[63]

FEWS advocates, in meetings with the panel, pointed to such missiles as a reason for replacing DSP.* But while the Everett group saw a need for a new system, FEWS was not it. The group concluded that the FEWS "design and cost is intertwined with requirements descended from SDI and Nuclear Warfighting," that FEWS "provides marginal advantage over stereo DSP for counterforce options," and that while FEWS would provide significantly better impact point prediction than stereo DSP, radars would provide more accurate impact point predictions for missiles with a range under 620 miles. Onboard processing, the group concluded, was a requirement "derived from nuclear warfighting," while such processing and cross-links "are not needed initially—if ever." Further, they judged that direct downlinks to users were "not worth [the] technical risk and cost." In addition, the group supported

---

*Also used as a justification by FEWS advocates was a Russian rocket with a 250-mile range, the Sphera. However, the rocket would burn out only a second or two after breaking through the clouds, making detection virtually impossible for any system, including FEWS. (Interview.)

the view that without Satellites 23 through 25 there would be danger of a coverage gap during a transition to FEWS.[64]

In addition to evaluating alternatives proposed by others, the group described what a system that satisfied its view of the necessary requirements would look like. The main body would be 6.5 feet by 7.0 feet by 7.7 feet, and the satellite would have an overall length of 88 feet. The spacecraft would carry a medium-wave infrared staring sensor as well as a short-wave infrared sensor. The medium-wave infrared sensor would cover 620-by-620-mile areas, with a 0.5-second revisit time and a 1.2-mile footprint, and would have high sensitivity. The short-wave infrared scanner would have a 3.0-second revisit time, a 1.2-mile footprint, and moderate sensitivity. The launch vehicle would be an Atlas-I/Centaur.[65]

The following month, a study by the Institute for Defense Analyses (IDA) also questioned the wisdom of the FEWS system, while being more favorable to DSP-II. DSP-II, concluded IDA, is a "technically sound, low risk concept" that "represents what a good program manager would come up with for fixing known problems and reducing life cycle costs if follow-on programs were not in the pipeline." The study also concluded that "DSP with Talon Shield/TSD is a very capable theater sensor."[66] IDA agreed with the authors of the DSP-II report that "current requirements are not justified" and were "developed when policy was nuclear warfighting." Further, it concluded that the "requirements [are] difficult to justify even under this policy."[67]

But before the end of the year, FEWS, or at least the name, was dead. Following the SBIR Technical Support Group panel report, the Air Force and Defense Department changed course, or at least acronyms. FEWS became, for a brief time, the Space-Based Early Warning System (SBEWS). What General Horner envisioned was a new competition for a less expensive follow-on to DSP, the half-size version of FEWS. Such a version, Horner said, would "cost less than buying an obsolete DSP and provide the accuracy and sensitivity we need."[68]

The Air Force's apparent decision to pursue a less costly, less ambitious system did not end the controversy that had resulted from the debate that began, or at least became public, with the Aru and Lunde DSP-II report. The controversy continued at a public hearing held by the House Committee on Government Operations, chaired by Congressman John J. Conyers, on February 2, 1994. Among those testifying at the hearing on "Strategic Satellite Systems in a Post–Cold War Environment" were Aru; Colonel Edward Dietz; Colonel Sanford Mangold; Thomas Quinn, deputy assistant secretary of defense for C^3I acquisition; Major General David Kelley of the Joint Staff, and Colonel Steven Sadler, deputy chief of staff for requirements of the Air Force Space Command.

In addition to dissenting from the Defense Department/Air Force position on FEWS, Aru, Dietz, and Mangold told of what they believed were reprisals in response to their work concerning a DSP successor. According to Mangold, after serving for a year as resource allocation team chief of the Air Force's Space Command, Control, Communications, Intelligence and Nuclear Deterrence Team, he was "extended for an unprecedented second year as team chief on June 1, 1993." However, three days later, allegations about his personal integrity resulted in his being summarily relieved of his position. He reported to the committee that he had been warned that his questioning of the need for FEWS and the MILSTAR communications satellite had made him "the most hated man in Space Command" and that he "was going to be taken out in some way, any way possible." A 210-day investigation followed his being relieved of duty, which established no guilt but resulted in an administrative reprimand, Mangold told the committee.[69]

Dietz did not deal with the question of harassment or retribution in his short statement to the committee. But in response to questions he answered that he "was fired and moved over to another job because I did my job." Aru told the committee that his coauthor received a negative performance review after ten years of positive reviews.[70]

In a subsequent written response to a question submitted after the hearing, Dietz pointed to six areas where he believed General Horner had given misleading statements concerning DSP or FEWS or both. To the claim that "FEWS makes launch location more targetable," he wrote, "Not true! Unless we use nuclear weapons" and noted the Everett committee's conclusion that FEWS was only marginally more useful than DSP for counterforce. As to the claim that "FEWS is 10X more accurate," he wrote, "Wrong! Current Army and USAF experience with DSP/Talon Shield has shown performance almost equal to FEWS specification." "Not true!" was his response to the claim that FEWS is cheaper than DSP, unless "you ignored the possibility of improving the current system or if you allowed all the specious cost additions to the DSP side of the cost account." "Sorely misleading" was his characterization of the claim that "FEWS detects with high cloud cover"; he then elaborated that infrared detection through clouds is not possible. As for FEWS detection of cruise missiles, Dietz characterized this as "Unlikely. Except for 1970 style weapons, under all but the most fortunate circumstances." As for the DSP life expectancy assumed by Horner, this was, according to Dietz, "Sorely overstated!"[71]

Aru dealt at length with many of the issues and claims in the DSP versus FEWS debate. He reported that at the end of May 1993, when the Office of the Secretary of Defense had instructed Major General Schnelzer to examine another DSP alternative beside DSP++, he ignored DSP-II in favor of "DSP-Forever," which had higher costs but poorer performance than DSP++. In regard to the assertion that FEWS could locate a launch site to within an area the size of RFK Stadium (in contrast to a DSP accuracy equivalent to the size of Washington), Aru stated that DSP performance during Desert Storm was significantly better than that, and that the Army's Tactical Surveillance Demonstration (TSD) program would provide missile launch point estimation comparable to FEWS.[72]

He also raised a point mentioned by other FEWS critics concerning such claims as FEWS ability to detect objects under cloud cover. He quoted George Paulikas, an Aerospace executive vice president, as telling him that some of the FEWS requirements "violate the laws of physics and thermodynamics simultaneously." In a similar vein, he noted that some targets were simply unobservable by any space-based system and that "some of the targets FEWS was required to detect would have been like trying to see a match in front of a floodlight—it cannot be done."[73]

Aru also charged that an internal Aerospace report did not support Aerospace president Edward Aldridge's statement, in a June 22, 1993, letter to General Horner, that "FEWS is the only system that will give us confidence in providing launch warning and tactical missile defense tip-off." He charged that every internal semiannual technical report to the Board of Trustees, had stated:

> FEWS designs have been driven by strategic requirements and the strategic concept of operations . . . Concerns have been raised by some users (e.g., the Navy and Army) that FEWS may not be configured to fully support their future needs. . . . The military war-fighting added value of enhanced surveillance information has been somewhat difficult to quantify, as clear metrics have not been delineated. An understanding of how end-users will and can take advantage of accurate and timely surveillance data must be established, so that tradeoffs of military utility can be performed.[74]

In addition, Aru accused Colonel Jeffrey Quirk, who had been the assistant director of engineering for the Space-Based Early Warning System program office, of including in DSP cost estimates, but not FEWS estimates, the cost of any augmentation or adjunct system to provide polar coverage. Further, he said that Quirk kept raising the cost of the system—from $1.1 billion to $1.7 billion to $2.1 billion to $3.3 billion to $4.3 billion. Further, Aru noted that the Everett group had concluded that the system was unnecessary.[75]

Aru also challenged the claim of General Horner that FEWS was required to detect extremely short-range tactical ballistic missiles such as the SS-21, which has a range of approximately 100 miles. Aru pointed to the findings of several review groups, including the Everett and IDA groups, that warning of missiles with ranges less than 186 miles should be handled by theater systems. He noted that "space-based infrared warning would not be timely enough to provide any significant utility due to the short flight time of these missiles."[76]

In an answer to a question for the record, Aru also charged that many of the requirements represented by the Air Force as driving FEWS and disqualifying DSP-II had not been validated by the JROC but had simply been part of an unapproved Air Force Space Command requirements document. He charged that unvalidated requirements included additional onboard satellite processing capabilities, satellite-to-satellite cross-links, full satisfaction of all threat detection requirements, and worldwide collection and distribution of wideband data back to the CONUS Ground Station.[77]

In his testimony, General Kelley referred to the conditions under which DSP operated during Desert Storm as ideal, observing that "we can't always be assured of such ideal conditions in future warfighting environments." Similarly, in his testimony, Colonel Sadler observed, "With DSP we got lucky in the desert. Certain very positive environmental conditions and things like that enabled us to use the DSP system against the Scuds."[78]

Challenging this assertion was a posthearing submission from Aru (a memo-gram from Jim Creswell, Aerospace Corporation's Talon Shield principal director) that quoted Air Force Secretary Donald Rice to the effect that "it turned out during Desert Storm to be the worst weather in the fourteen or fifteen years of recorded weather history in that part of the world."[79]*

----

*The weather over Iraq during the Persian Gulf War was a subject addressed by Richard P. Hallion, presently the Air Force historian, in his *Storm over Iraq: Air Power and the Gulf War* (Washington, D.C.: Smithsonian Institution, 1996). On pp. 176–77 Hallion wrote that: "the weather over Iraq was the worst in fourteen years, and twice as bad as the climatological history for the region would have suggested. The conditions, in fact, approximated a rainy European summer, and not the kind of blue-skies conditions one normally associates with desert warfare—Cloud cover exceeded 25 percent at 10,000 feet over central Iraq on 31 days of the 43-day war; it exceeded 50 percent on 21 of those days, and 75 percent on 9 days." In the Defense Department study of the Persian Gulf War, it is noted that "The worst weather in at least 14 years

Creswell's memo also noted that "apart from weather, viewing conditions that contribute to the complexity of a detection include background, geometery . . . , aspect angle (the direction of the missile's flight relative to a satellite's line of sight), and the complexity of the situation presented to the operator in the loop to support him in his real/false discrimination task." Complicating factors in Desert Storm, Creswell continued, included radiance from ordnance and target explosions when the absence of cloud cover and water vapor allowed early missile detection, which was particularly heavy during night hours.[80]

The hearings, which focused on both the DSP and the MILSTAR communications satellite systems, represented an exceptional airing of the internal battles over the two systems. Those battles soon became known to more than the viewers of C-Span and readers of a few major newspapers. The charges of potentially wasted billions and Air Force attempts to surpress dissenting views were too much for CBS's *60 Minutes* to pass up. On May 22, 1994 several million households saw the segment entitled "A Billion Here, a Billion There . . . ," featuring interviews with Mangold, Dietz, Aru, and General Horner.[81]

Neither the bad publicity from such a review of the recent past nor the decision to terminate FEWS prevented the Air Force from continuing its search for a DSP follow-on system. But coming up with a viable alternative would take time. The Air Force would face skeptical reactions from Congress and would experience yet another aborted program before hitting upon an acceptable solution.

(the time the USAF has kept records of Iraqi weather patterns) was a factor in all phases of the war. . . . Cloud ceilings of 5,000 to 7,000 feet were common." See Department of Defense, *Conduct of the Persian Gulf War: Final Report* (Washington, D.C.: U.S. Government Printing Office, 1992), p. 227.

THIRTEEN

High Now, Low Later

In 1994, while senior policy makers were focusing on the next-generation system, operators were consumed with employing the DSP constellation at hand to monitor missile and space launches and detect aircraft and space objects. As of March 1994 the constellation consisted of five satellites—the four primary satellites and one Limited Reserve Satellite—some of which also had limited capability. The oldest, DSP-12, which occupied the European slot at 5E, had suffered power supply failures in 1992. In a July 6, 1992, message Charles Horner, the U.S Space Command commander in chief, informed Colin Powell, chairman of the JCS, that he recommended a launch of DSP Flight 17 no later than July 1993. Inasmuch as there were three competing payloads for a July 1993 launch, Horner requested Powell's support "to ensure Flight 17 is launched as soon as practical with a goal of NLT [no later than] July 93." By March 1994 the target date had long passed.[1]

F-13, stationed at 105E, was capable of tracking both strategic and tactical ballistic missiles but had experienced a failure in the onboard electronics assembly for its station-keeping thrusters. As a result the spacecraft required constant control by either the CONUS Ground Station or the Nurrungar facility.[2]

F-14 and F-15, the Pacific (165W) and Atlantic (35W) satellites, were also experiencing problems: the thermal control systems for both of the satellites had failed. As a result, according to Air Force Space Command Vice Commander Moorman, "We cannot see certain classes of tactical missiles that don't burn bright or long."[3]

The newest satellite, F-16, covering the Eastern Hemisphere from its position at 65E, was in the best shape and capable of detecting both strategic and tactical ballistic missiles. At the time the satellite that it was hoped would become F-17 was awaiting a fall launch, while DSP 18 through 22 were in storage. Work on what was intended to be the last DSP—DSP 23—had begun.[4]

Meanwhile, the Air Force was expecting that the TALON SHIELD program would become operational, with the designation ALERT (an acronym for Attack and Launch Early Reporting to Theater), on October 1. An Aerospace Corporation briefing on TALON SHIELD/ALERT summarized its genesis, program objectives, and characteristics. The briefing noted that requirements for enhanced theater missile defense early warning included faster reporting, increased robustness, increased user confidence, improved launch parameters, and better impact point prediction.[5]

The centerpiece of the system would be the Central Tactical Processing Element (CTPE) at the National Test Facility, Falcon Air Force Base, Colorado. The CTPE would employ the full DSP constellation, having already demonstrated the ability to process data from five satellites and to detect and report tactical ballistic missile launches, SLOW WALKER events, and "Special Events"—ignoring ICBM launches, nuclear detonations, and space objects.[6] The data would arrive in the theater via the Theater Intelligence Broadcast System (TIBS) and Tactical Related Applications (TRAP) networks. The messages would provide event type information, launch data, burnout data, an impact estimate, and information on the number of viewing sources.[7]

The state of the DSP constellation in the spring of 1994 was of particular concern because of developments on the Korean peninsula. Reports that North Korea had crossed or was on the verge of crossing the nuclear threshold raised the prospect of a crisis that could escalate into war. At the time, in addition to the Taepo-Dong 1 and 2 missiles that were in the development stage, the North Koreans had an operational arsenal that included Scud-B missiles with a range of 186 miles and Nodong-1 missiles with a range of 620 miles.[8]

Concern that the situation might deteriorate to the point that North Korea would employ its missiles led to a decision to send Patriot missiles to South Korea along with the Army's prototype TSD ground station, which had been located in Germany for the past year. The transfer of the TSD prototype would permit U.S. military commanders in South Korea to receive data from two DSP satellites, allowing for stereo processing and more accurate estimates of launch points. The six Patriot batteries arrived in Pusan, South Korea, on April 17 and 18, to be deployed to major ports and airfields vulnerable to North Korean missile attacks.[9]

While senior officials were trying to avoid a crisis on the Korean peninsula, Air Force officials were trying to persuade various congressional committees that they had hit upon a reasonable successor to DSP—one that would satisfy the necessary requirements at a reasonable price. In a Febru-

ary 28 letter the Under Secretary of Defense for Acquisition officially informed Congress that the Defense Department planned to cancel FEWS and
DSP and start a new program. The new program came with a new acronym:
ALARM, for Alert, Locate, and Report Missiles. Plans called for ALARM
deployments to begin in 2004.[10]

One of the first briefings had occurred on February 25, when Air Force
Brigadier General Sebastian Coglitore, director of Air Force space acquisition
programs, and Darleen Druyun, deputy assistant secretary of the Air Force
for acquisition, met with congressional staffers. At least one staffer was not
impressed, telling *Space News*, "I thought it was an unmitigated disaster. . . .
There were very few details to be had." According to the staffer, the Air Force
representatives did not provide an estimate of ALARM's cost, other than to
say that it would be less than the projected cost of FEWS.[11]

According to Air Force spokesman Major Dave Thurston, the absence of
details was due to the fact that ALARM was in the conceptual stage. However, he said that "we all know that it's going to be lighter, smaller, and
cheaper than FEWS," partly because it would be launched on a medium-
sized rocket instead of a heavier Titan IV.[12]

Over the next couple of months, Defense Department and Air Force officials appeared before the relevant House and Senate committees—appropriations, armed services, and possibly intelligence—to explain their 1995 fiscal
budget requests and their vision for the future of space-based early warning. One part of such briefings was a comparison between the ability of DSP,
FEWS, and ALARM to meet stated requirements, as presented in Table
13.1. Given the recent history of program cancellations and changes from
AWS to BSTS to AWS to FEWS to ALARM, some congressmen were skeptical. Congressman Norman Dicks (D-Wash.) asked Lieutenant General
Moorman, if "we are ever going to make a decision and stay with it?" He went
on to say, "I think it is embarrassing that we get change after change on these
issues and I am not yet convinced that you know what you are doing."[13]

Answers to questions, oral and written, from members of the House of
Representatives indicated that DSP-23 would be the last DSP purchased.
DSP-24 and DSP-25 were canceled, against the Everett's panel's recommendation, based on a model that indicated at least a 90 percent probability
that the first ALARM satellite would be available before DSP 24 would be
needed. Thus, DSP-23 would be launched in 2003, with the two preceding
launches taking place in 2000 and 2001.[14]

As part of the sales pitch, several Defense Department officials claimed
that a number of problems existed with present DSP satellites that limited
their utility for spotting tactical missiles—problems whose specifics and im-

TABLE 13.1. *ALARM Requirements Comparison*

	DSP	ALARM	FEWS
Strategic detection	Yes	Yes	Yes
Theater detection capability	Does not meet	Limited to two MRC coverage	Worldwide
Number of areas of interest	Does not meet	2 worldwide	175 worldwide
Onboard data processing	No	No	Yes
Satellite cross-links	No	No	Yes
Messages direct to users	No	No	Yes
Nuclear detection system	Yes	No	No
Onboard attack detection	No	TBD	Yes
Full nuclear hardening	Yes	No	No
Requires Titan launch vehicle	Yes	No	Yes
MLV-Compatible	No	Yes	Yes
Separate ground segment	N/A	No	Yes
Mobile ground segment	Yes	Yes	Yes

plications would normally be classified. Moorman told a hearing of the House Appropriations Committee that overheating of the satellite focal planes was causing detectability problems, so that DSP "cannot see certain classes of tactical missiles that don't burn as brightly or as long."[15]

A written response from Thomas Quinn, Deputy Assistant Secretary of Defense for C³I acquisition, stated that DSP would be unable to provide warning to U.S. troops in South Korea if the North were to launch a Scud at them (which raises the question of why the TSD was being deployed to Korea if that were so). Quinn offered several reasons why DSP would not provide adequate warning, including the predominantly cloudy weather in Korea and the limited time during which the missiles would emit infrared radiation.[16]

Testimony and written responses from several witnesses also dealt with differences between ALARM, DSP, and FEWS. The written statement of Emmett Paige Jr., Assistant Secretary of Defense for C³I to the House Appropriations Defense Subcommittee, stated that ALARM "would have less onboard data processing than FEWS, but better detection performance than DSP . . . especially against tactical ballistic missiles."[17]

The absence of additional onboard data processing was only one of several capabilities that ALARM would not have, at least initially, that were to have been "essential" parts of FEWS. Also gone from plans for the initial ALARM satellites were cross-links and direct-to-the-theater event message capability, although Moorman stated that they would be introduced on the fifth and subsequent ALARM satellites. In response, Congressman Bill Young (R-Fla.) observed, "One of the major things we learned in Desert Storm about this type of activity was that on-board processing and better communication to the field was essential and we were planning to improve that. Now with ALARM, you are telling us that we are going to take those capabilities off."[18] Later in the hearing Moorman explained, "We are dealing with that by making warning improvements to the ground processing system to get warning more quickly to the field. We are talking about the difference in terms of less than two minutes."[19]

Another difference between the FEWS and ALARM plans was that a new ground system had been part of the FEWS package, while ALARM was to rely on an updated DSP ground system. Among the most dramatic differences was that FEWS was required to have the capability to simultaneously detect Scud-class missile launches over almost 80 percent of the earth, which was equivalent to requiring FEWS to handle 175 areas of interest simultaneously within Major Regional Contingency areas.[20]

As for cost, Quinn, in a written response, estimated that through 2006, and including the cost of upgrading the DSP ground segment to support transition, the cost of ALARM would come to $7 billion. That would be considerably less, indeed almost half, of the minimum cost ($12 billion) that had been estimated for FEWS.[21]

But the image of ALARM as a cost-saving alternative to FEWS was undermined when Moorman told the House Armed Services military acquisition subcommittee on April 14 that, in the long run, ALARM would not save American taxpayers any money. He revealed that while ALARM would cost substantially less than FEWS over the next five years, approximately until the end of the decade, each system would cost about $11 billion to develop through 2015.[22]

A little over two months after that revelation, several Pentagon officials, including Moorman and James Carlson, acting deputy director of the Ballistic Missile Defense Organization, attended a meeting called by Keith Hall, the Deputy Assistant Secretary of Defense for Intelligence. The purpose of the meeting was to launch yet another study of space-based early warning. The agenda for the Office of the Secretary of Defense Space-Based Warning Summer Study was wide open, Hall told a conference sponsored by the National Space Club. Officials would consider whether to accelerate ALARM, build more DSP satellites, or turn to another system. Hall noted, "I guess in one respect it'll determine whether or not the ALARM system is the way to go or should we go with a different solution set like a Brilliant Eyes approach."[23]

While members of the executive branch debated among themselves, the various congressional committees sought to settle their differences. In August the House and Senate armed services committees agreed to increase ALARM funding from the Pentagon's requested $150 million for fiscal year 1995 to $169 million. At the same time they agreed to terminate DSP with DSP-23, canceled a planned demonstration-validation effort for ALARM that would have piggybacked a test sensor on DSP-19 or some other satellite, and mandated that the first launch be brought forward to 2002.[24]

Entering a September 1994 conference between the House and Senate appropriations committees, each committee held radically different positions. The House appropriators had recommended raising 1995 fiscal year funding to $330 million and advancing the first launch to 2000. The Senate committee wanted to authorize $225.5 million for space-based early warning efforts but called for a 1997 fly-off between the ALARM test sensor and Brilliant Eyes.[25]

The wisdom of some congressional proposals was questioned by Aerospace president Edward C. Aldridge. The fly-off would be, according to Aldridge, "a waste of money. . . . You could do a fly-off using detailed simulations on a computer." As for accelerating ALARM, Aldridge agreed with Air Force and industry officials that attempts to accelerate the first launch to 2000 would risk cost overruns and schedule slips, saying, "If you accelerate this kind of program you can really screw it up."[26]

By late September the defense appropriators had reached agreement. Part of that agreement resulted from a report that had recently been completed by the Defense Resources Board, which suggested adapting the HERITAGE sensor, used as part of the DSP-AUGMENTATION program, as the basis for the ALARM sensor.* The conferees called for a review of that proposal, to be conducted by "a party without excessive linkages to the Air Force; the Intelligence community; the Alert, Locate and Report Missiles (ALARM) program; or the Brilliant Eyes program." The target date for completion of the report was set as February 15, 1995.[27]

Consideration of an adapted HERITAGE sensor as the basis for the ALARM sensor was among the alternatives being considered by the Space-Based Infrared Review Panel, which included representatives of the NRO, as well as Army, Navy, and Air Force headquarters staffs, the Air Force Space

---

*The DRB also considered three architectures for a launch-detection system: a system of low-earth-orbiting satellites based on a modified Brilliant Eyes concept, a scaled-back FEWS concept relying solely on geosynchronous satellites, and a mixture of Brilliant Eyes and a scaled-back FEWS. The Hall panel examined the same possibilities. (Steve Weber, "Pentagon Eyes New Early Warning System," *Space News*, September 26–October 2, 1994, pp. 3, 21.)

Command, U.S. Space Command, and the DIA. Indeed, it was panel chairman Keith Hall who suggested consideration of HERITAGE.[28]

Among other options being considered for the sensor may have been the High Resolution Multi-Spectral Instrument (HRMSI), a sensor that has been unsuccessfully searching for a home since the early 1990s. The HRMSI, developed by Aerojet, began as part of an NRO satellite program, which was canceled in 1991. In December 1992 the Department of Defense, specifically the NRO-run Defense Support Project Office, and NASA agreed to place the sensor, with a projected five-meter resolution, on LANDSAT 7. However, the arrangement fell apart when NASA was unable to obtain funds to pay for necessary improvements in the ground stations that would receive the data.[29]

Another multispectral sensor that may have been the subject of discussion is the sensor carried on the Pentagon's Miniature Sensor Technology Integration (MSTI) satellites. The most recent of the MSTI satellites, MSTI-3, carried a camera capable of detecting objects ninety-two feet across; it had been proposed as an operational missile warning sensor to increase coverage over regional hot spots.[30]

Other alternatives included piggybacking infrared launch-detection sensors on other geosynchronous satellites, such as MILSTAR, and elliptically orbiting satellites—as HERITAGE was piggybacked on elliptical satellites. The panel possibly also considered a Navy proposal to place a HERITAGE-type sensor on a proposed Navy satellite system designated RADIANT AGATE, which would provide stealthy communications to Navy forces operating in the Arctic region.[31]

Sometime in October or early November, the panel's report and recommendations went to Deputy Defense Secretary John Deutch, who approved an extensive ALARM system. Initially there would be four satellites in geosynchronous orbit and two satellites in highly elliptical orbits. Several satellites in low-earth orbit would possibly be added at later stages of the program. One candidate for the low-earth-orbit satellites was the Brilliant Eyes design, devised to track missiles in the middle portion of their trajectories.[32] The Deutch-approved architecture also bore a resemblance to a TRW concept for a hybrid system consisting of geosynchronous and low-earth-orbiting spacecraft. Under the plan, the initial operational capability for the new system would be 2002.[33]

But within a little over a month the bell was tolling for the ALARM program. No one familiar with the stops and starts of the last several years could have been terribly surprised to read the following in the November 29, 1994, issue of *Aerospace Daily:* "Marking yet another turn in its torturous effort to procure a replacement for the Defense Support Program (DSP) infrared early warning satellites, the Air Force has cancelled the ALARM program born earlier this year and announced a competition for the new Space Based Infrared (SBIR) program."[34]

On January 13, 1995 key Air Force officials, including Darleen Druyun and Major General Schnelzer, briefed industry officials whose companies might compete for the SBIR contract. They outlined the purposes of the SBIR system they wanted to deploy. It would provide missile warning to be used for national and theater missile defense, permit battle-space characterization, and acquire technical intelligence.[35]

The industry officials were told of the multitude of options that had already been evaluated—networks as widely diverse as five geosynchronous satellites and twenty-four low-earth orbit satellites had been studied. Combinations such as three geosynchronous and eighteen low-earth-orbit satellites also were considered. But the concept that finally won approval, labeled "High Now, Low Later," was remarkably similar to that envisioned for ALARM. It called for a constellation with four satellites in geosynchronous orbit (the first delivered in 2002), with two NRO satellites in highly elliptical orbits (the first also delivered in 2002) carrying an infrared package as a secondary payload. In addition, there might also be a number of low-earth-orbit satellites. A deployment decision on the low-earth-orbit satellites, initially designated the Space and Missile Tracking System (SMTS) and now SBIR-Low, would be made in the year 2000, subsequent to a flight test.[36]

The low-earth-orbit satellites, probably numbering around twenty, would operate below 1,000 miles and be equipped with two sensors to track intercontinental and theater ballistic missiles—sensors that could also survey other space objects and gather imagery intelligence. For wide-area acquisition there would be a scanning sensor employing short-wavelength infrared detectors to detect missiles in their boost phase. A high-resolution staring sensor would be used for tracking and would have detectors in three wavelengths: mid-infrared, mid- to far infrared, and visual.[37]

As envisioned by the Air Force, the satellites would produce a variety of infrared readings, with the bulk of the data processing being done on the ground. The data produced would have applications for missile warning, missile defense (by providing information to defensive systems), technical intelligence (including threat performance and target signature data), and "battlespace characterization."[38]

The announced plan also called for the 1999 closing of the DSP ground stations in Australia and Europe, with ground processing being consolidated in the United States. The SBIR ground segment would, according to the plans, be completed in 2001.[39]

Unlike the previous proposed DSP-successor programs, SBIR will be the actual successor, having garnered strong support in Congress. Indeed, in the summer of 1995, House and Senate conferees approved adding $135 million to the administration's SBIR request to accelerate deployment of the SMTS/SBIRS-Low (although the administration would attempt to shift $58 million to pay for Bosnia operations). The acceleration that the conferees had in mind would include a first SMTS launch in fiscal year 2002 and a network of eighteen satellites during the following fiscal year—almost five years ahead of the original Air Force schedule. Such an accleration, Air Force officials claimed, would cost an additional $2 billion—a claim supported by a General Accounting Office study. In July 1996 Paul Kaminski, Deputy Under Secretary of Defense for Acquisition and Technology, informed Congress that technical, funding, and management problems made it impossible to accelerate deployment to 2002. However, in September he ordered, in response to urging from the Defense Science Board, that the first launch be moved up to 2004 from 2006.[40]

In anticipation of the new high and low warning systems, the Army and Air Force have been discussing development of a JTaGs follow-on designated the Multi-Mission Mobile Processor. In addition, the Army is considering adding a direct downlink to the low-earth-orbit (SMTS) and elliptical-orbiting SBIRS satellites to give the Army the capability to downlink directly from any of the SBIRS satellites.[41]

Shortly before the Air Force briefing on SBIR, on December 22, 1994, a Titan IV carrying DSP-17 lifted off the launchpad at Cape Canaveral—just in time to allow the Air Force to use the Christmas holiday to switch to a new operations center at the Cape.[42] While plans called for DSP-19 through DSP-22 to be launched in consecutive years, beginning in 1997, possibly employing the space shuttle rather than expendable launch vehicles, 1997 passed with only the launch of DSP-18 on February 23. That may reduce the planned three-year gap between the next-to-last and last DSP launches to two years, with the final DSP being launched in 2003—a year after the first delivery of the new geosynchronous and highly elliptically orbiting satellites of the SBIR system.[43]

The launch of DSP-17 was only one of three significant events related to DSP operations that took place in 1994. In March the Aerospace Corporation began using a pair of laser beacons to calibrate the position of DSP satellites, reducing their position location error to one-third and thus increasing the accuracy of their launch point estimates.[44]*

On October 1 a 60-person 11th Space Warning Squadron (11th SWS), which will grow to 120 persons, was activated at Falcon Air Force Base (since renamed Schriever Air Force Base in honor General Bernard Schriever) to handle the ALERT system, to allow near-real-time tracking of missiles with ranges less than 1,800 nautical miles. The key element of the system is a battery of Onyx mainframe computers that filter out the background "noise" on the earth and distinguish the faint heat footprints left by tactical ballistic missiles. What the strategic system "threw away as noise is our bread and butter," according to the 11th SWS commander, Darrell Herriges.[45]

According to Ed Struck, the deputy commander of the 21st Operations Group, which supervises the squadron, "ALERT is a marked improvement from where we were in Desert Storm. We can process more data, more quickly and are able to detect dimmer targets."[46]

A six-person crew, including intelligence analysts, keeps watch twenty-four hours a day. The intelligence analysts look at incoming data, including that from intelligence satellites, to help determine what is actually taking place and who is at risk in the theater. According to Herriges, "we train our crews to minimize the time to get data to the 'shooters' — the Patriot missile batteries, the ships at sea and the soldiers in foxholes. We want to get into the 'theater operations tempo.' There's no time to waste when a missile warning alarm goes off."[47]

The 11th SWS represents one of three elements of the U.S. Tactical Event System. A second element is National Systems *Tac*tical *D*etection *a*nd *R*eporting (TACDAR). As its title indicates, TACDAR relies on information from one or more of the NRO's constellation of reconnaissance satellites to provide tactical warning information. The third component is the JTaGs, which, unlike the fixed ALERT system, is mobile. In April 1995 the Naval Space Command Detachments BUCKLEY and ECHO (which had been relocated to Dahlgren, Virginia, from Australia on September 1, 1993) relocated from Buckley ANGB, Colorado, and Dahlgren, Virginia, to Chesa-

---

*The calibration effort concluded in May 1996, when all active DSP satellites had been measured. In early 1997 it was reported that officials were considering whether to station pairs of laser beacons needed for the calibration efforts in potential conflict zones to permit real-time calibration, since the satellites' position will change slightly over time. As noted earlier, the LLYNX Eye system is being developed for such calibration. (Michael A. Dornheim, "In-Orbit Calibration Sharpens DSP," *Aviation Week & Space Technology*, February 10, 1997, p. 22.)

peake, Virginia. From there the members of the unified Detachment ECHO can be deployed to locations throughout the world.[48]

As of the spring of 1995, two Army JTaGs demonstration units, the Tactical Surveillance Demonstration (TSD) and the Tactical Surveillance Demonstration Enhanced (TSDE), were deployed to Stuttgart, Germany (to support the European and Central Commands), and Osan, Korea (to support U.S. Forces, Korea). Subsequently, the Stuttgart unit was temporarily transferred to Korea, where the two prototype stations were "doing the operational mission 24 hours," the Army's space integration chief Colonel H. C. Hostetter reported in April 1996. The two stations formed JTaGS Detachment Korea and could receive data from two or more DSP satellites, compute flight path parameters, and communicate the information to Patriot missile systems using theater communications networks such as the Tactical Information Broadcast Service.[49]

In early 1996 full production of five units was approved, with the two prototypes to be upgraded and another three units built. In February 1997 one of the five planned operational JTaGS stations was deployed in Europe, while an upgraded South Korean station entered service in May. Two additional stations are to be based in the United States and deployed when needed, while the fifth station will be used for training.[50]

FOURTEEN

Future Missions

Barring dramatic and unforeseen developments, the next decade will see the beginning of the transition from DSP to SBIRS. In space the four primary DSP satellites in geostationary orbit, augmented by the HERITAGE sensors on elliptically orbiting NRO satellites, will eventually be replaced by the SBIRS mixture of four primary geostationary satellites (and one spare), and two infrared-equipped, elliptically orbiting NRO satellites at high altitude, and the constellation of low-earth-orbiting detection satellites.

The geostationary satellites will be built by a team that reunites two of the earliest partners in the infrared satellite warning field: Lockheed and Aerojet. In November 1996 they, along with Grumman, were announced as the victors over a TRW-Hughes team. The prize was a $1.8 billion order to build five of the SBIRS-High satellites as well as the sensors for two satellites in highly elliptical orbit and the associated advanced ground processing system.[1]

The winners had proposed that their satellites carry a pair of new sensors to improve detection of Scud-class missiles—sensors that would integrate the technology developed over the past fifteen years. One sensor will be a high-speed scanning sensor that will sweep Earth and provide the first detection of a missile launch. The other will be a complementary staring sensor, which would focus on the launch and provide a more detailed view. In contrast, TRW and Hughes proposed a single HERITAGE-type sensor.[2]

Data from three experimental spacecraft presently in orbit will provide data to assist the sensor designers. In April 1996 the Midcourse Space Experiment (MSX) spacecraft, designed by the Johns Hopkins University Applied Physics Laboratory, was placed in orbit, carrying three sensors. Its primary sensor, an infrared payload designated SPIRIT III, operated until late February 1997. Two others, ultraviolet and visible-light sensors, were employed until that time to identify ballistic missile signatures during the period between booster burnout and missile reentry. Subsequently, MSX has been employed to detect and track objects in space.[3]

223

Overlapping MSX is the third Miniature Sensor Technology Integration (MSTI) satellite, which completed its first year in orbit in June 1997, collecting short-wave and midwave infrared imagery in support of the SBIRS program. The multispectral MSTI-3 collected more than 1.2 million short-wave and midwave infrared images to be used in analyzing the effect of atmospheric conditions on missile tracking from space.[4]

The third, lesser-known, contributors are the COBRA BRASS space experiments, which are conducted via infrared payloads on NRO satellites. The COBRA BRASS sensors were developed by Sandia National Labs, and the program is managed by the DIA's Central MASINT Office. As with MSTI, COBRA BRASS provides multispectral/infrared data in support of the SBIRS program.[5]

On earth, the makeup of the ground station network will also change significantly. That change began in the United States with the disestablishment of the 4th SWS, as the mobile ground terminal unit at Holloman had been designated since 1993. In late 1995 the mission was transferred to the 137th SWS at Greeley Air National Guard Station in Colorado, with the intention of manning the squadron with 173 full-time and 123 weekend guardsmen.[6]

Before another four years have passed, the United States will close the stations at Nurrungar and Kapaun as part of the ground station consolidation that represents Increment 1 of the transition to SBIRS. The United States and Australia have agreed that a Relay Ground Station for the Eastern Hemisphere DSP satellite, and for its SBIRS successor, will be established at the Joint Defence Facility at Pine Gap. The station will be the sort of "bent-pipe" facility envisioned in the Aru and Lunde report. The end of the American presence at Nurrungar may please some of those who converged on the site for yearly protests, many of whose cars were probably adorned with the Anti-Bases group's "Radomes Go Home" bumper sticker while they sported "Kick the Base" or "Cut the Lease" T-shirts.[7]

As part of the consolidation the processing and analysis activities conducted at Nurrungar and Kapaun will be relocated to the Mission Control Station (MCS) at Buckley, which will rely on some of the existing DSP architecture. Mission data will be transmitted, via bent pipes, from multiple satellites viewing the same event to the MCS, where they "will be fused to improve the detection and estimated target parameter accuracies." Groundbreaking for the Buckley MCS began in September 1997. The backup station for the MCS will be located at F. E. Warren Air Force Base, Wyoming.[8]

Increment 2 of the transition will take place both in space and on the ground. The development, launch, and operations of the geostationary and highly elliptical orbit SBIRS (SBIRS-High) satellites will be one part of Increment 2, with those satellites being launched as DSP satellites are retired. On the ground, the MCS and Warren facilities will be upgraded to allow them to process the data from the SBIR-High satellites. They will also be able to process DSP data, although redundant facilities will be deactivated.[9]

The final increment will involve addition of the SBIRS-Low component. According to one official involved in the SBIRS program:

> When Increment 3 becomes operational all of the SBIRS threshold requirements will be fully satisfied. The low-altitude satellites provide better viewing geometry for dim post-boost and mid-course objects. The low altitude also supports the need for very accurate trajectory state vectors on these objects. Because of the unique attributes of the low-altitude orbit, the operational requirements for midcourse tracking have been allocated to the Low Component. SBIRS Low will also support improved performance for other mission areas.[10]

Exactly which contractors will build SBIRS-Low will be determined in a competition in the year 2000. Originally, the competition was expected to be a contest between two major aerospace teams, with a TRW/Raytheon team launching two flight demonstration satellites to an altitude of about 1,000 miles and a Lockheed-Martin/Boeing/Aerojet team launching a single satellite into a 40-degree inclined orbit of about 450 miles. According to a Boeing brochure concerning the Low-Altitude Demonstration Satellite (LADS), approximately 400 experiments will be conducted in six months. The LADS will weigh 1,430 pounds and will carry medium-, short-, and long-wave infrared sensors. The brochure states that "all facets of midcourse tracking will be demonstrated": autonomous detection, acquisition, and tracking; handover from acquisition to tracking sensor; precise tracking to burnout; and ballistic cold-body tracking and discrimination. The Lockheed-Martin/ Boeing team states that its SBIRS-Low satellites will be capable of battle-space characterization, technical intelligence, integrated tactical warning/ attack assessment, space surveillance, missile defense support, and reentry vehicle tracking and discrimination.[11]

Then, in April 1998, at a speech before the National Space Symposium, the president of the 270-employee Spectrum-Astro company of Gilbert, Arizona, which built the MSTI satellites, announced its plans to bid for the SBIRS-Low contract. At the time Spectrum-Astro had begun a search for partners, and a month later it was reported that the company was finalizing a team.[12]

There are some uncertainties about the SBIRS system. The exact number of SBIRS-Low satellites is yet undetermined, with numbers ranging from eighteen to twenty-four often being mentioned. The exact status of the SBIRS nuclear detonation detection function also has yet to be determined. Department of Defense skepticism on the need for an electromagnetic pulse (EMP) sensor may limit SBIRS to carrying X-ray and optical sensors. A Space and Atmospheric Burst Reporting System (SABRS) package could become standard on the SBIRS satellites if the Defense Department agrees. The SABRS combines the fast-gamma and spectroscopy-class gamma-ray detectors now used on DSP into a single crystal.[13]

Despite the dramatic changes ahead, the missions and issues confronting the operators of SBIRS are likely to be similar to those that confront DSP's operators now and in the next decade—in terms of targets, information sharing, and scientific applications. Although Russia is not a likely military adversary, its missile development and operational testing programs, as well as China's, are still active and are closely monitored by the United States. On June 8, 1995, Russia conducted a test launch of an SS-N-8 missile, followed by test launches of a ten-warhead SS-N-20 on August 25 and an SS-25 "Topol" ICBM in early September. In July 1997 the Strategic Rocket Forces conducted the fourth test flight of the SS-27/Topol-M, which is being developed as a supplement and potential successor to the SS-25. The U.S. intelligence community believed that about 400 of the missiles, each with a 550-kiloton warhead, will be built. The first two were put on combat duty in late December 1997. On October 3, 1997, Russia concluded the final day of a strategic command post exercise, which included SLBM launches from the Barents Sea and ICBM firings from Plesetsk. An SLBM firing from a Delta I submarine followed the next day. Weather conditions permitting, one or two DSP satellites would have picked up the test failure of an upgraded SS-N-20, near Archangel, in November 1997. The SLBM exploded about 600 feet above its launchpad at the Central Naval Range at Nenonska. The new SLBM is scheduled to be carried on the new "Borey" (Arctic Wind) class of Soviet SSBN, which the United States expects to go into service in 2005.[14]

The United States and many of its allies in Asia are particularly concerned about the ongoing Chinese military buildup. Among the strategic missiles China has in development are the JL-2 SLBM, which carries a single warhead and has a range of almost 5,000 miles; the DF-31, which will carry one warhead more than 4,500 miles; and the DF-41, which is projected to carry a single warhead over 7,000 miles. In addition, China has been seeking to acquire foreign technology that could enhance its ability to deploy missiles with independently targetable warheads.[15]

In March 1996 China demonstrated its willingness to use military exercises, including repeated missile launchings into the sea near Taiwan, for foreign policy ends. The perpetually aggrieved Chinese leadership—angered by Taiwan's higher international profile, fearing a possible move toward independence, and probably terrified that Taiwan was about to demonstrate that free democratic elections and Chinese heritage were not incompatible—began a campaign of military intimidation by conducting an exercise based on a contingency scenario for the invasion of Taiwan. On March 12 Chinese warships and fighter aircraft began live-fire military exercises in the Taiwan Strait, practicing bombing runs and drills. In addition, the PRC also fired four DF-15 (also known as the M-9) surface-to-surface missiles, which landed in the test zones off the ports of Kao-hsiung and Chi-lung. The exercise ended on March 25, two days after the Taiwanese election.[16]

While the satellite launch detection program originated from fears of Soviet attack employing strategic missiles, the spread of tactical ballistic missiles throughout the world, along with the collapse of the Soviet Union, has resulted in a dramatic shift in emphasis. Monitoring the development and testing of IRBMs may be of more significance than monitoring Russian missile developments, and possibly even Chinese developments.

One application of intelligence on such programs is as a guide to U.S. diplomatic, trade, and export-control actions that seek to inhibit or stop such programs—as U.S. pressure helped stop the Argentinian Condor II program. Here imagery, communications intelligence, and human intelligence are likely to be far more useful than any data provided by launch-detection systems, since success in stopping a program is far more likely before the first test missile is fired.[17]

However, there will be a number of cases in which the United States may choose not to exert pressure to stop a missile program, may have no hope for success, or may simply fail. Thus, many nations are presently developing short- or intermediate-range missiles, including many who are hostile to the United States and its allies. Development programs by nations that are friendly to the United States are also of concern for at least two reasons: they may export those missiles, or the introduction of the missiles might be viewed by the United States as a destabilizing force in a particular region.

Among the missiles under development that most concern the United

States are the North Korean Taepo Dong 1 and 2. A mock-up of the Taepo Dong 2 was spotted by a U.S. imagery satellite in early 1994 at the Sanum Dong Research and Development Facility. The imagery indicated the missiles will be 105 feet long, have a 59-foot first stage with a diameter of about 8 feet, and a 46-foot second stage with a diamater of 4.3 feet—dimensions that led analysts to conclude that the missile will, when equipped with a 2,200-pound warhead, have a range far in excess of the 1,240 miles predicted for the Taepo Dong 1. Estimates of the Taepo Dong 2's range have varied from 2,170 to 3,720 miles. With the Taepo Dong 2, North Korea's offensive missile capability would extend beyond the Korean peninsula, permitting an attack on Guam. The Taepo Dong would be added to an arsenal that already includes Scud-B, Scud-C, and No-Dong missiles with ranges of 186, 310, and 620 miles.[18]

Nations developing intermediate-range missiles include India, Pakistan, Iran, North Korea, and Iraq. The nuclear tests conducted in May 1998 by India and Pakistan will aid those countries in developing warheads for their various missile programs. In December 1996 India had announced that it was halting development of the medium-range AGNI. However, in May 1997 a parliamentary panel urged the government to deploy the AGNI as a counter to potential threats from China and Pakistan. In July the Indian Minister for Defence announced, "It has been decided to accord high priority to the next phase of the AGNI program." A report by a congressional commission indicates India also has under development the AGNI-Plus and AGNI-B medium-range missiles as well as the SAGARIKA submarine-launched ballistic missile. In July 1997 Pakistan tested its Haft-3, reputed to have a range of 370 miles. In early April 1998 Pakistan tested a new missile, the GHAURI, with a range exceeding 900 miles and believed to be capable of carrying a nuclear warhead. The missile, named after a medieval Afghan king who defeated the Hindu ruler of Delhi, was apparently developed with technology obtained from North Korea.[19]

Prior to 1995 Iran was involved in the development of five intermediate-range missiles: the Iran 700 (or Scud-C), the Al-Fatah (in conjunction with Libya), the Nodong-1 (in partnership with North Korea), and the Tondar-68/M-18 (with China), and the DF-25 (also with China). The missiles were projected to be able to carry payloads ranging from 880 to 4,400 pounds, with ranges varying from 434 to 1,054 miles. In late 1997 reports from Iranian dissidents claimed that Iran's Revolutionary Guard was also developing the Shahab-3 missile, a Nodong derivative, with a range between 800 and 930 miles, as well as the Shahab-4, with a range of up to 1,240 miles. (Claims that a Shahab-3 had actually been tested were not supported by DSP data.)

Several months later, reports about the existence and range of the two missiles were confirmed in an Air Force publication. But then, in July 1998, a DSP satellite detected an Iranian test of a Shahab-3 that traveled for a minute and forty seconds and 625 miles before exploding. And UN inspectors believe that Iraq is attempting to develop an advanced version of the Al-Hussein, with a range of more than 2,000 miles.[20]

In addition to detecting missile launches associated with developmental programs, detection systems will probably continue to experience a growing number of "live" targets—an inevitable result of the spread of ballistic missiles and the continuation of war. Between 1973 and 1993 Scud missiles were employed in five regional conflicts: the 1973 Yom Kippur War, the Iran-Iraq war (1980–88), the U.S.–Libya conflict of 1986 (during which Iraq fired a Scud at the U.S. facility at Lampedusa, Italy), the Soviet intervention in Afghanistan, and the Persian Gulf War.[21]

During May 1994 those live targets included missiles fired in Yemen during the civil war there. At least some of those launches were monitored by the ALERT system at Falcon Air Force Base. Within one week in November 1994, missiles were fired in anger in at least two areas of the world. On November 3, apparently to retaliate for an offensive by Muslim-led government troops, Bosnian Serbs attacked the town of Bihac in northwestern Bosnia. According to a UN spokesman, two SA-2 missiles, converted for use against ground targets, landed near a school in Bihac, damaging thirty to forty buildings.[22]

On November 6 Iran carried out a missile attack on a base of the exiled Peoples Mujahideen (Mujaheddin-e Khalq) near Baghdad, after the rebels tried to bomb Iranian oil installations in the Mousian region. The mujahideen claimed that the Revolutionary Guards launched three Scud-Bs at their camps and had another nine poised for launch from western Iran. However, DSP satellites picked up only one firing, leaving U.S. officials skeptical of mujahideen claims.[23]

The data provided by DSP and SBIR satellites about such firings will have several applications. Aside from any additional technical intelligence that may be gathered about the missile systems themselves, the infrared data will constitute valuable intelligence about the war in progress—who is firing missiles, what missiles are being fired, who or what are the targets, how many missiles are being fired, and where the missiles have landed. The data may also be used to help manage the tasking of imagery and signals intelligence assets—for example, instructing a satellite to return imagery of an area that DSP returns indicate has been a launch or impact point.

★　★　★

Of course, if the targets of the missiles are U.S. forces or allied forces, the most valuable use of the data will be for warning purposes. The value of that data will vary with the nature of the missile's warhead, the time between launch and impact, accuracy, and the countermeasures available to U.S. or allied forces. If those countermeasures include one of the many actual, planned, or proposed ballistic missile defenses, the advanced warning that can be provided from space may be particularly valuable.[24]

The Army has been seeking to upgrade the Patriot defense system, whose Desert Storm performance became, as noted earlier, a matter of great controversy. The upgrade involves a number of phases, concluding with the replacement of the PAC-2 missile, used during Desert Storm, with a "highly-agile" PAC-3. The PAC-3 system, which was in its engineering and manufacturing development stage in mid-1998, will include a new missile, along with radar enhancements, communications upgrades, and increased computer capability. It is hoped that the PAC-3 will be more capable of dealing with missiles that break up in flight, as the Scuds did during Desert Storm.[25]

Use of DSP satellites to support Patriot tests has occurred on several occasions. The ability of JTaGs to cue a Patriot battery was tested, in November 1993, at White Sands, New Mexico. According to the Ballistic Missile Defense Organization, "JTaGS cued Patriot repeatedly, including some single beam acquisitions." In February 1997 DSP satellites were among the warning systems employed in a test in which Patriot missiles fired from Kwajalein Atoll destroyed two Scud-B missiles fired from Aur Atoll. The Scud-Bs were obtained from governments in Eastern Europe under a program designated WILLOW SAND.[26]

The Army has also been seeking to develop a Theater High Altitude Area Defense (THAAD) system, intended to neutralize ballistic missiles with ranges up to 2,170 miles and to protect a wider area than the Patriot batteries. The system envisioned, which has experienced significant developmental problems, would give the THAAD interceptors two shots at an incoming missile, the second after an assessment of the first interceptor's performance. Patriot batteries would get the third shot, if necessary.[27]

According to official testimony, THAAD would

provide a capability for in-depth defense against tactical ballistic missiles with increased range and carrying chemical, biological and nuclear warheads. The most effective defense against hardened warheads is to intercept them at higher altitudes by a non-nuclear hit-to-kill missile interceptor. High altitude interception of chemical and biological warheads destroys these warheads at extended ranges from the defended area. High altitude interception of threat nuclear warheads also negates the nuclear blast produced at lower altitudes and dilutes detrimental nuclear radiation effects against targeted civilian populations.[28]

Two Navy systems are under development or consideration. The Theater Ballistic Missile Defense System (TBMDS), or "Lower-Tier," is meant to provide protection of ports, coastal airfields, and landing areas from Scud-type missiles for "tens of miles." It is intended to be deployed beginning in 2002 on Ticonderoga-class guided missile cruisers and Arleigh Burke–class guided missile destroyers with AEGIS combat systems, shipboard weapons, surveillance, and communications systems. DSP or SBIRS data would, after ground processing, be transmitted via Navy communications satellites to AEGIS-equipped ships, which would use the data in targeting their defensive missiles.[29]

The second system is in a very early stage of development. Formerly designated "Upper Tier" and now called "Theater Wide," it would attempt to deal with longer-range missiles, intercepting missiles at heights of up to 310 miles and distances of up to 744 miles.[30]

In addition to theater defenses, a national missile defense system to protect the United States may also be deployed early in the next century. The current notional architecture for such a system would include a single set of ground-based interceptors at Grand Forks, North Dakota; a phased radar at Grand Forks; command centers at NORAD and the U.S. Space Command; upgraded early warning radars; DSP and SBIRS spacecraft; and a battle management C^3I system to link all the elements together.[31]

Data from DSP and SBIRS would be used to cue the upgraded early warning radars, which would gather tracking and threat assessment data. The radars would then hand off that information to the radar and interceptors at Grand Forks. Final release authority for the interceptors would be held by the Commander in Chief of the U.S. Space Command.[32]

The Clinton administration has adopted a "three plus three" policy with regard to national missile defense. It has announced that, if in three years projected threats justify building such a system, it would begin work toward deployment, with the expectation that it would be up and running within another three years.[33]

In addition to providing cuing data to a missile defense system, the SBIRS-Low constellation may also be able to provide data to help distinguish between actual warheads and decoys. Scientists plan to monitor the signatures of warheads and decoys in space across several infrared and visible-light wavelengths to distinguish the differences. A warhead that weighs several hundred pounds will cool at a different rate than a balloon that weighs only a few pounds.[34]

As the ability of the United States to provide more timely early warning grows, whether in the DSP or the SBIRS era, that ability becomes a more valuable commodity, not only to the United States but also to its allies. Some may seek access to the data for use with any antimissile system or may be induced to help develop such a capability by the offer of warning data. Israel has already scored a successful hit on a Scud-type target in a test of its Arrow-2 ABM system.[35] Thus, despite the closure of the overseas ground stations and concentration of regular processing activity in the United States, DSP and its successor will still play a role in the United States' relations with a number of countries.

In April 1994 General Charles Horner, chief of the U.S. Space Command, told a breakfast meeting that data sharing is the first step toward wider cooperation on ballistic missile defense between the United States and its allies and with other selected countries. He said that Japan and France were very interested in having access to U.S. launch-detection data. However, while the French are discussing ballistic missile defense with the United States, they may find simply being a subscriber to the U.S. system to be unacceptable—just as they have found being a subscriber to the U.S. satellite imagery system to be unacceptable.[36]

Among those whose desire for full-time access to the information is predictable is Israel, particularly in light of its Gulf War experience. On April 28, 1996, the Clinton administration announced that it would help Israel develop a ballistic missile defense system and share the data from DSP satellites. At a joint press conference in July with visiting Prime Minister Benjamin Netanyahu, President Clinton said the new capability would be operational by the end of 1996. The arrangement will be permanent and, according to U.S. officials, will provide warning information "of any ballistic missile launch that in any way could threaten Israel."[37]

Shortly after the April announcement, it was reported that Japan would be a recipient of launch detection data, in "yet another attempt to encourage Japan to move more rapidly toward deployment of anti-missile defenses." In 1994 the Pentagon had proposed four options for missile defense systems designed to address the threat from North Korean and PRC ballistic missiles that are expected to be deployed in the next decade. Whereas in January 1997 the Japanese declined to join a U.S.–sponsored plan for theater ballistic missile defense, in early 1998 the Japanese government asked Diet approval to contribute $650,000 toward ballistic missile defense research.[38]

In April 1996, at the time of General Horner's speech, the United States began providing most of its NATO allies with access to DSP data through 270 Linked Operations–Intelligence Center, Europe (LOCE) terminals. Nations plugged into LOCE include the United Kingdom, Canada, Norway, Denmark, Belgium, France, Spain, Italy, and Greece.[39]

The United States has also decided to share SBIRS data only with allies who are already cooperating on missile defense programs, such as the Medium Extended Area Defense System, a planned $3 billion replacement for the Hawk air defense system. Italy, Germany, and the United States have already committed to the project, and access to SBIRS data was intended as an inducement to get other European nations to participate.[40]

Another nation that will receive U.S. early warning data is Russia—the successor to the very nation whose missile activities required the creation of the DSP program. The prospect of data sharing had been raised even before the formal demise of the Soviet Union, as noted in chapter 9.

General Horner subsequently observed, with respect to Russia, that "it's important that we examine how we can share more ballistic missile warning [data]. . . . First of all, it's a vital step in building trust; and second of all, you really don't give up anything if you do not intend to conduct first-strike operations on another country." It was reported in early 1995 that an agreement had been reached to share data from the DSP system and the Russian PROGNOZ satellites.[41]

In 1995 the U.S. BMDO budgeted $5 million for a five-year study of a Russian-American Observation Satellite (RAMOS) program that could improve missile detection. The first phase of the program was to involve joint observation of backgrounds and "targets of opportunity" using Russian space-based sensors and U.S. airborne sensor platforms.[42]

In late 1995 a different opportunity for cooperation arose when it became clear that a Russian lunar module test vehicle, Cosmos 398, would reenter the atmosphere and hit earth. In the four days preceding reentry, Russia and the United States exchanged radar data via an electronic mail link in an attempt to determine where the spacecraft might land. When it did reenter the atmosphere and land about 200 miles north of the Falkland Islands at 20:40 Greenwich Mean Time, its reentry was observed by the DSP-Atlantic satellite.[43]

Subsequently, in light of limited Russian resources, the agreement was revised. Data from the MSX satellite will be provided to the Russians, as well as to the SBIRS program, for analysis.[44]

DSP and its successor will also continue to have a scientific role. As noted earlier, all DSP satellites since F-14 have been equipped with a variety of scientific sensors to help determine the environment in geostationary orbit. DSP data have also been used to address a question of growing concern—the frequency and potential result of asteroids entering earth's atmosphere.[45]

Beginning in 1975 the magnetic tapes that recorded data from the sensor were examined, although on a less than systematic basis, for indications of meteoroids slamming into the earth's atmosphere. When such meteoroids enter the atmosphere, they are heated to several thousand degrees. If they are about six inches or smaller, they burn up and are seen as "shooting stars" in the night sky. Some of the largest ones may actually reach the earth's surface, while most explode under the pressure of air resistance.[46]

One motivation for beginning the process of looking for evidence of such explosions was to help those within the U.S. intelligence community responsible for monitoring foreign nuclear weapons efforts to discriminate between nuclear explosions and natural explosions. Since 1975 there have been a number of instances in which it has not been immediately clear whether an "event" of nuclear proportions was caused by nature or was the result of a nuclear explosion. As discussed earlier, in 1979 a mysterious flash registered on VELA sensors that occupied numerous goverment panels.[47]

In December 1980 another flash in the vicinity of South Africa set off another round of debate. After two months the consensus was reached that the flash was caused by a meteor. On April 15, 1988, a large meteoroid exploded over Indonesia, giving off energy equivalent to at least 5,000 tons of TNT. On October 1, 1990, an explosion with the force of 2,000 tons of high explosives took place over the Pacific Ocean. Only after several months, and a collection effort that began with the assumption that a nuclear explosion had been involved, was the conclusion reached that a 100-ton asteroid rather than a nuclear bomb had exploded.[48]

Another major impact occurred on February 1, 1994, near the island of Kusaie. The meteor carried the kinetic energy of an atomic bomb, was the brightest in the nineteen years that such data had been analyzed, and triggered sensors on six DSP satellites. Some reports suggested that President Clinton was woken for fear that the explosion was nuclear.[49]

The data on impacts from the DSP satellites are also of value to scientists. For that reason an Air Force officer, Colonel Simon P. Worden, who also holds a doctorate in astronomy, exerted considerable effort to get the Air Force to declassify the data. His effort proved successful in October 1993 when the Air Force declassifed data about the 136 blasts, ranging from the equivalent of approximately 500 to 15,000 tons of high explosives, that had been detected between 1975 and 1992.[50] Eugene M. Shoemaker, an astronomer at the Lowell Observatory in Arizona and a pioneer in the field of earth-impact studies, referred to the data as "a unique source of scientific information." Such data can be used in attempting to estimate the frequency with which meteoroids enter the atmosphere and the distribution of meteoroids of different size. That information is relevant to studies that seek to estimate the probabilities of the earth being the victim of a catastrophic collision with a meteoroid.[51]

At least one other application of DSP and its successors for environmental purposes has been noted by the Director of Central Intelligence's (DCI's) Environmental Task Force, a group of scientists chartered in 1992 by then DCI Robert Gates to examine the potential environmental applications of the data that had been or could be gathered by U.S. technical collection systems. A summary report noted what had been apparent from the earliest days of the DSP program: that "satellite systems designed to detect missile launches can also be used to detect and monitor large fires—which generate the greenhouse gases carbon monoxide and carbon dioxide—in remote areas." Conceptual development of a program that would include data from DSP and other sensors for fire detection began in the 1995 fiscal year, and the demonstration stage began in the 1997 fiscal year. Prototype operations are scheduled to begin in the 1999 fiscal year.[52]

Whatever environmental missions DSP and its successor system are employed for represent one of the many unexpected bonuses that have come from the program. The system proved far more capable of detecting and providing data on missile launches than even its advocates could have expected. And DSP's ability to monitor aircraft and spacecraft, and to provide measurement and signature intelligence on a variety of events in addition to missile launches, made it a valuable asset to additional consumers in the military services and intelligence community.

Of course, its key contribution during the Cold War was to superpower stability. One cannot say for certain what would have happened without DSP. But if one could travel to an alternative universe in which the world was identical, except for the absence of DSP, it is not hard to imagine that the risks of war, whether by aggression or miscalculation, would have been greater. It is certain that U.S. leaders would have had only radar for detecting an attack, no means of checking on the validity of radar signals, and half the warning time if an attack did take place (or none if the Soviets used a southern route). Awareness of these facts would have meant a greater susceptibility to false alarms and drastically reduced decision time. Certainly, the false alerts of the late 1970s and early 1980s might have had more serious consequences in the absence of DSP.

The DSP program is one of the most evident legacies of the Cold War. MIDAS and its successors were products of the Soviet missile threat, and those systems continued to evolve in response to the changes in the Soviet missile program and U.S. nuclear policy. The DSP system has already evolved to address the new international environment. And the continued existence of ICBMs and the spread of theater ballistic missiles make a continued space-based detection program essential for the foreseeable future. Thus, while DSP and its successor may be legacies of the Cold War, they are not relics.

Launch of DSP-23.

Epilogue: A Long, Hard Road: From DSP to SBIRS

The Air Force's November 1996 award of a contract for the development and production of SBIRS-High, along with the launch of DSP Flight 18 (F-18) in 1997, suggested that the end of the DSP program and the advent of its successor might be just over the horizon. But there could be no certainty that the transition would be an easy one: problems might arise with future DSP launches or with on-orbit satellites, and the development of the Space Based Infrared System (SBIRS) might run into obstacles. In addition, the decision to cancel production of DSP-24 and DSP-25 reduced the margin for error.

On April 9, 1999, a Titan IV rocket blasted off from Cape Canaveral to take Flight 19 into geosynchronous orbit. But when its first and second stages failed to separate, the $250 million spacecraft was stranded in its 28-degree, highly elliptical transfer orbit, making it useless for launch detection. The failure left the Air Force contemplating moving up the launch of the spacecraft that was intended to be DSP Flight 22 by eighteen months (from August 2002), since DSP-19 and DSP-21 were too far along in the launch process to reschedule.* Doing so would also require using the space shuttle, since Titan boosters were already committed to other launches in early 2001. If turning Flight 22 into Flight 20 was not feasible, then, said Lt. Col. Craig Lowery, the chief of the warning operations branch at Air Force Space Command, "we'll just have to cross our fingers and hope the current satellites continue to function properly."[1]

---

* A DSP flight number indicates the number of DSP spacecraft that have been launched up to that point in time. However, spacecraft have not always been launched in the order produced; thus the spacecraft employed for F-18 was the twentieth produced, DSP-20. The spacecraft used for Flight 19 was DSP-22. There may also be a disparity between the sequence in which the sensors are produced and the order in which they are flown. See *History of the Space and Missile Systems Center, 1 October 1998–30 September 2001*, vol. 1 (El Segundo, Calif.: Air Force Space and Missile Systems Center, n.d.), p. 195.

There never would be another DSP launched from the shuttle, but in early January 2000 the Air Force was preparing to launch DSP Flight 20, when there was another problem. On December 22, about five weeks before the anticipated launch, an overhead crane used to hoist the rocket's nose cone into place was discovered to have dripped oil, causing concern that the spacecraft may have been contaminated. As a result, on January 6, workers began inspecting the satellite for possible oil damage. Ultimately, DSP Flight 20 was launched from Cape Canaveral into geosynchronous orbit, apparently to become the Pacific satellite. But that did not occur until May 9, 2000. An undoubtedly relieved Maj. Todd Granger, Air Force Space Command deputy chief for missile warning, claimed it was "the most beautiful sight I've seen in 16 years in the business."[2]

"Beautiful" was used again, in August 2001, following the launch of DSP Flight 21, which was also aboard a Titan IVB booster and ultimately deposited into geosynchronous orbit on its way to becoming the Eastern Hemisphere satellite. This time it was the turn of the chief of Air Force Space and Missile Systems Center launch programs, Mike Dunn, to use the word, when he asked in the immediate aftermath of the successful launch, "wasn't that beautiful?" The August 6 launch did come after a short delay, caused by the discovery of a defective part in the rocket's upper stage. As a result, in a twenty-month period, two DSP spacecraft made it into orbit to replace aging members of the constellation.[3]

Along with the changes in space, the DSP program was also experiencing change on the ground. On October 12, 1999, the Overseas Ground Station—the Joint Defence Facility, Nurrungar—closed, with the Relay Ground Station–Pacific (at Pine Gap) serving as a bent pipe and relaying data from the relevant DSP satellites to the Continental Ground Station at Buckley Air Force Base in Colorado. By the end of September 2001, the European Ground Station at Kapaun, Germany, had also closed, with Menwith Hill serving as the European Relay Station for the European DSP satellite. At that time, the Buckley DSP ground station was in the process of being replaced by the new SBIRS Mission Control Station (MCS), also at Buckley.[4]

DSP launches from 1999 to 2001 bracketed the year 2000 and the feared Y2K impact on computers, including those employed at ground stations and on board a host of satellites. While one National Reconnaissance Office (NRO) satellite system suffered a Y2K problem, DSP satellites apparently made the transition to the new millennium without incident. Of course, there was no way that the Air Force could be certain that America's space sentinels—or Russia's—would be immune to some computer malfunc-tion. Russian authorities couldn't help but worry about the impact of Y2K on U.S. warning systems and had good reason to worry about its impact on their own warning capability, which was far from robust at the time.[5]

An August 1999 message from the U.S. Defense Attaché Office in Mos-cow noted that Russia was lagging in its efforts to identify and fix problems that could threaten its command and control capabilities, including warning systems. In July, in an effort to prevent any misadventures due to warning system malfunctions at the beginning of the new year, the Pentagon invited Russia to send a delegation to what would be known as the Center for Stra-tegic Stability and Y2K. An earlier invitation had been canceled by Moscow to protest the NATO air campaign over Kosovo.[6]

But in September, the Russians agreed to permit ten to twenty officers to spend their New Year's Eve and beyond at the center, which would be lo-cated at Peterson Air Force Base, Colorado. That agreement was formalized when Defense Secretary William Cohen and his Russian counterpart, Igor Sergeyev, signed an agreement during Cohen's September visit to Moscow. It specified that Russian officers would sit next to Americans "to read data flowing in from American radar sites and satellites from around the world." An Air Force Space Command official explained that they would be "be feeding data straight back to Russia and the Russian officers [would] be in direct contact with the leadership back in Moscow." On December 21, and continuing until January 16, 2000, nineteen Russian officers were at the Y2K center, with access to warning data that included the time, position, and tra-jectory of any missile launched anywhere in the world.[7] *

---

* In June 2000, the United States and Russia signed a memorandum of agreement to establish a Joint Data Exchange Center in Moscow in order to "set up an uninterrupted exchange of information on launches of ballistic missiles and space launch vehicles from the early warning systems of the United States of America and the Russian Federation," which built on a Sep-tember 1998 agreement between President Bill Clinton and President Boris Yeltsin. However, the effort ground to a halt during the administrations of their successors. In June 2010, a joint statement by President Barack Obama and President Dmitry Medvedev stated that the two na-tions "would continue their efforts to share-early warning data on missile launches." See "Text: Clinton/Yeltsin On Exchange of Info on Missile Launches," September 2, 1998, www.fas.org; The White House, "Memorandum of Agreement Between the United States of America and the Russian Federation on the Establishment of a Joint Center for the Exchange of Data from Early Warning Systems and Notifications of Missile Launches," June 4, 2000; Viacheslav K. Abrosi-mov, "Landmark U.S.-Russian Data Exchange Deal," *Jane's Intelligence Review*, January 2001, pp. 51–53; Peter Baker, "Nuclear 'Milestone' Divides U.S., Russia," *Washington Post*, June 13, 2001, p. A23; Tom Z. Collina, "Russia, U.S. Working on Joint Launch Notification," *Arms Control To-day*, July/August 2010. Another possible joint venture in launch detection and notification—the

In the eighteen months leading up to Y2K, a variety of countries provided the Air Force Space Command with the opportunity to verify that DSP satellites were working properly. On July 22, 1998, Iran tested the 800-mile-range Shahab-3 ballistic missile, whose launch *The New York Times* reported had been detected by a "United States spy satellite." The missile, believed to be either a close variant or an exact copy of the North Korean Rodong missile, apparently blew up in the later stages of its flight—an explosion that undoubtedly would also have been detected. Then, on August 31, North Korea fired what was first identified as a Taepodong-1 missile, whose first stage fell into the sea before reaching Japan, with the remainder of the missile flying over Japan and landing in the Pacific, about 1,000 miles from its launch site on North Korea's east coast. The North Korean regime claimed that the rocket's purpose had been to place a satellite in orbit, a claim met initially with great skepticism by U.S., Japanese, and South Korean experts. But there was "tons of data"—including telemetry, communications intelligence, and DSP data—for U.S. intelligence analysts to examine, an effort that ultimately led to the conclusion that the Hermit Kingdom had attempted and failed to place a spacecraft into orbit.[8]

While an anticipated December test of China's mobile Dong Feng-31 (DF-31), reported to have a range of 5,000 miles, was postponed until the next year, Russia did test-fire a new ICBM that month. The missile was known to NATO as the SS-27, but bore the Russian designation Topol-M. Russia's strategic nuclear forces described the missile as one "which has no parallels in the world" and claimed that the first operational Topol contingent would be combat ready before the end of the year.[9]

In early 1999, India and Pakistan contributed targets for DSP's infrared sensors. On April 11, India test-fired a nuclear capable, solid-fueled, Agni-II intermediate range missile, which flew for eleven minutes and more than 1,250 miles before splashing down in the Bay of Bengal. Pakistan responded three days later with a test of its Ghauri-II, which lifted off from a site in northeastern Pakistan and reportedly hit a target approximately 700 miles away. Then, in August, China finally conducted the first test of the mobile DF-31.[10]

Russian-American Missile Observation System—has also floundered. See Leonard David, "U.S.-Russia Working on Satellite Missile Watching System," October 24, 2001, www.space .com; Victoria Samson, "Prospects for Russian-American Missile Defence Cooperation: Lessons from RAMOS and JDEC," *Contemporary Security Policy* 28, no. 3 (December 2007), pp. 494–512.

The remainder of the year included a number of Russian missile launches. On October 2, Russia test-fired a Topol (SS-25) missile from Plesetsk as well as a submarine-launched ballistic missile. Far more deadly Russian missile launches followed in late October and November, when Russia fired, and a DSP satellite detected, the firings of SS-21 and Scud-B missiles from the Russian airbase at Mozdok, about sixty miles northeast of Grozny. An attack on October 21 involving two missiles hitting a Grozny marketplace and a maternity ward resulted in at least 143 deaths. In early November, SS-21s landed in and around Grozny. Less deadly was the test of two missiles on November 17, which were fired from a submarine in the Barents Sea and hit their targets on the Kamchatka Peninsula, about 3,100 miles away. Then on December 14, Russia conducted its ninth Topol-M test, which took twenty-four minutes from the time the missile left Plesetsk to the time it hit the target area on Kamchatka.[11]

On September 21, 2000, Iran conducted the third Shahab-3 test, which proved to be a failure for Iran's missile program when the rocket exploded shortly after liftoff, but another success for the DSP program when one of its satellites detected the launch. Precisely one week after the Iranian failure, Russia proved more successful when it conducted the twelfth Topol-M test, again departing from Plesetsk and landing on Kamchatka. In early November, another DF-31 was test-fired from China's Wuzhai Missile and Space Center, about 250 miles north of Beijing. Another test followed in mid-December.[12]

As expected, testing continued in 2001, including an Indian test of the Agni-II on January 17. A month later Russia conducted several tests in a single afternoon. At 1:28 P.M. Moscow time on February 16, a submarine in the Barents Sea launched a missile toward the Kamchatka Peninsula, while fifteen minutes later a Topol-M was fired from Plesetsk, also headed to Kamchatka. According to official Russian sources, all the missiles hit their targets. In mid-October China launched a JL-2, the submarine version of the DF-31, firing it from a specially modified Golf-class diesel submarine off the coast of north-central China. That test was followed in December by a test of CSS-7 (M-11) missile, launched from a facility in northwest China.[13]

While monitoring and reporting on such tests represented DSP's primary mission, the satellite constellation also continued to provide data on other events of both military and civil/scientific significance. In the aftermath of the September 11 attacks, the United States began, before the end of the year, bombing raids on Taliban facilities in Afghanistan. Among the infrared signals detected by the DSP constellation were the secondary explosions that followed attacks on those facilities, which often contained weapons and ammunition. Some of the resulting signatures were the largest ever detected by the launch detection satellites.[14]

One event that was both a military event and a tragedy occurred on October 4, 2001, when a Russian TU-154 airliner, which had left Tel Aviv en route to Novosibirsk, exploded over the Black Sea, killing its seventy-six passengers and crew. It was soon reported that a U.S. satellite, undoubtedly a DSP satellite, had detected the launch of a S-200 (SA-5 GAMMON in U.S. terminology) surface-to-air missile at the same time that the aircraft exploded (an explosion probably also detected by DSP spacecraft).[15]

The constellation also continued to detect the detonations of meteors and assorted fires. On April 23, 2001, a DSP satellite detected "a blinding flash of light over the Pacific several hundred miles southwest of Los Angeles," a flash that was matched by seismic signals that were detected halfway across the world. While the initial concern was that the flash and seismic signals indicated a atmospheric nuclear detonation, it soon became evident that DSP had performed one of its civil functions, the detection of meteors that crash into the earth's atmosphere and explode, creating an intense fireball.[16]

As noted earlier, DSP has been detecting fires since the first satellite was orbited in November 1970 and its use was noted in the 1994 Environmental Task Force report. Such data was apparently employed in response to large wildland fires in California and Nevada in 1999. During July and August 2000, the most active fire season for the United States in fifty years, the "Federal fire community benefitted greatly [deleted]," according to a report by the interagency Civil Applications Committee, whose functions include passing on requests for the use of "national systems" in support of civil missions (i.e., those of nonmilitary, nonnational security agencies).[17] The material deleted from the committee's declassified 2000–01 activity report undoubtedly refers to data by DSP spacecraft that allowed identification of the intensity and breadth of fires.

While DSP satellites in orbit were busy detecting missile launches and other infrared events, back on earth the Air Force and some of its contractors were busy anticipating and working on the next generation of warning satellites. In mid-November 1998, at the Combat Air Force Commander's Conference, the Air Force Space Command presented an overview of the SBIRS program, which noted four key mission areas—ballistic missile warning, provision of data to missile defense systems, technical intelligence, and "battle space characterization"—that included both the SLOW WALKER and FAST WALKER missions.[18]

Attendees were told of the planned SBIRS architecture (four geosynchronous satellites, two highly elliptically orbiting payloads, twenty-four low-earth orbiting satellites, bent pipes to relay the data to a Mission Control Station or its backup, and mobile processors). It was also noted that the SBIRS-High sensors—a scanning sensor like the one possessed by DSP and a staring sensor to look at a selected territory—would cover a wider portion of the electromagnetic spectrum than the DSP sensor, and the SBIRS-Low sensors would cover an even wider range. The Air Force Space Command also specified the distribution of missile warning and defense missions between the two types of satellites. SBIRS-High satellites would be performing global and theater coverage, detecting missiles during their boost phase, and then handing over the tracking mission to SBIRS-Low. One SBIRS-Low sensor would acquire the target, while another would track the missile through its postboost, midcourse, and reentry phases. In performing the technical intelligence and battle space characterization missions, SBIRS-High satellite staring sensors would detect missile launches and other "bright" events while SBIRS-Low spacecraft would focus on smaller areas to detect "dim" events.[19] In addition, the audience learned of different concepts of how SBIRS-Low would fit into national and theater missile defense schemes, including missile defense radars being cued by SBIRS-Low or interceptors being fired in direct response to an SBIRS-Low detection. The presentation also noted the success of DSP satellites in detecting the explosions of strike aircraft, a subject relevant to the Air Force's occasional need to launch search and rescue missions. Four events were noted: the crash of an F-15E during Operation Desert Shield, the shootdown of Sean O'Grady and his aircraft during operations in the Balkans, the crash of an F-117 stealth fighter in western New Mexico, and an A-10 crash near Eagle Colorado. The presentation promised that SBIRS would also detect such events but would "provide better information on location and time." Also promised was the ability to detect artillery flashes and explosions.[20]

Over the next several years, articles in military-oriented magazines occasionally focused on the promise of SBIRS. The October 2000 issue of the Navy Space Command's *Space Tracks* contained a review of the SBIRS program that discussed its expected ability to support naval missions, including theater missile defense. Its readers also were told that the initial launch of SBIRS-High would take place in fall 2004 and the constellation of four geosynchronous satellites would be in place in 2008. Another article, in the July 2001 issue of *Air Force Magazine* (a publication of the private Air Force Association), noted both the potential benefits of SBIRS and some of the problems of the low-altitude portion of the program.[21]

But while both the high and low segments of SBIRS held great promise, it was also clear that there were concerns about the feasibility of realizing those promises. Problems with the software for the SBIRS-High ground segment led to a halt in testing in October 1999, and resumed in July 2000 to determine if the problem had been fixed. By late in the year, the Air Force had decided to alter the design of the SBIRS-High satellites to allow them to detect dimmer missile exhaust plumes after testing indicated that the sensors (for the geosynchronous satellites) might be collecting more sunlight than expected. In early 2001 it also became apparent that there were problems—technical issues along with testing failures—with the infrared payload for the highly elliptically orbiting portion of SBIRS-High. In October, the House Appropriations Committee, citing major technical problems, added $30 million to the SBIRS-High development account while cutting $94 million that had been earmarked for purchase of satellite components.[22]

Not surprisingly, as software and hardware problems appeared, projected costs rose and kept rising. In mid-November 2000, *Space News* reported that the original price tag for the SBIRS-High satellites and ground station—which had been $1.8 billion when the contract had been awarded to Lockheed Martin—had risen to $2.9 billion, mostly due to the Defense Department decision to stretch out the SBIRS-High development program by two years. It also reported the Air Force's claim that the first SBIRS-High launch remained on schedule for 2004, although that schedule represented a two-year delay from the original plan for a 2002 first launch. But by the end of the year—in an effort to save some more money—the Defense Department's comptroller proposed delaying, by one year, the launches of the second through fourth satellites. That proposal was rejected, but in March it was reported that the first launch had been delayed until after September 30 so that the costs incurred would be part of the 2005 fiscal year budget. About a year later, government officials reportedly described the program as being "as much as $2 billion over cost and running almost three years behind schedule," with possible further cost growth and schedule delays in the future.[23]

By December 2001, in the wake of the problems that had plagued SBIRS-High over the previous three years, as well as a classified General Accounting Office (GAO) audit of the program, there was talk in the Pentagon of killing the program entirely, although that option was considered unlikely.[24]

A more likely component of SBIRS to be a casualty of technical difficulties and escalating costs was SBIRS-Low. Missile launch detection from high-altitude orbit had been part of America's national security space effort for over three decades and was still considered vital. In contrast, the

SBIRS-Low constellation represented a new capability, and one that might be of relatively little value if a viable national missile defense system was not established. In addition, the very size of the SBIRS-Low constellation might make it appear to be an unaffordable luxury.

And certainly, the SBIRS-Low program would experience its share of problems, delays, and cost growth. In early 1999, with initial deployment slated for 2006 and the tenth satellite to be orbited in 2010, the Air Force canceled a pair of SBIRS-Low flight demonstration projects involving Boeing and TRW because both efforts experienced major delays and cost overruns. Then there was a delay in awarding the two study contracts that were preliminary to picking a contractor, awards that had been scheduled for mid-June. The problem was a dispute between the Air Force and the Defense Department's operational test and evaluation office. While the Air Force planned to include space-based experimentation as an option in the studies, the test and evaluation office director, Philip Coyle, wanted some type of on-orbit test to be mandatory. The dispute was resolved with an agreement that the first satellites in the constellation would be test satellites, used to determine whether certain satellite components—including the filters used to screen out certain portions of the electromagnetic spectrum—were adequate for the mission.[25]

With that problem resolved, in August the Air Force awarded the contracts, worth $275 million and extending thirty-eight months, to Spectrum Astro (and subcontractor Northrop Grumman) and TRW (and subcontractor Raytheon). Instructions to the competitors made it clear that their proposals should focus on designing systems to support both national and missile theater defense, and that capabilities for collecting technical intelligence and performing battle space characterization should be secondary considerations.[26]

But long before those study efforts were due to conclude, there were suggestions from official sources that the program's projected cost—$10.6 billion according to the Congressional Budget Office—was more than the Defense Department was willing to pay. In a September 2000 interview with *Space News*, Secretary of the Air Force Whitten Peters reported that "the whole program is under review again" and that some Pentagon officials were questioning whether SBIRS-Low was needed, affordable, or technically feasible. It was also possible that the Air Force would lose responsibility for the program, since Congress had mandated its transfer to the Ballistic Missile Defense Organization. In October, Deputy Secretary of Defense Rudy de Leon deferred a decision on whether to transfer the program or request that the mandate be voided, partly because of uncertainties about its ultimate cost.[27]

In February 2001, the GAO produced a report on the program that concluded there was a high-risk of SBIRS-Low missing its initial deployment date. According to the report, the SBIRS-Low program office judged that there was a high risk that five of the six critical technologies—including the scanning and tracking infrared sensors as well as the satellite communications crosslinks—would not be ready when needed. In addition, the watchdogs complained that Air Force plans for flight-testing crucial satellite capabilities would not be available until five years after the start of production.[28]

The findings troubled Rep. Jerry Lewis (D-Calif.), who had requested the report. He announced, "I will urge my colleagues to review this report and weigh its findings as we consider these programs this year." Also disturbed, but for a far different reason, was Col. Michael Booen, the Air Force's program director for SBIRS-Low. In a letter to his superiors, Booen wrote that the report was "inaccurate, and its conclusions, recommendations and basic premise . . . are largely overcome by events," which included a plan to stretch out the schedule for SBIRS-Low launches so they could be more thoroughly tested. In addition, Art Money, the assistant secretary of defense for command, control, communications, claimed that GAO auditors misstated the risks associated with the software development strategy and other technologies, a critique rejected by the GAO.[29]

Whether in response to the GAO, congressional concerns, or their own internal evaluation, the Pentagon announced plans in summer 2001 to add nine months and $230 million to the TRW and Spectrum Astro study contracts to, it was hoped, resolve complex issues that would be more expensive to deal with later in the program. Coyle, by now the former Pentagon operational test and evaluation chief, questioned whether nine months would make a difference in solving the most serious problems. More significant to the Air Force were the reservations of the Senate and House of Representatives, both of which cut program funding. The Senate characterized the proposed extensions as "premature," given an ongoing Defense Department study on alternatives to SBIRS-Low. Then, by late October, the House Defense Appropriations Committee, charging that an internal Pentagon study "call[ed] into question the effectiveness of a SBIRS Low system to provide target discrimination," in light of a projected cost of $23 billion for the complete constellation, wanted to kill the program entirely—while allocating $250 million for satellite sensor technology.[30] But by the end of the year, it was reported that Rep. Lewis had agreed to consider changing the language in the House version of the 2002 defense authorization act so that SBIRS-Low would continue to be a procurement program rather than being downgraded to a research effort.[31]

★   ★   ★

In late 1999 the first test of a kill vehicle for a national missile defense system proved successful. On October 2, 1999, a Minuteman missile was launched from Vandenberg Air Force Base, California, followed by the launch of an interceptor from the Marshall Islands, about 4,300 miles away. The Minuteman released a dummy warhead and decoy balloon, which was followed by the intercept of the warhead over the Pacific Ocean.[32]

That test did not spare the program from severe criticism from a Pentagon-chartered review group, the Panel on Reducing Risk in BMD Flight Test Programs, headed by retired General Larry Welch, whose report was written before the October test. That report charged that the missile defense effort was plagued by overoptimism, inadequate testing, shortages of spare parts, and management failures. The next two tests helped validate the panel's critique. In mid-January 2000, an interceptor launched from the Marshall Islands failed to hit a mock warhead deployed from a Minuteman fired from Vandenberg when the interceptor's infrared sensor failed to direct the kill vehicle to the target. The next test, the following June, also resulted in a failure.[33]

In the midst of the failures, there were charges that the missile contractor tests were doctored or rigged by using simpler and fewer decoys, making it easier for the interceptors to find their targets. During this same time period, the GAO noted that only three of the nineteen planned intercept attempts were scheduled before a July deployment review and that "none of these attempts will expose the interceptor kill vehicle to the higher acceleration and vibration loads of the much faster, actual system booster." That report was also followed by another report by a Welch-led group, the National Missile Defense Independent Review Team. It concluded that while there was the technical capability to develop and deploy a limited system by 2005, "meeting [that] goal with required performance remains high risk. The team also noted that due to flight test restrictions on overflight, impact area, and space debris, "current plans provide flight tests in only a limited part of the required operating envelope." In addition, their report warned that advanced decoys were "likely to escalate the discrimination challenge." Subsequently, President Clinton decided to leave the decision for national missile defense to the next administration.[34]

Such a decision was only one of a number of decisions that needed to be made concerning missile defense, given the number of military service–theater missile defense programs—programs that could also use the data provided by SBIRS satellites, both high and low. A May 2000 GAO report recommended that the Navy Theater Wide program—whose goal was to defend military and civilian targets from medium- and long-range missile attacks—be revised to reduce the risk. The report expressed concern about the rapid test schedule, which included a test every two and a half months rather than the standard six months, and that there would not be enough time for the evaluation of test results and the incorporation of solutions into the following test. In addition, the GAO found that the funds the Defense Department had budgeted for the program were less than what the Navy believed it required to fund the program.[35]

Also in the mix of missile defense systems were the Navy Area Defense, the Army's Theater High Altitude Area Defense, the Air Force Airborne Laser, and the Patriot PAC-3, whose earlier configuration benefitted from DSP data during the 1991 Gulf War. The Patriot program's mission remained the same: protection of U.S. troops and bases from short- and medium-range ballistic missiles as well as cruise missiles and aircraft.[36]

The 2001 launch of DSP Flight 21 was followed by two years of DSP launch inactivity. But there were developments in the network on the ground for receiving and processing the satellite's infrared data. On December 18, 2001, the Air Force declared the SBIRS Mission Control Station (MCS) at Buckley Air Force Base, operated by the 2nd Space Warning Squadron, which had begun performing primary command and control operations for the DSP satellites in June 2001, to have reached its initial operational capability. The declaration represented the formal replacement of the three DSP ground stations—at Buckley, Nurrungar, and Kapuan—as well at the ALERT facility at Falcoln Air Force Base with the MCS.[37] The data that was once received at Nurrungar and Kapuan was now relayed via "bent pipes" at the Relay Ground Station–Pacific (at Pine Gap, Australia) and the Relay Ground Station–Europe (at Menwith Hill, United Kingdom).

By 2004 at the latest, the ground network for DSP consisted, in addition to the MCS, of the SBIRS MCS Backup, the SBIRS Interim MCS Backup, and three operational and two contingency JTaGS sites. The three operational JTaGS sites were JTaGS-Pacific Command (JTaGS-PAC, Osan, Korea), JTaGS-European Command (JTaGS-EUR, Stuttgart, Germany), and JTaGS-Central Command (JTaGS-CENT). There were two additional Contingency Units remaining within the United States: JTaGS-COS at Colorado Springs, Colorado, and JTaGS-TX at Ft. Bliss, Texas.[38]

With the delays and problems plaguing the SBIRS program, launching the remaining DSP spacecraft on schedule and successfully became even more important. Col. Robert S. Reese, the SBIRS deputy system program director at the Space and Missile Systems Center, called the final two DSPs the "cornerstone" for extending DSP operations sufficiently far into the future to guarantee that there would be no gap in America's ability to detect missile launches. The Titan IVB launch of February 14, 2004, from Cape Canaveral, successfully carried DSP Flight 22, nicknamed "Eagle Eye," to its geosynchronous orbit and represented the first of two steps.[39]

When it was anticipated that DSP Flight 23 would be launched in March 2005, it was also expected that its launch would be one of the initial launches on a Delta IV heavy rocket, which was to replace the Titan IVB as the way heavy satellites were placed in geosynchronous orbit. But it would not be until May 2005 that the last DSP spacecraft would arrive at Cape Canaveral. By that time, there were questions about the reliability of its new launch vehicle, resulting from its initial launch in December 2004, which failed to place a test payload into the proper orbit. Thus, a year after being shipped to the launch site, DSP-23 was not expected to launch before 2007.[40]

The satellites that were in orbit continued, of course, to report on the multitude of targets that generated infrared signals of sufficient strength to be detected and reported. Predictably, China and Russia contributed a number of missile launchings. Chinese tests included a test of a DF-31 reentry vehicle on January 3, 2002, that failed when the booster blew up in midflight, a July 2002 launch of a CSS-5 medium-range missile to test warheads designed to defeat missile defenses, and a late August 2002 test of the 4,340-mile range DF-4. During the last quarter of 2003, Russia conducted three ICBM tests, including a December 5 launch of a SS-19 STILETTO.[41]

In early 2002, Iran also conducted an engine test, possibly for a Shahab-3, that resulted in an explosion and fire that was probably detected by the DSP satellite that covered that region of the world. Of course, in 2003, the short-lived second war with Iraq required DSP to be employed to detect any Iraqi missile launches intended to hit allied targets, including those in Saudi Arabia and Kuwait. With the beginning of the war on March 19–20, Iraq began firing those missiles at targets in Kuwait, including Camp Doha, the Army's main base in the country, with thirteen launches between March 20 and March 29. Against targets in Iraq, the Iraqis began with a March 27 launch (which was either an Ababil-100 or an extended range FROG), followed by a single launch on April 1, along with double and single launches of FROG-7 missiles on April 2.[42]

However, DSP was not able to provide the same level of information as in the 1991 Gulf War because most Iraqi missiles launched were Ababil and

Al-Samoud missiles, which flew between sixty-two and a hundred miles, longer-range missiles having been prohibited and destroyed after the 1991 war. While the infrared sensors on DSP apparently detected twenty-six launches, there was insufficient time between launch and impact to verify that a short-range missile had been fired (rather than that some infrared event occurred). An Army briefing attributed only one instance of missile attack warning to DSP: the March 3 launch of a short-range ballistic missile. (And that was attributed to DSP-Augmentation.) Instead, the Army had to rely on ground-based and sea-based radar systems to provide warning, including cuing of the Patriot missile defense batteries.[43]

DSP contributed to the war effort in another way: helping with bomb damage assessment, specifically by cataloging the explosions resulting from cruise missile strikes. Space operations (i.e., DSP) were identified as having detected 1,493 "static IR events" and 186 "high explosive event[s]," which included both detonations of arms facilities and shot-down aircraft.[44]

With the war over, DSP detections of missiles again usually involved missile tests such as the mid-February 2004 Russian test that resulted in the missile's self-destruction in midair. The following month Pakistan fired a Shaheen-2 with a range of up to 1,250 miles. It was also reported that month that China had carried out at least five missile tests since January, including tests of the short-range DF-11 and DF-15 missiles and the longer-range DF-21 (with a range of about 1,116 miles) and DF-31 missiles. On August 11, Iran tested its Shahab-3 while Russia test-fired a SS-19 STILLETO missile that traveled from Kazakhstan to its target on Kamchatka, a journey of about 4,154 miles.[45]

Missile tests over the next two years that were probably or definitively noticed by DSP infrared sensors include a late April 2005 short-range missile test by North Korea and Syria's test-firing of three Scuds (with ranges between 185 and 435 miles) in late May, including one that broke up over Turkish territory. Between July and October, Iran conducted three or four tests of an upgraded version of the Shahab-3 that traveled between 930 and 1,240 miles and were detected by one or more DSP spacecraft. On November 1, Russia tested a Topol-M (SS-27) missile with its maneuverable warhead, a test that involved a flight from Kapustin Yar.[46]

In the early years of the decade, DSP satellites also detected a number of explosions. On January 28, 2002, a series of explosions at the Ikeja munitions depot in Lagos, Nigeria, produced fireballs and rattled Nigeria's commercial district, destroying homes and businesses over a significant area and killing at least 500 people.[47]

DSP satellites also picked up their share of asteroids, usually about thirty per year. At least two of the spacecraft detected the infrared signal generated by a large meteorite (fifteen to thirty feet in diameter) that dove into the Earth's atmosphere over the Mediterranean Sea on June 6, 2002. Later that year, in late September, DSP tracked an asteroid above a remote region of Siberia. That satellite monitored the rock from thirty-nine miles above the earth down to nineteen miles.[48] One particularly tragic detection involved the crash of the space shuttle orbiter *Columbia,* which, engulfed in flames, crashed in central Texas on February 1, 2003. About the time the shuttle reentered, a DSP sensor detected a heat spike. DSP data, along with imagery, was offered to NASA to aid in the investigation of the crash. Another occurred on February 28, 2004, when a DSP satellite detected the explosion of the *Bow Mariner,* a tanker carrying 3.5 million gallons of ethanol, fifty-five miles off the coast of Chincoteague, Virginia, a blast that killed eighteen crewmen. First considered an unknown event, additional sources of information made it clear that the infrared signal came from an incident in the shipping lanes off Virginia.[49]

Two foreign blasts were also, given their intensity, probably detected by DSP during the first half of 2004. A February 23 explosion at the Solid Propellant Booster plant at the Satish Dhawan Space Center, just off India's east coast, flattened Building 117, killed six workers, and injured three. Then, on April 22, a collision of two trains loaded with fuel produced a massive explosion just north of the Ryongchon train station in North Korea. The blast, which was first estimated to have killed 3,000 people (although a later estimate was 160 killed and 1,300 injured), destroyed homes and other buildings and spread debris for miles. In September, another large explosion in northern North Korea, which may also have been detected by DSP sensors, resulted in a two-mile-wide mushroom cloud. It was quickly determined not to be a result of a nuclear detonation, and North Korea stated it was the result of excavation for a hydroelectric project.[50]

DSP data was employed in 2002 by a JTaGS unit to guide firefighters toward the hottest spots in the Hayman Forest Fire that burned over large portions of Colorado. During the following years, there was significant fire activity, including the 2004 fire season, during which 65,878 fires burned 8,094,531 acres, the details of which DSP satellites provided to the Department of Agriculture's Forest Service. Two years later, there were 96,385 large fires and 9,873,429 acres burned, and DSP spacecraft apparently again provided data to the Forest Service.[51]

In 2003, DSP also apparently provided data in response to the May 2003 eruption of the Anatahan Volcano in the Commonwealth of the Northern Marianas, an eruption that blanketed Anatahan Island with ash. Such information was not an uncommon occurrence since the Volcano Hazards Program was established by the U.S. Geological Survey in 1980. DSP detection of volcanic activity sometimes was due to the thermal signatures of lava flow, although more frequently it resulted from the sunlight reflected from the ash plume rising high in the atmosphere.[52]

While the successful launch of DSP Flight 22 helped replenish the constellation, the need to solve the problems with the successor program, particularly the SBIRS-High component, remained urgent. One sign of concern was the report in early 2002 that the National Reconnaissance Office was working on possible alternatives to the SBIRS program. Peter Teets, the Under Secretary of the Air Force and NRO director, stated that the Pentagon would consider one of the reconnaissance office's "creative" solutions if the Air Force could not solve the problems plaguing SBIRS-High.[53]

A few months later, in April and May, there were continued warning signs that the future of SBIRS-High might be problematical. *Space News* reported the program had "slipped into disarray," and that an independent review team had found that the program had been "allowed to move through programmatic milestones before the technology was ready, that certain program requirements were not considered in designing certain components, and that there was a general breakdown in program management." *Aviation Week & Space Technology* headlined an article "It's High Noon for SBIRS-High" and informed its readers that the Pentagon would soon face the choice of sticking with SBIRS or seeking some alternative program.[54]

In early May, the Defense Department "certified" the SBIRS-High program, a requirement of the Nunn-McCurdy amendment to the 1982 defense authorization act, which specified that program unit cost increases of 15 percent or more trigger a requirement for detailed reporting to Congress about the program, while increases of 25 percent or greater require the Secretary of Defense to certify the program and stipulate that the system is essential to national security, that there is a lack of alternatives that provide equal or greater capability at less cost, that costs had been brought under control, and that program management was sufficient to manage and control costs. The certification attributed the high costs to "increased complexity and software for the [highly elliptically orbiting] payload . . . and increased complexity and weight for the [geosynchronous] payload." Despite that recertification, there was no guarantee that the program would survive for long. On the day of certification, Under Secretary of Defense Edward Aldridge told reporters that a recent look for cheaper, better alternatives to SBIRS-High had come up empty. But he also told his audience that "if we find that six months from now, the program is going south, I have no hesitation to pull the plug." Seven months later, the plug had not been pulled but SBIRS-High and what had been SBIRS-Low were among the space programs that Air Force Secretary James Roche characterized as being "in trouble" and with "costs billions of dollars more than expected."[55]

Early 2003 brought a report that radio waves emitted by the first SBIRS-High sensor earmarked for highly elliptical orbit might disrupt control systems on the host satellite, which would require thirty-nine modifications to the first two sensors and delay delivery of the first sensor by ten months. Tests also revealed problems with the first sensor's ability to maintain earth coverage and missile tracking. In May, the Defense Science Board and the Air Force Scientific Advisory Board issued a joint report on national security space program acquisition. It observed that "SBIRS High has been a troubled program. It could be considered a case study for how not to execute a space program." It specified a variety of problems, including underfunding, an optimistic contractor proposal, uncontrolled requirements, limited program manager authority, as well as funding and program manager instability. It did recommend continuing with the program as it had been restructured.[56]

That fall, the GAO cast a critical eye on SBIRS-High, as well as other space programs. In October, the office reported that the cost of the program had risen by over $2 billion to $4.4 billion, and noted the recent restructuring of the program, which included tighter management controls to minimize further changes in requirements and to improve management of software development. However, the office concluded that "the restructuring . . . did not fully address some long-standing problems in the development of SBIRS-High." Thus, the program was "at substantial risk of cost and schedule increases."[57]

Among the continuing problems cited was that "delays in the development of the first sensor have had a cascading effect," with continuing design and software development work on the first sensor diverting personnel and other resources from follow-on developmental work. Thus, it was unlikely that the second highly elliptically orbiting satellite and first geosynchronous satellites would adhere to their revised development and launch schedules. The report also noted a continuing problem with changing requirements, reporting that there had been ninety-four requirement changes before restructuring, and thirty-four actions that required contract modifications after restructuring. It also noted the recent delays that had already become part of the revised program, with the launch of the first geosynchronous satellite pushed back from September 2004 to October 2006, and the second launch pushed back from September 2005 to October 2007.[58]

The report also noted that delivery of flight software for the highly elliptically orbiting satellite had been delayed, as had its ground software, which missed its revised delivery date of August 2003. There was also concern about the remaining computer memory margin on board the satellites. In addition, the GAO expected that on-orbit performance information about the elliptically orbiting sensor would not be available in a timely manner for geosynchronous sensor development efforts.[59]

October and November also brought further reports of SBIRS troubles, although they were coupled with Air Force claims that the problems were not serious. *Space News* reported that "technical difficulties continue to hamper progress on the U.S. Air Force's next-generation satellite system for missile warning," but Col. Mark Borkowski, the SBIRS program manager, claimed that the problems were manageable and would "only" add tens of millions of dollars to the overall cost. The additional costs would be "a drop in the bucket next to the estimated $8 billion total cost of the SBIRS High program," an almost fourfold increase from the original expected cost of $2.1 billion. But in November it was reported that the first SBIRS-High sensor would not be ready for operations until summer 2005, rather than the previously expected November 2004, due to a delay in the expected delivery of the sensor and the satellite that would carry it in highly elliptical orbit.[60]

Early 2004 brought at least a glimmer of good news for the troubled program—Lockheed Martin had successfully completed testing on the first geosynchronous satellite's sensor—which involved simulating the harsh conditions of space. But February also brought a report that the Pentagon's acting acquisition chief, Michael Wynne, had directed the department to examine alternatives to SBIRS-High. That report was followed by another, which provided more specifics: the first two SBIRS-High launches, scheduled for late 2006 and late 2007, were likely to be delayed by a year, and the Air Force was exploring the possibility of halting the SBIRS-High program at two geosynchronous satellites. A new program would then be established to provide further high-altitude sensors and spacecraft.[61]

While the Air Force was reconsidering the future of SBIRS-High, it also wanted more money to pay for whatever future it did have. In late March, two months after submitting its 2005 budget request, Air Force Under Secretary Peter Teets informed Congress that the $508 million it had requested wouldn't be enough, explaining that the Air Force was anticipating "another cost problem" and additional delays. That cost problem helped push the expected bill for the program up to $9.9 billion, an increase of more than 15 percent from the previous year's estimate.[62]

The summer did bring some good news when, on August 5, Lockheed Martin delivered the first sensor for highly elliptical orbit, whose electromagnetic interference problem had accounted for about $1.5 billion in cost growth. SBIRS also received a vote of confidence when Thomas Young, whose 2003 report was critical of the program, observed that "a lot of critical technical issues [had] been resolved." As might be expected, that news was followed in October by news that the first SBIRS-High launch might not take place until fall 2008. There was one further piece of good news for the year when a review of the SBIRS-High signal processing software, which allows the infrared signatures of events such as missile launches to be distinguished from background noise, established that it would function properly within the satellite assembly.[63]

The end of 2004 also found Congress directing the Air Force to provide a detailed report on "the cause of the most recent SBIRS cost increases, schedule delays, and technical problems; the most recent Defense Support Program gap analysis and any effect that further delays will have on U.S. early warning, technical intelligence, and missile defense capabilities; steps taken to address the most recent SBIRS technical difficulties; . . . remaining risk areas; and an assessment of the confidence level in the SBIRS schedule and cost estimates current as of October 1, 2004." That obligation was fulfilled in March when the Office of the Secretary delivered its *Report to the Defense and Intelligence Committees of the Congress of the United States on the Status of the Space Based Infrared System Program.*[64]

The report assessed the risks of fielding the first elements of a new system, software-related risks, and other high-priority risks (including theater probability of warning and the probability of collecting technical intelligence on target areas). It assessed all of the risks as either medium or medium-low. It also provided a schedule risk assessment involving twenty-two items in three different categories: system effectiveness, the space segment, and the ground segment. Specific items included message certification, delivery of sensors, payloads, satellites, and launch dates. In every case but two, the confidence level was medium. With regard to cost, the report's authors concluded that there was "75% confidence" that the program could be carried out within its current budget.[65]

Along with whatever optimism might have been engendered by the Department of Defense report came the, by now predictable, notification of an increase in the taxpayers' bill for the program. Pete Teets, now Acting Secretary of the Air Force, in a letter to Senator John Warner, reported that he had "reasonable cause" to believe that it was likely that the average procurement unit cost of the SBIRS program had exceeded the 15-percent threshold and possibly the 25-percent certification threshold. That increase came from a per-unit cost increase for the third through fifth geosynchronous satellites. Teets also informed the Senate Armed Services Committee chairman that he directed an independent program assessment to review the technical and cost components of the program, an assessment that would consider alternative acquisition strategies. He also reported, in subsequent congressional testimony, that the first sensor for the highly elliptically orbiting portion of SBIRS had been delivered in August 2004 and "we are on track for delivering our second payload in June of this year."[66] March also brought news that the second infrared sensor destined for the highly elliptically orbiting portion of SBIRS-High was having problems, including a software "anomaly that required additional analysis and rework," according to an Air Force spokesman. It was hoped that the anomaly would be resolved by the end of the month. Complicating matters further were the technical problems plaguing the NRO signals intelligence satellite that would host the sensor.[67]

Later that month Teets, who retired on March 25, tried to put the SBIRS problems in context. He told *Space News* that the Air Force had time to solve the remaining technical difficulties with the SBIRS-High program without risking a coverage gap before the DSP constellation began blinking off. However, it would be a concern if the SBIRS satellites were not on orbit early in the next decade—that is, in the 2010–2012 time frame. He further estimated that a hard deadline for launching the first SBIRS-High spacecraft without facing a gap was 2015. The leeway that did exist, he explained, was the result of the longer-than-expected life of the DSP satellites (about ten years). Of course, there was one DSP satellite yet to be launched and Gen. Lance Lord, the commander of the Air Force Space Command, noted that it was vital that the final DSP be launched successfully and join the operational constellation.[68]

About a month after Teets's departure, the Air Force revealed that the cost of getting those satellites into orbit had gone up again, by another $1 billion. Options for ending the SBIRS effort with two geosynchronous satellites and moving on to a new program were still being reviewed. More bad news followed in July and August. First, it was that the second infrared elliptically orbiting sensor, which Teets predicted would be delivered in June, was not likely to be delivered by August. Then, on July 28, Acting Secretary of the Air Force Michael Dominguez, informed Congress that the pricey SBIRS-High program had gotten even pricier, with the price tag going over 25 percent, triggering another Nunn-McCurdy review. To add insult to injury, in early August the Air Force revealed that the current expected first launch date of June 2008 was in jeopardy.[69]

September and October did bring some good news. In late September, Lockheed Martin shipped the SBIRS sensor for the second elliptically orbiting satellite to the Air Force. In early October, Northrop Grumman announced that it had completed the mechanical and electrical integration work on the payload for the first elliptically orbiting satellite, which would then be shipped to Lockheed Martin.[70]

Then, in December, key congressional leaders, including the chairmen of the appropriations, intelligence, and armed services committees, received a two-page letter from Under Secretary of Defense for Acquisition, Technology, and Logistics Kenneth Krieg. One aspect of the letter was no surprise, since it certified the necessity of a restructured SBIRS program. What was different were the specifics of the restructured, and scaled-back, program. It would consist of the guaranteed purchase of two geosynchronous satellites, two highly elliptically orbiting payloads, and the associated ground system. But rather than buying three more geosynchronous satellites, only one more would be purchased, and even then only once Krieg was "confident that the first . . . geosynchronous satellite can perform its mission."[71]

In addition, Krieg noted "the continued importance to support strategic and theater missile warning and missile defense" and his conviction of the "need to develop a competitor capability, in parallel with the SBIRS program." Thus, he wrote that the Defense Department would "work with Congress to initiate a new program for space-based Overhead Non-Imaging Infrared (ONIR)."[72]

That new program, or prospective new program, became known as the Alternative Infrared Satellite System (AIRSS). Its objective, Air Force Deputy Under Secretary for Space Programs Gary Payton said in February 2006, was "to keep it simple—a simple spacecraft, a simple sensor." But simple, for an infrared launch detection system, still required money, and the Air Force asked for $103 million for the 2007 fiscal year for the program, with the expectation of increased requests of $350 million, $535 million, and $700 million for the 2009, 2010, and 2011 fiscal years, respectively.[73]

In March, the Government Accountability Office (as the GAO had been renamed in July 2004) issued its yearly summary of the projected capabilities and status of selected major weapons programs. The report noted that "program officials . . . believe[d] a more realistic program schedule" for the geosynchronous SBIRS satellites had been developed, with "the first [geosynchronous] satellite delivery having been delayed an additional five months until late 2008." The next month a GAO official concisely summed up the history of the trouble-plagued program: "With unit cost increases of more than 315-percent over the 1996 initial estimate, the program has undergone four Nunn-McCurdy unit cost breaches. Total program costs have increased from about $4 billion to more than $10 billion. The launch schedule has slipped over six years; the first satellite is currently scheduled to be delivered in September 2008. Defense Department officials recently called for initiating planning efforts for the development of a new missile warning system, parallel to SBIRS."[74]

In April, Lt. Gen. Michael Hamel, commander of the Air Force Space and Missile Systems Center, told a national space symposium that the Air Force would take a lower-risk approach with the alternative system, while hoping to take advantage of advances that had taken place since the SBIRS program had begun. One possibility, raised by Deputy Under Secretary Gary Payton, was combining the wide-area coverage and staring capabilities into a single sensor.[75]

But, despite the Air Force's focus on an alternative program to SBIRS, in late November, Col. Roger Teague, Commander, Space Group of the SMSC's Space Based Infrared Wing, was still briefing the full-scale version of SBIRS—its two elliptically orbiting payloads and four geosynchronous satellites. He reiterated the view of SBIRS as having four missions—missile warning (including data on launch origin, missile type, trajectory, and impact point), missile defense (detecting missile tracks and cuing missile defense system), technical intelligence (characterizing missile and other infrared events, signatures, and phenomena), and battle space awareness (providing situational awareness, targeting, and assessment for command, control, and execution of joint operations).[76]

Teague also explained how the SBIRS system would improve on DSP. It would, his briefing asserted, have higher sensitivity, allowing it to see dimmer objects more often. Its staring sensor could be tasked, so while the SBIRS scanning sensor would perform the traditional worldwide launch detection mission of DSP, its staring sensor could be directed to focus on a particular region of the world. It would be able to provide more accurate estimates of launch points and impact points. Another comparison made in the presentation showed, in the form of a figure, the expanded capabilities of SBIRS. That figure is reproduced as Figure E-1.[77]

In early 2002, there were still two competitors for the SBIRS-Low contract, TRW (with subcontractor Raytheon) and Spectrum Astro (with subcontractor Northrop Grumman). But the original plan for the Ballistic Missile Defense Organization (which had assumed responsibility for the program in 2001 and was renamed the Missile Defense Agency in 2002) to choose a contractor that coming September had been scrapped because the program was being restructured, a restructuring that could result in the initial launch date being moved to 2011 from 2006 and being treated as a research and development effort, with the results from the first launch being used to redesign subsequent satellites. Under that scenario, a truly operational constellation might not be in place until 2015.[78]

Then, in April the Pentagon awarded TRW the SBIRS-Low contract, although Spectrum Astro would remain involved. But the initial work would be restricted to the production of two research and development satellites that would be used to determine if there was an actual need for a full constellation with long-wave infrared sensors and, if so, what particular technologies were required. And by August, the Pentagon's concept of a full constellation had shrunk dramatically. Under the new plan, SBIRS-Low would consist of no more than eight satellites and receive a new name, to avoid confusion with SBIRS-High.[79]

That new name would be the Space Tracking and Surveillance System (STSS). In March 2004 Northrop Grumman was completing construction of the first two satellites, with the first launch expected in June 2007. The satellites would then be employed during ballistic missile system defense tests. In May 2005, the Missile Defense Agency was tentatively planning to begin design work on an operational STSS constellation, which was projected to cost $2.7 billion through 2011.[80]

The George W. Bush administration brought a number of significant changes to the nation's various missile defense programs. One change, early in the administration, was the cancellation of the Navy Wide Area missile defense program, intended to be a sea-based system to defend ships and coastal areas against short-range ballistic missiles. The Navy's determination in fall 2001 that the program was 32 percent over budget led the administration to terminate the program.[81]

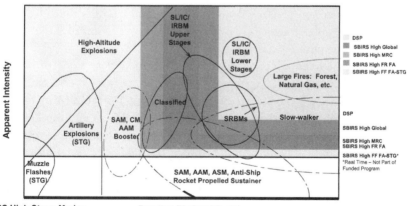

SBIRS High Starer Modes   Duration (Notional Scale)
• Step-Stare - Theater Major Regional Conflict (MRC)
• Step-Stare - TI Fast Revisit Focused Area (FR FA)
• Dedicated Stare – Fast Frame Focused Area (FF FA)*
• Step-Stare - TI High Sense Focused Area (HS FA) – not shown

E.1. DSP vs. projected SBIRS capabilities.

Another change was conceptual. Rather than thinking of prospective missile defense in terms of what was to be defended (e.g., U.S. territory, ships at sea, the theater, and so on), the new framework focused on where attacking missiles or their warheads were to be intercepted—in the boost phase, during the midcourse of their flights, or at the final (terminal) portion of their flight.[82]

Thus, programs for boost-phase interception included the Air Force's (now-defunct) Airborne Laser program, which was envisioned in 2005 as consisting of seven aircraft, five of which would be deployed to a theater to support two twenty-four-hour combat air patrols. Other possibilities for boost-phase intercept included a space-based laser, space-based kinetic weapons, and sea-based interceptors deployed on Aegis ships.[83]

What had been known as the National Missile Defense program, and involved the use of ground-based interceptors to defend the United States, became the Ground-Based Midcourse component of the Midcourse Defense Segment. Cuing for the system could come from a variety of sources, including DSP, SBIRS-High, STSS, an X-Band radar to be located on a sea-based platform, the COBRA DANE radar on Shemya Island, upgraded Ballistic Missile Early Warning System (BMEWS) radars in the United Kingdom and Greenland, and the Aegis SPY-1 radar. The interceptor missiles were to be based at Ft. Greeley, Alaska, and Vandenberg Air Force Base.[84]

Another potential option for midcourse intercept was the Sea-Based Midcourse component, which succeeded the Navy Theater Wide program (which was also called the Navy Upper Tier program). The interceptors would be based on Aegis ships. As of 2005, a number of options, some based on the SM-3 missile, were being considered as ways to improve the system so that it could be used against ICBMs.[85]

Also scheduled to benefit from the infrared detections of DSP and SBIRS satellites was the Terminal Defense Segment of the missile defense architecture. Since the use of the Patriot in the 1991 Gulf War, advanced versions of the Patriot Advanced Capability-2 (PAC-2) and the Patriot Advanced Capability-3 (PAC-3) had been developed. Patriots were intended to work in conjunction with the longer-range Terminal (formerly Theater) High Altitude Area Defense.[86]

Patriot batteries, in addition to being deployed to the Middle East, where they were employed during the 2003 Iraq War, would also be deployed to South Korea (in 2003) and Japan (in 2006) to protect those nations against missile attacks emanating from North Korea. At least in South Korea, the Army, in 2004, was integrating the JTaGS downlink for DSP data into the Patriot system.[87]

Of course, the development of plans and the deployment of interceptors did not mean that there had not been questions raised, spurred by test failures, about the capabilities to fulfill the actual mission. Just as questions were raised about the effectiveness of the Patriots during the 1991 Gulf War, there were challenges to claims of their effectiveness during the 2003 Iraq War. And there were challenges to the credibility of the tests as well as clear failures (in which the interceptor failed to hit its target), such as the one of December 15, 2004, when an interceptor fired from the Marshall Islands in the Pacific failed to hit a mock warhead fired from Kodiak Island in Alaska. There were also Patriot PAC-3 test failures in March, April, and May 2004.[88]

        By early 2007, the launch of DSP Flight 23 was on the horizon, but not on the immediate horizon. The delay had been caused by the problems with the original Delta IV test launch, as well as other launch requirements. In mid-January, the launch was scheduled for April 1, but was expected to slip significantly. And it did, but for at least one reason not anticipated in January. A liquid-oxygen leak during a countdown dress rehearsal in March damaged the launch pad for the Delta IV heavy rocket that was to place the DSP in orbit. Repairs required disassembling the launch vehicle, pushing the expected launch date to the summer.[89]

        By late July, the satellite was being stored at Cape Canaveral's Large Processing Facility, with August 28 as the anticipated launch date. But that date would represent only one of many that would pass without the launch of DSP 23, as would dates in October and early November. But at about 9 P.M. on November 10, the Delta IV lifted off from Cape Canaveral and eventually placed it in geosynchronous orbit, apparently to serve as the European satellite. Along with its infrared telescope and its usual nuclear detonation detection package, the spacecraft also carried an experiment, the Space and Atmospheric Burst Reporting System (SABRS) Validation Experiment (SAVE), which was intended to demonstrate new technology for subsequent nuclear detonation payloads.[90]

        After the numerous delays with DSP Flight 23 and the virtually endless problems with SBIRS, the Air Force, the contractor, and those responsible for early warning were elated and relieved by the apparent success of the final DSP mission. A Northrop Grumman official declared the launch to be "the beginning of Flight 23's operational service." A company spokeswoman also noted that the Air Force "[had] been telling the contractors . . . that this DSP-23 spacecraft [was] essential for the transformation to SBIRS." Northrop Grumman also took out an ad in trade publications, part of which read, "Congratulations DSP. The final satellite has launched and the mission continues." Gen. Robert C. Kehler, the commander of the Air Force Space Command, told reporters that the successful launch had boosted the Air Force's confidence that it would not face a missile warning gap before SBIRS became operational, as well as that the DSP satellite was "waking up the right way" and that the issue of a warning gap "looks a little bit different today than it did last Friday."[91] But a little over a year later, that optimism would evaporate. In July, the Air Force decommissioned DSP Flight 19, which after failing to arrive in its geosynchronous orbit had been employed to collect radiation data about the Van Allen belt and test command and control features. But then in November, reports from amateur satellite observers and Air Force sources indicated that there was a problem with Flight 23. Reuters reported that it had stopped working in September, and by December the satellite was drifting into the vicinity of other geostationary satellites.[92]

While the Air Force Space Command was able to avert any collisions with other satellites, it continued to investigate why such a crucial satellite failed. One means of the investigation was the use of two 500-pound Defense Advanced Research Projects Agency spacecraft, which were designated Mitex. The spacecraft, launched in 2006, maneuvered around and inspected each other while in geosynchronous orbit. Their photographs of the Flight 23 spacecraft might have revealed a bent antenna or the impact of a micrometeorite hit.[93]

The Mitex investigation effort was underway by mid-December, but in early January a U.S. defense official said, "There's not that much data available," and "You have to go back and recreate what might have been going on." He added that there might never be "great certitude" about what happened. Possibilities included defective parts, software problems, a natural phenomenon such as a solar flare, space debris, and even an intentional attack, although the last two possibilities were considered unlikely.[94]

Whatever the cause for the failure, the inability to employ DSP Flight 23 in the operational launch detection constellation meant that concerns about a possible gap reemerged and the requirement for SBIRS to remain on schedule became more pressing.

By the time of DSP-23's apparently successful launch, there had been both good and bad news concerning the SBIRS program. On June 27, 2006, about five years late, the first SBIRS highly elliptically orbiting sensor, the 500-pound HEO-1, was carried into space on a NRO signals intelligence satellite launched from Vandenberg Air Force Base on a Delta IV rocket. In April 2007 it was reported that the Air Force was contemplating declaring the sensor—which was already providing data on satellite activity—operational. The sensor was being used, according to one Air Force official, to track satellites from their launch to their deployment in orbit. It may also have provided data on China's antisatellite test of January 2007. In June, the SBIRS-High program manager reported that he expected that the sensor would be ready to contribute to other missions by the end of the month, characterizing it as "well-behaved" and stating, "We're very happy with the performance we are getting out of it." That judgment was based, at least in part, on HEO-1's use against operations at U.S. launch facilities and the Eglin Air Force Base test range, where the United States develops and tests air-to-air and air-to-ground weapons, as well as its monitoring of activity in the Middle East.[95] In March 2007, the GAO reported that the SBIRS-High program's three critical technologies, including the infrared sensor and onboard processor, were mature and that the launch of the first SBIRS geosynchronous satellite, GEO-1, was expected in late 2008, with another geosynchronous satellite expected to be launched in late 2009. But it also observed, "given the high probability of design flaws, costly redesigns that further delay GEO delivery are possible." It also reported that the Office of the Secretary of Defense had instructed that the priority was to maintain the schedule, even if that meant launching with a reduced capability.[96]

That same month, it was reported that the Air Force intended to award a contract in August for a launch-detection demonstration satellite that was to be used in the AIRSS program, whose deployment was envisioned to occur in 2015, and would rely on advances in focal plane technology. At the same time, there were rumors that the Air Force would return to its original plan to purchase five dedicated SBIRS satellites—buoyed by the success of the HEO-1 sensor—with AIRSS becoming a true next-generation replacement program, and the A standing for Advanced rather than Alternative. In June, the Air Force did decide to buy a third SBIRS spacecraft, while the Senate Armed Services Committee killed AIRSS funding in its version of the 2008 defense authorization bill, explaining that while AIRSS began as an alternative to SBIRS, it was now its successor.[97]

By July, one trade publication headlined an article, "SBIRS returns," and asked, "Is America's missile warning network back on track?" It noted a "blizzard of contractor press releases" that gave the impression that the SBIRS-High satellites were "finally taking shape." But it also reported that the director of the GAO's acquisition and sourcing management team stated that SBIRS managers were "just approaching the really hard part of it," adding, "Until they get through that, I wouldn't say this program is back on track."[98]

In mid-August it was reported Northrop Grumman had delivered the payload for the first geosynchronous SBIRS spacecraft to Lockheed Martin for integration into the satellite bus and final testing. Integration was completed by mid-September. In September, *Aviation Week & Space Technology* revealed that, due to Air Force confidence in the SBIRS effort, the service was changing the direction of the AIRSS program, designating it as the successor to SBIRS rather than as an alternative, and renaming it the Third-Generation Infrared Surveillance Satellite (3GIRS) program. The confidence was the result of the performance of the HEO-1 payload as well as the results of testing the geosynchronous payload.[99]

But two days later, a less rosy picture came from Secretary of the Air Force Michael W. Wynne, in a memo for the Under Secretary of Defense for Acquisition, Technology, and Logistics. The memo began, "This is about [the] Space Based Infrared Satellite (SBIRS) system, and is not good news." "The problem," Wynne wrote, "is with the GEO birds," specifically "a safe hold that did not work on a current satellite, causing mission termination."* The design similarity to the geosynchronous satellites caused a "no-fly condition." He also noted the fragility of the computer architecture and that the fourth geosynchronous satellite option might be at risk.[100]

Some viewed the memo skeptically, wondering whether Wynne was trying to lay the groundwork for restoring the AIRSS effort from a follow-on/technology demonstration program to an alternative to SBIRS. A congressional aide told a trade publication, "It's either really bad or they are trying to make it look really bad." AIRSS was not the only investigation of an alternative to SBIRS, with the Office of the Secretary of Defense commissioning, in early February, the Sandia National Laboratory to conduct a study, designated ARGUS, that explored the possibility of quickly building and launching a missile warning satellite to fill any gaps in coverage that might occur.[101]

A different reaction to the Wynne memo appeared in a *Space News* editorial, "Same Old Story on SBIRS," and included the opinion that "SBIRS long ago earned the ignominious distinction as the poster child for program mismanagement." However, it also stated that while it was "tempting to declare that the time has come to put SBIRS out of the American taxpayer's misery once and for all . . . the Pentagon has no choice but to dip into the

---

*The satellite was a classified NRO satellite whose safe-hold failed and rendered it useless only seven seconds after it was turned on. The safe-hold system, which controls basic satellite functions such as positioning and communications, engages when an on-orbit satellite encounters anomalies. That allows ground controllers to evaluate the problem. See "More Sbirs Trouble," *Aviation Week & Space Technology*, October 8, 2007, p. 27; "Sbirs Slip," *Aviation Week & Space Technology*, November 12, 2007, p. 25.

public till for SBIRS yet again," for the "alternative would be to bet billions of those same taxpayer dollars that the Air Force can develop a brand new system and have it ready to enter service around 2009."[102]

In December, it was reported that the Defense Department was exploring whether GEO-1 could be launched without major software modifications, which were projected to cost between $300 and $550 million and delay launch by twelve to eighteen months beyond its planned 2009 launch, which was already late. The problems were discovered in January testing, and Pentagon officials were concerned about whether there was a link between the safe-hold problem and the test results. The possibility of relying on AIRSS/3GIRS was discounted, given the expectation that it would not be ready in time, which made the purchase of a fourth SBIRS satellite likely.[103]

Through 2008 there was continued work on the successor as well as a mixture of frustration, caution, and optimism with respect to SBIRS and, at least until September, optimism with regard to the health of the U.S. early warning satellite constellation. In March, the GAO noted that while the SBIRS-High program's critical technologies were mature, the program office's confidence in the planned December 2009 launch date was only moderate. In addition, the assessment reported that the Defense Contract Management Agency judged cost and schedule variances as "high risk, and worsening." The Government Accountability Office also concluded that "the program continue[d] to have problems with its flight software system and the pointing and control assembly software."[104]

In March, the Defense Department announced that Lockheed Martin would be building the third geosynchronous SBIRS spacecraft. The following month it was reported that the safe-hold problem had been fixed, as part of an upgrade in command, data handling, and fault management for the geostationary spacecraft. At the end of April, Lockheed Martin announced that the GEO-1 spacecraft was ready for environmental testing "in preparation for launch in late 2009." In June, good news came with regard to the HEO-2 payload, which had been placed in orbit that March on another NRO signals intelligence satellite launched from Vandenberg Air Force Base. The Air Force reported that it had successfully completed the initial checkout of the payload (expected to become operational in 2009). The expectation was that the sensor would be certified to dispatch missile warning messages by September, about a year after HEO-1 was certified.[105]

June also brought Lockheed Martin's announcement of the on-orbit handover of HEO-1 and its ground system to the U.S. Air Force "in preparation for the start of certified operations later this year." In late August, Lockheed Martin announced that the Air Force had certified the readiness of HEO-1 for operational utility evaluation, which would involve live HEO-1 data being injected into warfighter operational networks providing warning and intelligence data. The end of the process would be the U.S. Strategic Command's final certification of HEO-1 and its ground processing elements later that year when the HEO-1 sensor and its data was declared operationally proven.[106]*

In September, the GAO reported, "DOD has estimated that the SBIRS program will be delayed by 15 months and cost $414 million in funding to resolve flight software problems, but these estimates appear optimistic." Further, the office said, "if DOD should need additional time or encounter problems beyond what was planned for, more funds will be needed and launch of the first satellite in December 2009 could be jeopardized." The situation was summed up by a *Space News* headline, "Nagging SBIRS Software Issue Tempers Optimism on Program."[107]

September also brought news that the Air Force was going to order the fourth SBIRS satellite. In October, *Aviation Week & Space Technology* informed its readers that while the U.S. Strategic Command had hoped to certify HEO-1's ability for warning message certification by August, it was expected by the end of December.[108]

The year had also seen a number of developments with regard to alternative or successor programs. In early March 2008, there was the news that Raytheon had demonstrated a fully integrated infrared sensor that had been funded under the AIRSS program, while near the end of the month another prototype sensor was delivered by Science Applications International Corporation (SAIC). It was also expected that the sensor developed by SAIC, which incorporated four telescopes, would be carried into space by a commercial SES Americom satellite in 2010, a project designated the Commercially Hosted Infrared Payload (CHIRP) Flight Demonstration Program.[109]

A replacement for AIRSS/3GIRS as a stopgap measure for SBIRS in 2014 appeared in the Air Force 2009 budget as the Geosynchronous Earth Orbit Infrared Gap Filler System for $117 million. It was not long before the unwieldy potential acronym associated with the effort, GEOGFS, was replaced with a simpler one, IRAS, which stood for Infrared Augmentation

---

* Certification that a sensor can be employed for operational purposes is the responsibility of the Strategic Command. There is no specific sequence among the different SBIRS missions and "it is possible to receive certification for all, none, or only some of the end-user systems." During the testing phase, "a variety of scenarios, [recorded] historical events, and Targets of Opportunity (TOO), foreign and domestic [are] used to evaluate the system's mission processing and reporting capabilities." Information provided by Air Force Space and Missile Systems Center, January 27, 2012.

Satellite. But in 2009, the Pentagon killed the plan, despite concerns about a potential coverage gap in 2014 between the expected launch of the second and third SBIRS satellites. But that didn't kill the idea of a gap-filler project for long. By May, the Pentagon's Operationally Responsive Space Office was examining options for a lower-cost solution.[110]

In 2009, there was some progress in the SBIRS program, but not enough to actually put a satellite in orbit. At the beginning of the year, it was hoped that the first geosynchronous SBIRS satellite would be launched in early 2010, and that the first satellite would pass a number of tests on the way to deployment over the course of the year. In April, Northrop Grumman delivered the payload for the second geosynchronous satellite to Lockheed Martin, and by mid-June it had been integrated into the second satellite.[111]

In April there was also news that the Pentagon planned to order four additional geosynchronous SBIRS spacecraft (numbers 3–6) and two more HEO sensors. The Air Force had been ordered in late 2008 to procure the third and fourth dedicated satellites and hosted sensors, as well as to begin the process of acquiring the fifth and sixth satellites. And by June, Lockheed Martin had received $1.5 billion to build HEO-3 and the third geosynchronous orbit spacecraft (GEO-3). There was also good news in July about the second HEO payload: it had entered an operational trial period in advance of being put into full operational use.[112]

But there were also continued reasons for concern. That month, an Air Force official acknowledged that the process of fixing the flight software, which had consumed twenty months, had cost $750 million. And, in September, Air Force Space Command chief Gen. Robert Kehler told reporters that SBIRS hardware and software problems had not been solved and that it was imperative to start a follow-on program. One consequence of those continued problems was the Air Force's announcement in December 2009 that the first SBIRS geosynchronous satellite would not be delivered until 2011.[113]

In 2010 there was the usual mixture of good and bad news. The first two geosynchronous satellites as well as the ground system passed a number of tests. There was also good news about the SBIRS elliptically orbiting payloads that were already on orbit. On May 10, the Air Force Space and Missile Systems Center announced that the second elliptically orbiting payload had been certified for operational technical intelligence missions. The first had been certified for the technical intelligence mission in August 2009. But while the chief of the Air Force Space Command, Gen. Robert Kehler, was optimistic that SBIRS was on track and that the first geosynchronous satellite would arrive in orbit early the following year, Gen. Kevin Chilton, the head of the Strategic Command, was still concerned about a future missile warning gap, because of the gap in purchasing the third satellite.[114]

The year began with $73.4 million having been provided for the program that had evolved from the remains of the third generation infrared satellite effort—for the fiscal year that would end in September. By that time, with growing optimism about SBIRS, it had evolved into a two-pronged effort: the planned launch of the CHIRP sensor built by SAIC and work by Raytheon on an experimental sensor. In late January, Gen. Robert Kehler, the commander of the Air Force Space Command, announced that the third generation effort would change significantly, telling reporters, "I think we're going to find it significantly different," and "We'll still continue with some demonstration activities." It soon appeared that difference meant an end to separate funding for the third generation program, cancellation of Raytheon's work, and funding the CHIRP effort out of the SBIRS budget.[115]

In 2011, nine years after the first geosynchronous SBIRS satellite was expected to reach orbit, GEO-1 actually made it into space. In mid-February the spacecraft was poised for a May 4 launch. Brig. Gen. (select) Roger Teague, who had become head of the Space and Missile Systems Center Infrared Space Systems Directorate, said, "There have been a lot of doubters along the way," and "We had a lot of issues to overcome, but we've done that." He also promised that there would be "a system on orbit that [would] demonstrate that success."[116]

In late February, the spacecraft arrived at Cape Canaveral, and was fueled on April 11. On the afternoon of May 7, an Atlas V rocket blasted off from Cape Canaveral, beginning its payload's journey to geosynchronous orbit, which it was expected to reach on May 16. On May 18, the satellite deployed its solar arrays, antennas, and a light shade to protect the infrared sensors, with their three infrared bands, from the sun. Two days later, an Air Force official acknowledged that the spacecraft had reached its intended orbit, apparently at about 116W, which placed it west of Latin America and south of the southwestern United States. In June, the satellite began transmitting photographic data to ground operators. Jeff Smith, Lockheed Martin vice president, released a statement that observed that the satellite was "performing flawlessly" and that "the first image sent from the satellite is outstanding."[117]

The success in getting an operational space-based infrared sensor into orbit was followed, in September, by success in getting an experimental one into orbit. On September 21, an Ariane 5 rocket blasted off from Kourou, French Guiana, carrying the SES 2 satellite with the CHIRP sensor onboard. In early November, the Air Force announced that the payload had been turned on and was functioning as expected.[118]

★   ★   ★

While the SBIRS-High program had a rocky road over the fifteen years from the time it was selected to the first launch in 2011, the STSS program, formerly SBIRS-Low, had an even harder time. In April 2007, the Missile Defense Agency (MDA) was considering purchasing a missile tracking satellite to bridge the expected gap between the yet-to-be launched STSS satellites and a future operational constellation, which the MDA had postponed from 2012 to 2016. Also postponed was the expected launch of the STSS pair. In March the GAO reported that the launch was planned for December. But by mid-June, the missile defense agency had postponed the STSS launch until spring 2008, although the agency was planning to launch a related classified experimental payload by the end of 2007.[119]

But it would not be until 2009 that any of the three spacecraft would leave earth. On May 5, 2009, a Delta II rocket blasted off from Vandenberg Air Force Base, carrying the Space Tracking and Surveillance System Advanced Technology Risk Reduction spacecraft. Then, in late September 2009, the Air Force launched the two STSS spacecraft on the same rocket, which resulted in their orbiting in formation, in a 58-degree inclination, at an altitude of 840 miles, and 15 degrees apart. The satellites, each about the size of a compact car, are controlled from the MDA operations center at Schriever Air Force Base in Colorado.[120] Once they arrived in orbit, the two STSS craft experienced more than their share of problems. According to the GAO, the demonstration satellites experienced over twenty-two on-orbit issues. Those issues included the inability, during launch deployment, of the satellites to stabilize themselves, which could have resulted in the loss of the satellites if not for the multiple mobile command sites set up to monitor and control the satellites during postlaunch operations.[121]

The STSS spacecraft reached full capability in November 2010. Between June 2010 and February 2011, the spacecraft were employed by the MDA to track at least six targets, including ground-based interceptor terminal area defense, the subsequently canceled airborne laser, the Aegis missile, and Minuteman readiness tests. In September 2010, the satellites tracked a Minuteman III missile test through the boost and postboost phase, "demonstrating their ability to follow a cold-body object through space," according to Northrop Grumman. Subsequently, another Northrop Grumman official reported that the satellites, in March 2011, had detected and tracked a ballistic missile launch through all phases of its flight.[122]

Ultimately, the MDA decided not to pursue the STSS program, instead shifting its focus to the development of an alternative low-earth orbiting infrared detection system, the Precision Tracking Space System (PTSS).

According to the MDA's plans, PTSS satellites will "track ballistic missiles shortly after launch and throughout their midcourse flight." Launch of the first two development satellites is planned for summer 2016. Thus, it will be well over a decade after the initial anticipated deployment of the SBIRS-Low constellation that a PTSS constellation might begin operations.[123]

From 2007 to 2011, DSP (and, from May 2011, SBIRS) had the usual array of foreign and domestic infrared events to monitor. On April 12, 2007, India tested an 1,860-mile-range Agni-III missile from its Wheeler Island test range. The next month, Russia tested a RS-24 Yars ICBM, which was launched from Plesetsk, with the missile's multiple reentry vehicles landing on the Kura testing range on Kamchatka peninsula. In late December, the Russians tested another RS-24 as well as a very different type of missile, a submarine-launched RSM-54 Sineva, a hybrid missile that in its later stages becomes a cruise missile.[124]

Among the tests that undoubtedly lit up DSP sensors in 2008 was the barrage of missiles fired by Iran that July. On July 9, Iran claimed that it fired a combination of nine long- and medium-range missiles, including a new version of the 1,200-mile-range Shahab-3; in addition, a state-issued photograph showed the simultaneous firing of four missiles. Additional tests took place the following day. However, some of the Iranian claims may have been attempts at deception, with a three-missile rather than four-missile salvo having been fired on July 9, and a 900-mile range Shahab-3 having been employed in the July 9 test. And on July 10, one of the missiles tests may have been a Scud-C rather than the Shahab, as was claimed.[125] A minimum of two, and possibly three, DSP spacecraft detected the missile launches, and the infrared data they collected would have been used to determine the truth or falsity of Iranian claims. It is also possible that one of the SBIRS highly elliptically orbiting payloads also detected some of the launches. A month later, one or possibly two DSP satellites monitored the failed flight test launch of an Iranian space launch vehicle.[126]

Iran and Russia would be among the countries providing missile launches for DSP and SBIRS to monitor during the last months of 2008. In November, Iran launched what it reported to be a new generation of long-range missiles, with a range of 1,200 miles. Iranian television identified the missile as the solid-fueled Sejil. In late August the Russians test-launched a Topol (SS-25) intercontinental ballistic missile, which flew 3,700 miles from Plesetsk to Kamchatka. In October, the Russians tested three submarine-launched missiles and one land-based missile. A Sineva SLBM was fired from a submarine in the Barents Sea, northeast of Norway, and traveled more than 7,145 miles. Other submarine launches were from the Barents Sea and the Sea of Okhotsk, while a Topol ICBM was fired from Plesetsk. In late November, the Russians tested a RS-24 Yars, whose deployment was expected to begin in 2009. Just as the Yars flew 4,000 miles, so did a Bulava missile fired from the White Sea, with their warheads each landing in the Kamchatka impact area.[127]

Early 2009 featured tests by Iran and North Korea. In early February, Iran launched its first spacecraft, designated Omid, into a 150-by-237-mile orbit, with an inclination of 55 degrees. Not only is it likely that one or more DSP satellites detected the launch, but the orbiting spacecraft would have been a FAST WALKER target. Only a few days later, the U.S. intelligence analysts were anticipating a North Korean missile test, while North Korea claimed it would be orbiting a communications satellite. In early April, the North Korean launch materialized and gave DSP and other intelligence assets what turned out to be a brief exercise, since the missile and payload fell into the ocean 2,390 miles from the launch site. Then, on May 20, Iran successfully tested a solid-fueled, two-stage Sejil-2, believed to have a range of 1,200 miles.[128]

In June, DSP data was being mined for a very different, and far more tragic, event than a missile test. The objective was to find any clue to what happened to Air France Flight 447, carrying 228 passengers, which disappeared en route from Rio de Janeiro to Paris early on June 1. About three hours after takeoff, the plane's telemetry system transmitted data indicating electrical problems and possible depressurization of the aircraft. A few days later, the first debris was recovered, a process that continued into 2011.[129]

The next month DSP data was again contributing to the U.S. detection and understanding of foreign missile tests. On June 25, a long-range missile, fired from a Russian submarine, hit its target "and the pieces of the missile landed in the designated area," according to Russian President Dmitri Medvedev. That test was followed by the December 10 test of a Topol (SS-25), which traveled from Kapustin Yar to its impact site in Kazakhstan. Another of the older Russian missiles, known by Western intelligence agencies as the SS-18 SATAN, was tested two weeks later and, according to a Russian strategic forces spokesman, "hit the intended target area on the Kamchatka Peninsula with astounding accuracy."[130]

From July to December in 2009, there were also launches by India and Iran. An Agni-2 launch, from Wheeler Island in late November, failed when the missile went off course after a problem arose with the separation of the second stage and missile. In mid-December, Iran test-fired another Sejil-2, with Iran's defense minister claiming that the missile was faster and more accurate, and thus harder to intercept, than earlier versions.[131]

In 2010, China, Russia, and India all conducted launches that were probably monitored by one or more DSP satellites. On January 11, 2010, China launched an SC-19 missile from its Korla Missile Test Complex, which successfully intercepted a CSS-X-11 medium-range missile fired from the Shuangchengzi Space and Missile Center. According to a State Department cable, "US missile warning satellites detected each missile's powered flight as well as the intercept, which occurred at 1157:31Z at an altitude of approximately 250 km." On October 7, Russia successfully tested a Bulava submarine-launched ballistic missile, the twelfth in the series of Bulava tests, most of which were failures. Then, in late November, India was also successful in a test-launching of an upgraded version of a 435-mile-range Agni missile.[132]

The following year, Iran, Russia, and India all tested missiles. Between January 21 and February 19, Iran launched a pair of medium-range missiles at target areas in the Indian Ocean. In late April, from a location in the Barents Sea, Russia fired a Sineva submarine-launched ballistic missile, fifty feet long and capable of traveling more than 6,200 miles, from the nuclear submarine Yekaterinburg. Variants of the missile can carry between four and ten warheads. Then, in late September, India successfully completed a trial launch of another Agni-2 missile, which was launched from the Integrated Test Range in Orissa; this test followed two tests that had taken place the previous week. A month earlier, in mid-August, the Libyan dictator Moammar Qadhafi fired a Scud missile at opposition forces from a location about fifty miles east of Sirte; it landed in the desert, east of Brega, which was an area the opposition fighters were attempting to seize.[133]

By the beginning of 2011, the U.S. missile defense system included components that had already been fielded and others in various stages of development. One component consisted of the assorted warning systems, including DSP and SBIRS, and a variety of radars, including the SPY-1 on the Aegis warships, COBRA DANE, the upgraded BMEWS radars (at Beale Air Force Base, California; Fylingdales, United Kingdom; and Thule, Greenland), the sea-based X-band radar operating in the Pacific, and seven transportable X-band radars (with one deployed to Japan and one to Israel).[134]

Deployed interceptors included the thirty ground-based, midcourse interceptors deployed at Ft. Greeley, Alaska (26), and Vandenberg Air Force

Base, California (4), with the mission of defending the United States from limited attacks from Iran or North Korea, although how well they would perform continues to be a matter of dispute. In addition, Aegis Ballistic Missile Defense cruisers were equipped with SM-3 interceptors to defend U.S. deployed forces and regional targets against short- and medium-range missiles. In addition, Patriot Advanced Capability-3 missiles are deployed in a number of locations around the world, including South Korea and Israel, to intercept short-range missiles such as Scuds or Scud-variants.[135]* In its advanced flight-testing stage at the beginning of 2011 was the Army's Terminal High Altitude Area Defense system, with three to six launchers, interceptor missiles, and an X-band radar. However, in March 2011 the GAO reported that "the program [was] still experiencing design and production issues" and that delivery of the first two batteries had been delayed by at least a year. Also in development was the Navy's Aegis Sea-Based Terminal Defense System.[136]

At the beginning of 2012, the DSP/SBIRS network consisted of the satellites, the MCS at Buckley and its backup, the JTaGS stations (which since 2008 included a station at Misawa Air Base in Japan), the DSP Mobile Ground System consisting of the six Mobile Ground Terminals (MGTs) at Holloman Air Force Base, and the relay ground stations in Australia and the United Kingdom. Through the SBIRS Survivable/Endurable Mobile Ground Terminal program, the MGTs are being upgraded to allow them to be compatible with the SBIRS satellites, an effort necessitated when it was concluded that the planned replacements, the Mobile Multi-Mission Processors, could not be funded. The DSP constellation and SBIRS GEO-1 is operated by the 2nd and 8th Space Warning Squadrons at the MCS, while the 11th Space Warning Squadron at the MCS Backup at Schriever Air Force Base is responsible for the SBIRS high-elliptically orbiting payloads. The issue of when, if ever, there will be a low-earth orbiting constellation is still unresolved.[137]

The GEO-1 satellite was reported as delivering better than expected results during testing. The Air Force reported that the spacecraft scanning sensor was able to detect targets 25 percent dimmer than required and measure

---

* The Bush administration's plan of deploying ten ground-based interceptors to Poland along with a radar in the Czech Republic was scrapped by President Barrack Obama, who announced that he planned to deploy SM-3 interceptors on ships and then in Europe. See Peter Baker, "Obama Reshapes a Missile Shield to Blunt Tehran," *New York Times*, September 18, 2009, pp. A1, A6; Peter Spiegel, "U.S. to Shelve Nuclear—Missile Shield," *Wall Street Journal*, September 17, 2009, pp. A1, A20.

intensity more accurately (60 percent more accurately) than was specified. However, due to budget constraints the Air Force delayed work on aspects of the ground station and, as a result, it will not be possible, until 2016 or 2017 to fully exploit data from the satellite's staring sensor in real-time. Data is being sent for analysis to the National Air and Space Intelligence Center at Wright-Patterson Air Force Base.[138]

As of mid-2012 the Air Force was planning on launching GEO-2 in March 2013. While a GAO representative told Congress that the third and fourth geosynchronous SBIRS satellites were behind schedule, the Lockheed Martin official responsible for the SBIRS program asserted that they were on schedule to launch in late 2014 and late 2015 respectively. The Air Force was also expected to issue a contract for GEO-5 and GEO-6 in 2013.[139]

At the moment, the satellite constellation of geosynchronous and highly elliptically orbiting infrared missile launch detection satellites is apparently still sufficiently robust—despite the failure of DSP and the exceptionally long delay in getting even the first geosynchronous SBIRS satellite in orbit—that there has been no gap in providing missile launch detection as well as other varieties of infrared data obtained from those satellites. That situation was in part a tribute to the extra-long life of the DSP satellites orbited, a key trait of the program since its inception. Whether, given the decision not to fund DSP 24 and DSP 25, those long lifetimes prove sufficient to ensure the absence of a gap remains to be seen, and depends on whether future SBIRS spacecraft arrive in orbit on time and perform up to expectations.

A Geosynchronous SBIRS Satellite

Appendices

A. Space-Based Infrared Detection of Missiles

The ability of the Defense Support Program (DSP) system to detect missile launches and provide data on such systems is a function of two key factors. One is the ability to *detect* from space the infrared radiation emitted by missile plumes. The second is the ability to *distinguish* those emissions from those caused by fires, explosions, or other events as well as from the earth's natural background radiance.[1]

Ballistic missiles can be classified according to their range. Intercontinental ballistic missiles (ICBMs) have ranges of approximately 6,000 miles, while missiles with ranges of between 620 and 3,200 miles are characterized as intermediate-range ballistic missiles. Short-range ballistic missile have ranges of less than 620 miles. Together the latter two categories are often characterized as theater ballistic missiles.[1]

The flight of a ballistic missile from its launch point to its impact on its target involves several phases. Boost phase begins with launch and continues until the missile's motor stops burning. For an ICBM this phase typically lasts between one and two minutes, depending on the range and type of missile. For theater ballistic missiles the boost phase generally lasts between thirty seconds and two minutes—again depending on the range and type of the missile. Figure A.1 illustrates ballistic missile trajectories.[2]

The midcourse, or sustainer, phase begins after the missile motor burns out and ends when the missile begins to reenter the atmosphere. Since this phase occurs outside of the atmosphere, the missile simply coasts, subject only to the force of gravity. During midcourse, the missile reaches its maximum altitude. If the missile's range is so short that it does not leave the atmosphere, there really is no midcourse phase. Most theater missiles have maximum ranges short enough that they spend very little time outside the atmosphere. In contrast, an ICBM spends more than twenty-five minutes outside the atmosphere while traveling 6,200 miles.[3]

The final, or terminal, phase of the missile occurs when the warhead (and the missile if it is still attached) begins to reenter the atmosphere at an altitude of between forty-eight and sixty-two miles and heads toward its target. This phase typically lasts only a matter of seconds. During this portion of the flight, the warhead heats up as it is buffeted by an increasingly dense atmosphere.[4]

The key to detecting a missile is detecting the infrared emissions from its rocket plume—the moving formation of the hot rocket exhaust gases (and sometimes also entrained small particles) outside the rocket nozzle—gases that reach 2000 K. This gas formation is not uniform in structure, velocity, or composition. It contains several flow regions and supersonic shock waves. It is usually visible as a brilliant flame. The plume size, shape, structure, emission intensity of photons or sound pressure waves, visibility,

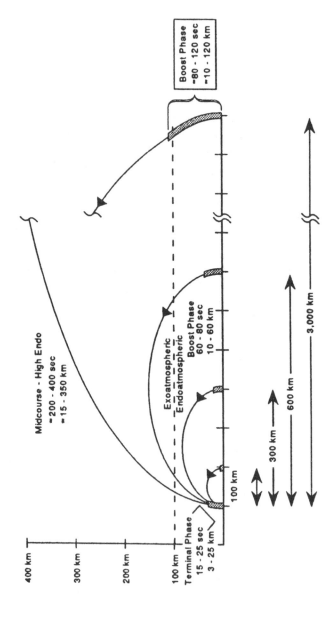

A-1. Ballistic missile flight trajectory. (From Congressional Budget Office, *The Future of Theater Missile Defense* (Washington, D.C.: CBO, 1994), p. 6.)

Boost Phase
≈80 - 120 sec
≈10 - 120 km

Midcourse - High Endo
≈200 - 400 sec
≈15 - 350 km

Exoatmospheric
Endoatmospheric

Boost Phase
60 - 80 sec
10 - 60 km

Terminal Phase
15 - 25 sec
3 - 25 km

400 km
300 km
200 km
100 km
100 km

100 km
300 km
600 km
3,000 km

electrical interference, or smokiness depend not only on the system or its propellants but also on the flight path, flight velocity, altitude, weather conditions (such as winds, humidity, or clouds), and the particular vehicle configuration.[5]

The plume emits radiation in the infrared, visible, and ultraviolet segments of the electromagnetic spectrum. The primary radiation emissions from most of the plume gases are usually in the infrared spectrum and to a lesser extent in the ultraviolet spectrum, with relatively little energy being emitted in the visible spectrum. The specific nature and intensity of the emissions are determined by the particular propellants and their respective exhaust gas compositions.[6]

The pattern of emissions, in terms of intensity at different wavelengths, for a solid-fueled missile varies from the pattern for a liquid-fueled missile, allowing the infrared data to be used to distinguish between such missiles. The earliest liquid-fueled missiles, the U.S. Atlas and Titan and Soviet SS-6 missiles, used an oxidizer such as liquid oxygen and a fuel such as kerosene (RP-1) or hydrogen. Such missiles burned brighter than either solid-fueled missiles or their liquid successors. However, subsequent liquid-fueled missiles used a fuel such as ammonia and an oxidizer such as nitrogen tetroxide, and they produced a less intense signature than either their liquid predecessors or solid-fueled missiles. Solid fuel, which consists of a combination of hydrocarbons, burns on its exposed surface to produce hot gases. The signature of solid-fueled missiles is boosted by the practice of injecting aluminum pieces into the fuel to increase mass, which results in increased thrust.[7]

Within the infrared portion of the electromagnetic spectrum are the short-, mid-, and long-wave categories (SWIR, MWIR, and LWIR). The hot exhaust gases from missile booster engines radiate primarily in the short- and mid-wave region. Colder bodies such as reentry vehicles, the booster body, and the earth radiate at much longer wavelengths.[8] The exact wavelength at which the exhaust gases radiate is determined by the particular composition of the gases, but the primary emission bands for gas plumes are near the water vapor and carbon dioxide lines at 2.7 microns in the short wave and 4.3 microns in the midwave.[9]

Atmospheric water vapor and carbon dioxide attenuate most of the infrared radiation from a missile plume in the early stages of flight. However, the higher temperature and pressure of the water and carbon dioxide in the plume produce a broader infrared signal than the atmospheric absorption bands. Infrared energy will therefore leak through on both sides of the 2.7- and 4.3-micron lines, even from rockets close to the surface of the earth. Hence, infrared detectors, which have peak sensitivity at 2.7 and 4.3 microns but are also sensitive to wavelengths around 2.7 and 4.3 (specifically in the 2.69 to 2.95 and 4.3 to 4.5 range), can detect missiles close to the surface of the earth, before they exit the atmosphere. They can therefore also detect other terrestrial events or objects, such as aircraft, industrial heat sources, fires, and explosions.[10]

The infrared detection mechanism on the DSP satellites consists of a twelve-foot-long Schmidt infrared telescope at a slight angle from the axis of spin of the satellite, which spins at six revolutions per minute. The telescope receives the infrared radiation and focuses it on to the focal plane, where the detectors reside.[11]

Infrared detectors are the transducers that convert infrared radiation to a measurable electronic signal. The earliest detectors employed on DSP satellites were lead sulfide detectors, which could best detect infrared energy from about 1 to 3 microns, when uncooled (at 300 K).[12]

Detecting radiation represents only the first step in satellite operations. Onboard signal/data processing is employed to sort out the signals of interest from the enormous number of signals detected by the sensors. As the satellite sweeps over its area of responsibility, each detector output is sampled up to 7,000 times per second. Each sample is converted to a five-bit binary value representing brightness. Brightness values *below a specific threshold* are then discarded. They are also compared with previous and subsequent values to determine their peak. Brightness values are then tagged to identify source detector. To prevent data channel overload, the algorithm selectively discards returns having low threat probability. Brightness ID data, combined with star, radiation, and health data, are encrypted and transmitted to the appropriate DSP ground facility.[13] The process is illustrated in Figure A.2.

Clearly, onboard signal processing requires use of data about the brightness expected from different types of events, including missile launches. That is true to an even greater extent with regard to ground processing. The first stage of ground processing involves reconstructing individual data from the decryption of the transmitted data. Successive data from the same area are compared, and constant-level data can be discarded as clutter. Areas where there is high noise, including reflections off clouds or ice-covered mountains, can be blanked. Data from a general scan area are then correlated to form prospective missile flight patterns. The patterns are then compared with known profiles, and "best fit" is used to identify missile type.[14]

The flight profile for an identified missile type is fixed to azimuth/elevation reports and projected backward to calculate launch point, heading, and launch time. Reports are then displayed for assessment and prepared for distribution to users.[15] The process is illustrated in Figure A.3.

Important factors in the ability to identify a missile flight profile, and thus its launch point and trajectory, are the presence or absence of cloud cover, the spin rate of the DSP satellite, and missile burn time. All three affect the number of detections that a DSP satellite can make of a missile plume's infrared radiation.

Since infrared radiation does not penetrate clouds, the presence of a cloud deck at a certain altitude precludes DSP satellite detection of infrared radiation below that altitude. Once a missile breaks cloud cover, or is launched in the absence of cloud cover, a DSP satellite can detect its presence once every ten seconds. Thus, the maximum number of detections is $b/10$, where b is the missile burn time (in seconds) while in view of the DSP sensor. Unless there are multiple detections of an object, a missile cannot be distinguished from a stationary object.

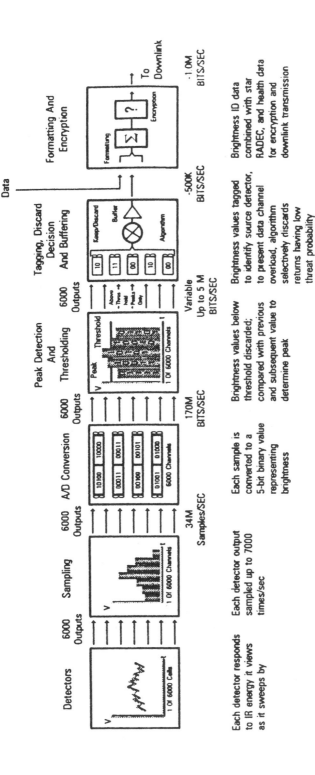

A-2. DSP onboard processing. (From Jonathan Kidd and Holly Caldwell, *Defense Support Program: Support to a Changing World.* AIAA 92–1518. AIAA Space Programs and Technologies Conference, March 24–27, 1992, Huntsville, Alabama.)

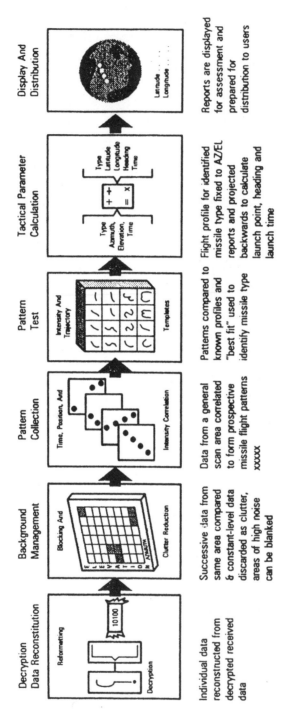

A-3. DSP ground processing. (From Jonathan Kidd and Holly Caldwell, *Defense Support Program: Support to a Changing World.* AIAA 92–1518. AIAA Space Programs and Technologies Conference, March 24–27, 1992, Huntsville, Alabama.)

B. Chronology

October 22, 1946: The Soviets begin the deportation of German missile scientists to the Soviet Union.

May 1954: The suggestion of Sergei Korolev to begin a program to build an intercontinental ballistic missile, is approved by the Soviet Politburo.

October 1954: The National Intelligence Estimate *Soviet Capabilities and Probable Programs in the Guided Missile Field* notes that "we have no firm current intelligence on what particular guided missiles the USSR is presently developing or may now have in operational use."

August 21, 1957: The first successful Soviet ICBM test takes place. The missile is launched from Tyuratam and impacts in the Pacific Ocean.

October 4, 1957: The Soviet Union orbits *Sputnik,* the earth's first artificial satellite, weighing 188 pounds.

November 3, 1957: *Sputnik II* is orbited, along with the launch vehicle's second stage, representing a total of 4,000 pounds placed in orbit.

December 17, 1957: The Intelligence Advisory Committee approves a Special National Intelligence Estimate, which concludes that the "USSR is concentrating on the development of an ICBM which, when operational, will probably be capable of carrying a high-yield nuclear warhead to a maximum range of about 5,500 nautical miles." The estimate also predicts that the Soviets would probably have an initial operational capability of up to ten prototype ICBMs between mid-1958 and mid-1959.

January 14, 1958: Secretary of Defense Neil H. McElroy approves construction of the Ballistic Missile Early Warning System (BMEWS)—three large ground-based radars located in Alaska, Greenland, and the United Kingdom—to detect Soviet ICBMs launched at the United States.

November 5, 1958: The Advanced Projects Research Agency (ARPA) announces, in a classified order, that Subsytem G of WS117L is to become an independent project, designated the Missile Defense Alarm System (MIDAS).

February 26, 1960: MIDAS 1 (Series I, Flight 1) is launched unsuccessfully from the Atlantic Missile Range. The second stage does not separate, and it and the payload land in the Atlantic Ocean.

May 24, 1960: MIDAS 2 (Series I, Flight 2) is successfully launched in a 33-degree, 300-mile orbit. Transmission of infrared data to a ground readout station lasts for only a short time, and the satellite falls silent permanently on the thirteenth orbit.

September 6–9, 1960: The President's Scientific Advisory Committee meets to review the MIDAS program.

July 12, 1961: MIDAS 3 (Series II, Flight 1) is launched from the Pacific Missile Range. It is the first of the Series II tests, for which Baird-Atomic of Cambridge, Massachusetts, provides the sensor.

October 21, 1961: MIDAS 4 (Series II, Flight 2), is launched from the Pacific Missile Range. The satellite proves to be unstable, and its main power source fails after seven days.

November 30, 1961: A study group chaired by Jack Ruina of ARPA completes its "Evaluation of the MIDAS R&D Program." It states that it cannot be predicted with confidence whether an operationally significant version of MIDAS could be completed in the coming decade.

April 9, 1962: MIDAS 5, the third and last of the Series II flights, is launched into a 1,500- to 1,800-nautical-mile polar orbit from the Pacific Missile Range. A power malfunction occurs after the first six orbits.

June 25, 1962: Director of Defense Research and Engineering Harold Brown warns the Assistant Secretary of the Air Force (Research and Development) in a memo that the Air Force should not attempt to launch an operational MIDAS system in the guise of a research and development effort.

July 31, 1962: An ad hoc group on ballistic missile warning, chaired by H. R. Skifter, turns in its report, "Evaluation of Ballistic Missile Warning Systems," dated July 10. The report states that MIDAS probably could not reach operational status until 1966.

November 27, 1962: The Air Force submits the *Air Force Tactical Warning Study* to the Department of Defense to refute the Skifter study.

December 17, 1962: Approximately eighty seconds after liftoff, flight control of MIDAS 6 (Series III, Flight 1) is lost. The booster breaks up and explodes.

December 28, 1962: Deputy Secretary of Defense Roswell Gilpatric, in a memo, informs Air Force Under Secretary Joseph Charyk that the Defense Department cannot accept the conclusions of the *Air Force Tactical Warning Study.*

May 6, 1963: In testimony before the House Appropriations Committee, Harold Brown states that half of the $423 million that had been spent on MIDAS had been wasted on "premature system-oriented development," and that "the way the program was going, it would never produce a reliable, dependable system."

May 9, 1963: MIDAS 7 (Series III, Flight 2) is launched from Vandenberg Air Force Base, California, carrying an Aerojet-designed sensor. It operates in a 87-degree orbit at an altitude of approximately 2,250 miles. During its six weeks of operation it detects nine U.S. ICBM launches, including those of several solid-fueled missiles. It represents the first time a missile is detected from space and represents a turning point in the program.

June 12, 1963: MIDAS 8 (Series III, Flight 3) suffers a booster malfunction that terminates the mission ninety-four seconds into the flight.

July 18, 1963: MIDAS 9 (Series III, Flight 4) is launched into a circular orbit of approximately 2,300 miles and performs satisfactorily for ninety-six orbits. It confirms the successful performance of the MIDAS 7.

June 9, 1966: The first of the Research Test Series/Program 461 satellites is launched from Vandenberg Air Force Base, but goes into a highly elliptical orbit. For that and other reasons the mission objectives are not achieved.

August 19, 1966: The second RTS/Program 461 satellite is launched into a polar orbit and performs successfully—establishing worldwide performance data, exhibiting detection capability against lower target levels and service lifetimes that were almost a factor of two over requirements.

October 5, 1966: The last of the Program 461 satellites is launched. Along with the second 461 satellite it detects forty-five launches, including launches of Soviet solid-fueled missiles.

November 1, 1966: The Air Force designates Program 949 as the successor to Program 461.

December 1966: TRW and Aerojet-General, as associate prime contractors, are awarded a contract for the production of four launch detection and nuclear detonation detection satellites.

August 1968: A U.S. Air Force site survey team arrives in Australia to examine possible sites for the overseas ground station for the Program 949 satellites.

April 23, 1969: Australian Prime Minister John Gorton informs the Australian House of Representatives that the government of Australia has accepted a U.S. proposal to establish a "joint United States–Australian defense space communications facility" near Woomera. The Prime Minister does not reveal the exact nature of the mission.

June 14, 1969: As a result of a security breach, Program 949 becomes Program 647. It also is given the unclassified title Defense Support Program.

November 10, 1969: The United States and Australia sign the agreement to establish the Joint Defence Space Communications Station (JDSCS). The station site is named "Nurrungar."

June 1970: Buckley ANGB, Colorado, is chosen as the Continental Ground Station for the Defense Support Program/Program 647.

October 16, 1970: *Aerospace Daily* publishes an article referring to a planned mid-November launch of a Program 647 (as Program 949 had been retitled) satellite to detect missile launches, which whose ground station would be located at Nurrungar.

November 6, 1970: The first DSP satellite is launched from Cape Kennedy on a Titan III rocket. Although intended to go into geosynchronous orbit, it actually goes into an elliptical orbit.

May 5, 1971: DSP Flight 2 is launched into geostationary orbit and is positioned to detect Soviet and Chinese missile launches.

May 19, 1971: Nurrungar assumes command and control of F-2, whose mission is to detect Soviet and Chinese missile launches.

March 1, 1972: DSP Flight 3 is launched into geostationary orbit, above the Atlantic, to detect Soviet SLBM launches.

July 12, 1973: DSP Flight 4 is launched into geostationary orbit, where it monitors the Pacific for Soviet SLBM launches.

October 6, 1973: Egypt launches an attack on Israel, beginning the Yom Kippur War. A six-man team from Aerospace, Aerojet, and the Air Force units is sent to Nurrungar to examine DSP-E detections, in an attempt to fully understand the detection capability of the DSP system against targets other than ICBMs and SLBMs.

December 7, 1973: A report by the Foreign Technology Division of the Air Force Systems Command confirms the ability of DSP satellites to detect Scud missiles.

October 18, 1975: The DSP-E satellite experiences intense illumination at 2.7 microns. Four similar occurrences take place in October and November. Initial speculation is that the illumination is the result of a Soviet laser. The Defense Department ultimately concludes that the cause was a gas pipeline fire.

December 14, 1975: DSP Flight 5, the first of the second-generation DSP satellites—the Multi-Orbit, Performance Improvement Spacecraft (MOS/PIM)—is launched. Five days after launch a catastrophic failure in the propellant system sends the satellite into an uncontrollable spin.

June 25, 1976: DSP Flight 6 is successfully launched into geostationary orbit. A modified sensor permitted an above-the-horizon experiment.

February 6, 1977: DSP Flight 7 is launched and positioned over the Pacific.

June 10, 1977: ADCOM publishes Required Operational Capability (ROC) 3-77 for a Simplified Processing Station (SPS).

June 1978: The Simplified Processing Station is shipped to Vandenberg Air Force Base, California, for testing.

December 1978: The Simplified Processing Station is shipped to its initial operational location: Cornhusker Army Ammunition Plant at Grand Island, Nebraska.

December 27, 1978: The Air Force informs the Defense Communications Agency that rather than purchase six Simplified Processing Stations it would buy six Mobile Ground Terminals.

June 10, 1979: DSP Flight 8 is launched from Cape Canaveral and positioned over the Pacific.

November 8–15, 1979: An ADCOM site survey team evaluates several possible sites for SPS overseas siting.

December 20, 1979: The Defense Systems Acquisition Review Council meets to review options for creating a more survivable DSP system.

February 1980: The DSARC announces its decision to add a 4.3-micron detection capability to DSP as protection against laser jamming.

March 8, 1980: Lieutenant General James V. Hartinger, commander in chief of the Aerospace Defense Command, and Assistant Secretary of Defense (C^3I) Gerald Dineen discuss an apparent increase in DSP-E ground station outages.

July 25, 1980: President Jimmy Carter signs Presidential Directive 59, "Nuclear Weapons Employment Policy," which requires the United States to be able to fight a prolonged nuclear war, lasting up to six months.

January 20, 1981: Ronald Reagan is sworn in as fortieth President of the United States.

March 16, 1981: DSP Flight 9 is launched into geostationary orbit to monitor possible Soviet SLBM launches from the Atlantic.

March 26, 1981: The Air Force informs Congress that the SPS has been moved to Kapuan Air Station, Germany.

March 30, 1981: President Reagan is shot during an assassination attempt.

April 5, 1981: Ground processing of signals from DSP Flight 9 at the CONUS Ground Station incorrectly indicates a Soviet SLBM launch.

April 13, 1981: False returns are again apparently received from Flight 9. An ADCOM memo noted that "the false reports of April 13 could have resulted in unacceptable posturing of SAC forces." It would later be discovered that the false signals were the result of software problems at the U.S. ground station.

March 6, 1982: DSP Flight 10 is launched into geostationary orbit and assumes responsibility for monitoring the Atlantic.

June 18, 1982: The Soviet Union conducts an integrated test of ICBMs, an SLBM, ballistic missile defense, and antisatellite weapons. The use of DSP to detect the tests becomes the subject of a classified video—"The Event of 18 June 1982: Executive Summary"—which is shown to President Reagan.

1983: A small contingent of U.S. Navy personnel arrive at the Overseas Ground Station at Nurrungar to investigate the possibility of employing DSP data to identify

Soviet naval aircraft and missiles. The program is code-named SLOW WALKER.

March 25, 1983: President Reagan signs National Security Decision Directive 85 (NSDD-85), "Eliminating the Threat from Ballistic Missiles."

May 3, 1983: The Navy briefs General James V. Hartinger, commander in chief of the Air Force Space Command, on its recommendations for the future of the SLOW WALKER program, which Hartinger approves.

1984: Mobile Ground Terminals are transported to their main operating base at Holloman Air Force Base, New Mexico.

The Strategic Defense Initiative Organization is created. It assumes responsibility for future missile detection programs. The Advanced Warning System program is transferred to SDIO from the Air Force and renamed the Boost Surveillance and Tracking System (BSTS).

January 19, 1984: Under Secretary of the Air Force Edward C. Aldridge informs the Defense Department of his decision to produce a laser cross-link system for DSP satellites, which would make it possible to send data from the Eastern Hemisphere satellite back to the United States via other DSP satellites, rather than through Nurrungar.

January 31, 1984: The National Reconnaissance Office launches a signals intelligence spacecraft, code-named VORTEX. Intended to go into geosynchronous orbit, it goes into a high-altitude elliptical orbit.

April 14, 1984: DSP Flight 11 is launched and placed in operation over the Western Hemisphere.

May 5, 1984: The NRO VORTEX satellite launched on January 31, 1984, passes within three kilometers of DSP Flight 7.

June 28, 1984: DSP Flight 7 is moved to replace Flight 9 as the primary Eastern Hemisphere satellite after Flight 9 develops a fuel leak.

December 22, 1984: DSP Flight 12 is launched into geostationary orbit above the Pacific. Flight 12 is a remodeled version of the sixth satellite produced. The satellite had been retrofitted to increase the number of infrared detectors from 2,000 to 6,000, improving its ability to determine the exact point from which a missile was launched. It also has an above-the-horizon detection capability.

August 28, 1985: A Titan IIID rocket, carrying a KH-11 reconnaissance satellite, malfunctions, requiring the destruction of the payload and the rocket.

January 26, 1986: The space shuttle orbiter *Challenger* explodes.

April 18, 1986: A Titan III rocket, carrying a KH-9 reconnaissance satellite, explodes shortly after liftoff.

October 1, 1986: The Mobile Ground Terminals are turned over to the Air Force Space Command for operational use.

August 1987: Desmond Ball, of the Australian National University Strategic and Defence Studies Centre, publishes a *A Base for Debate: The US Satellite Station at Nurrungar.* Ball argues that the United States should be told that Nurrungar must be closed in 1989, after the current agreement expires.

November 28, 1987: DSP Flight 13 is launched, the first DSP launch since December 1984, due to the Titan and *Challenger* disasters. F-13 is the second of two retrofitted DSP satellites equipped with the Sensor Evolutionary Development sensor. It becomes the DSP-W (Atlantic) satellite at 35W.

November 22, 1988: Prime Minister Robert Hawke of Australia informs Parliament that the United States and Australia have signed a new ten-year agreement covering the operation of Nurrungar and other U.S. bases in Australia. The specifics of implementing agreements remained to be determined.

June 14, 1989: DSP Flight 14 is launched. The launch represents the first use of the Titan IV launch vehicle. Flight 14 is the first of the new-generation DSP satellites, DSP-1, which carries the SED sensor as well as detectors sensitive to 4.3 microns and modifications intended to enhance DSP survivability. Missing from the satellite is the Laser Cross-Link, which had been scheduled to make its initial appearance on Flight 14. The satellite is placed over the Pacific at 165W.

July 5, 1989: A South African SRBM is launched from South Africa's Arniston Missile Range.

September 15, 1989: A Jericho-II missile is launched from the Israeli Air Force's Palmikhim research and development facility. Satellite imagery and DSP data indicate that the South African and Israeli missiles are virtually identical.

January 10, 1990: The SPS at Kapuan is renamed the DSP European Ground Station (EGS).

July 20, 1990: Negotiations involving the Air Force, TRW, and McDonnell-Douglas (the LCS subcontractor) conclude. The agreement calls for the second through eighth laser cross-links to be delivered between October 15, 1991, and February 15, 1994. It is also agreed that the first cross-link need not be delivered in flight-ready form.

August 2, 1990: Iraqi Republican Guard forces invade Kuwait.

November 12, 1990: A Titan IV places DSP Flight 15 into geostationary orbit. It becomes the Eastern Hemisphere satellite in June 1991.

December 2, 1990: Iraq conducts missile tests, launching three missiles from a base in southeast Iraq toward a remote area in western Iraq.

Late 1990 to early 1991: The BSTS is transferred back to the Air Force and is renamed Advanced Warning System. It subsequently becomes known as the Follow-On Early Warning System (FEWS).

January 15, 1991: President George Bush signs National Security Directive 54, "Responding to Iraqi Aggression in the Gulf," which sanctions the use of force to expel Iraqi troops from Kuwait.

January 17, 1991: U.S. and allied forces begin bombing raids on Iraqi targets. That afternoon Iraq fires two Scuds at Israel. DSP provides an average of 5.4 minutes' warning between January 17 and January 20. It averages 3.3 minutes from January 21 to the conclusion of the war.

January 20, 1991: Iraq fires eight Scuds at Saudi Arabia.

February 26, 1991: A Scud missile fragments in the atmosphere above Dhahran and tears into a warehouse housing American soldiers. Twenty-eight are killed and about 100 wounded. While DSP did detect the launch, a "worst-case combination of events" prevented a full warning from being issued. In addition, the Patriot battery defending the area was out of commission for a computer repair.

April 11, 1991: Deputy Secretary of Defense Donald Atwood informs the National Space Symposium that work on a DSP follow-on would begin in 1992 and predicts deployment within a decade. The system is intended to provide "increased accuracy and more timely warning" to American military forces" and to "enhance the nation's ability to deal with the increased short-range missile threat."

November 24, 1991: The space shuttle orbiter *Atlantis,* the first shuttle to carry a DSP spacecraft, lifts off from Cape Canaveral. The satellite is deployed at 1:03 A.M. on the 25th. The first unclassified DSP launch, it is placed over the Eastern Hemisphere at 65E.

April 24, 1992: The Army and Navy sign a memorandum of agreement to jointly develop a portable DSP ground station: the Joint Tactical Ground Station (JTaGS).

April 12, 1993: The *Air Force Space Sensor Study* is published, and makes the case for the need for the Air Force's envisioned DSP successor, the Follow-On Early Warning System (FEWS).

April 23, 1993: The Aerospace Corporation publishes *DSP-II: Preserving the Air Force's Options,* which suggests that 85 percent or more of the FEWS requirements could be attained with a modified DSP system, at considerably less cost.

May 24, 1993: U.S. Space Command chief General Charles Horner, in a letter to Aerospace Corporation president Edward C. Aldridge, asks him to "help me understand how such a document could be distributed," challenging "my top-priority program."

September 1, 1993: The Naval Space Command Detachment Echo is relocated from Australia to Dahlgren, Virginia.

1994: The laser cross-link program is canceled after ground test failures.

February 2, 1994: The House Committee on Government Operations holds public hearings on "Strategic Satellite Systems in a Post–Cold War Environment." The controversy over the DSP follow-on is one of the two topics discussed.

April 1994: A Tactical Surveillance Demonstration prototype is relocated to South Korea to provide additional warning to Patriot batteries placed there in fear of a North Korean attack.

October 1, 1994: The TALON SHIELD program becomes operational, with the designation Attack and Launch Early Reporting to the Theater (ALERT). The ALERT system is located at Falcon Air Force Base, Colorado. A sixty-person 11th Space Warning Squadron is established.

November 29, 1994: *Aerospace Daily* reports the Air Force has canceled the ALARM program.

December 22, 1994: DSP Flight 17 is launched on a Titan IV.

January 13, 1995: Air Force officials brief industry executives on the Space-Based Infared System they envisioned to replace DSP. Labeled "High Now, Low Later," it would initially consist of four geosynchronous satellites and two elliptically orbiting satellites, with a network of low-earth-orbiting satellites added in the future.

April 1995: Naval Space Command Detachments BUCKLEY and ECHO are relocated to Chesapeake, Virginia, to form a unified Detachment ECHO.

December 1995: The 4th Space Warning Squadron at Holloman Air Force Base is disbanded and its mission assigned to the 137th Space Warning Squadron at Greeley Air National Guard Base in Colorado.

April 28, 1996: The Clinton administration announces that it will help Israel develop a ballistic missile defense system and share the data from DSP satellites. At a press conference with Israel's Prime Minister Benjamin Netanyahu, President Clinton says the new capability will be operational by the end of 1996.

July 26, 1996: Ian McLachlan, the Australian Minister for Defence, announces that the Australian government has agreed in principle to a proposal from the U.S. Air

Force to establish a Relay Ground Station at the Joint Defence Facility at Pine Gap and to close the Joint Defence Facility at Nurrungar.

September 3, 1996: Paul Kaminski, Deputy Under Secretary of Defense for Acquisitions and Technology, directs the Air Force to prepare to begin deployment of SBIRS-Low (the Space and Missile Tracking System) in 2004, two years earlier than planned.

November 11, 1996: Lockheed, Aerojet, and Grumman are selected as contractors for the SBIRS-High system. The initial contract for $1.8 billion calls for them to build five satellites for deployment into geosynchronous orbit and the sensors for two highly elliptical satellites. Their winning proposal called for use of both scanning and staring sensors.

February 23, 1997: DSP Flight 18 is launched into geosynchronous orbit.

September 1997: Construction for the SBIRS Mission Control Station at Buckely ANGB begins.

April 9, 1999: Flight 19 is launched from Cape Canaveral but is stranded in a useless orbit when the launch vehicle's first and second stages fail to separate.

October 12, 1999: The Overseas Ground Station at Nurrungar, Australia, formally closes.

May 9, 2000: Flight 20 is launched and successfully arrives in geosynchronous orbit.

December 21, 2000: Nineteen Russian officers begin work at the Center for Strategic Stability and Y2K at Peterson Air Force Base, Colorado, as a precaution against Y2K-induced warning failures.

April 23, 2001: A DSP satellite detects an intense flash of light over the Pacific, which is subsequently determined to be the product of a meteorite entering the Earth's atmosphere.

August 6, 2001: Flight 21 is launched from Cape Canaveral into geosynchronous orbit.

December 18, 2001: The Air Force declares the Mission Control Station at Buckley Air Force Base to have reached Initial Operational Capability.

January 25, 2002: A DSP satellite detects a series of explosions at a munitions depot in Lagos, Nigeria.

April 2002: TRW is awarded the contract for SBIRS-Low (subsequently renamed the Space Tracking and Surveillance System).

May 2003: A joint report from the Defense Science Board and Air Force Scientific Advisory Board states that "SBIRS has been a troubled program. It could be considered a case study for how not to execute a space program."

February 14, 2004: DSP Flight 22, nicknamed Eagle Eye, is launched from Cape Canaveral and successfully reaches geosynchronous orbit.

February 28, 2004: A DSP satellite detects the explosion of the tanker Bow Mariner off the coast of Virginia.

March 10, 2005: Key congressional leaders are informed, by a letter from the Under Secretary of the Air Force, that there would be further cost problems and delays with the SBIRS-High program.

December 12, 2005: In a letter to congressional leaders, Kenneth Krieg, the Under Secretary of Defense for Acquisition, Technology, and Logistics, states that the SBIRS program might end with the purchase of two highly elliptically orbiting satellites, two geosynchronous satellites, and the associated ground system.

June 27, 2006: The first highly elliptically orbiting SBIRS payload, HEO-1, is carried into orbit on board an NRO signals intelligence satellite.

September 26, 2007: A letter from Secretary of the Air Force Michael Wynne informs the Under Secretary of Defense for Acquisition, Technology, and Logistics that there is a problem with the geostationary SBIRS spacecraft.

November 10, 2007: DSP Flight 23 is successfully launched into geosynchronous orbit.

March 3, 2008: HEO-2 is launched into orbit, from Vandenberg Air Force Base, on board an NRO signals intelligence satellite.

September 2008: DSP Flight 23 stops working.

May 5, 2009: The Space Tracking and Surveillance System Advanced Technology Risk Reduction satellite is launched.

September 5, 2009: A pair of Space Tracking and Surveillance satellites are launched.

January 11, 2010: One or more DSP satellites detect the test of a Chinese interceptor missile.

May 7, 2011: The SBIRS geosynchronous satellite, GEO-1, is successfully launched from Cape Canaveral.

September 21, 2011: The experimental Commercially Hosted Infrared Payload (CHIRP) is launched on an Americom SES-2 satellite.

C. Satellite Launch Listing

	Launch Date	Inclination	Perigee (miles)	Apogee (miles)	First Station
MIDAS 1	Feb. 25, 1960		[failed to achieve orbit]		
MIDAS 2	May 24, 1960	33.0	300	317	N/A
MIDAS 3	July 12, 1961	91.2	2,082	2,191	N/A
MIDAS 4	Oct. 24, 1961	95.9	2,168	2,329	N/A
MIDAS 5	April 9, 1962	86.7	1,745	2,097	N/A
MIDAS 6	Dec. 17, 1962		[failed to achieve orbit]		
MIDAS 7	May 9, 1963	87.4	2,234	2,282	N/A
MIDAS 8	June 12, 1963		[failed to achieve orbit]		
MIDAS 9	July 18, 1963	88.4	2,275	2,311	N/A
461-1	June 9, 1966	90.5	108	2,242	N/A
461-2	Aug. 19, 1966	90.1	2,282	2,294	N/A
461-3	Oct. 5, 1966	90.2	2,282	2,295	N/A
DSP F1	Nov. 6, 1970	7.8	16,151	22,249	N/A
DSP F2	May 5, 1971	0.87	22,104	22,221	EH
DSP F3	March 1, 1972	0.2	21,958	22,296	ATL
DSP F4	June 12, 1973	0.3	22,181	22,187	EH
DSP F5	Dec. 14, 1975	3.0	22,116	22,187	N/A
DSP F6	June 25, 1976	0.5	22,084	22,223	ATL
DSP F7	Feb. 8, 1977	0.5	22,084	22,223	ATL
DSP F8	June 10, 1979	1.8	22,141	22,229	PAC

(continued)

C. Satellite Launch Listing (continued)

	Launch Date	Inclination	Perigee (miles)	Apogee (miles)	First Station
DSP F9	March 16, 1981	1.99	21,987	22,027	ATL
DSP F10	March 6, 1982	1.97	22,022	22,070	ATL
DSP F11	April 14, 1984	1.27	22,028	22,028	PAC
DSP F12	Dec. 22, 1984	3.4	22,267	22,437	PAC
DSP F13	Nov. 28, 1987	2.92	22,018	22,046	ATL
DSP F14	June 14, 1989	3.1	22,080	22,133	PAC
DSP F15	Nov. 12, 1990	INA	INA	INA	EH
DSP F16	Nov. 24/25, 1990	INA	INA	INA	EH
DSP F17	Dec. 22, 1994	INA	INA	INA	INA
DSP F18	Feb. 23, 1997	INA	INA	INA	INA
DSP F19	April 9, 1999	28.8	391	21,515	N/A
DSP F20	May 9, 2000	3.4	22,185	22,191	PAC
DSP F21	August 6, 2001	6.1	22,180	22,196	EH
DSP F22	February 14, 2004	3.4	22,188	22,196	INA
DSP F23	November 9, 2007	8.4	22,177	22,195	EUR
SBIRS HEO-1	June 27, 2006	63.3	1,472	23,547	N/A
SBIRS HEO-2	March 3, 2008	63.3	1,197	23,824	N/A
SBIRS GEO-1	May 7, 2011	6.3	22,226	22,245	INA

Key: ATL = Atlantic; EH = Eastern Hemisphere; EUR = European; INA = Information not available; N/A = Not applicable; PAC = Pacific.

Notes

1. Bad News Travels Fast

1. Michael J. Neufeld, *The Rocket and the Reich: Peenemünde and the Coming of the Ballistic Missile Era* (New York: Free Press, 1995), p. 281; James Harford, *Korolev: How One Man Masterminded the Soviet Drive to Beat America to the Moon* (New York: Wiley, 1997), pp. 65–68.

2. On the recruitment of German scientists by the United States, see Linda Hunt, *Secret Agenda: The United States Government and Project Paperclip, 1945–1960* (New York: St. Martin's Press, 1991).

3. Neufeld, *The Rocket and the Reich*, p. 267; Peter Alway, "The Soviet Legacy of the V-2," *Quest*, fall 1995, pp. 20–24.

4. John Prados, *The Soviet Estimate: U.S. Intelligence Analysis and Russian Military Strength* (New York: Dial Press, 1982), p. 53.

5. Harford, *Korolev*, p. 67.

6. Ibid., pp. 1, 57, 70.

7. Steven J. Zaloga, *Target America: The Soviet Union and the Strategic Arms Race, 1945–1964* (Novato, Calif.: Presidio, 1993), pp. 65, 108–9.

8. Neufeld, *The Rocket and the Reich*, p. 267.

9. Central Intelligence Agency, *Scientific Research Institute and Experimental Factory 88 for Guided Missile Development*, March 4, 1960, p. 2; Alway, "The Soviet Legacy of the V-2."

10. Central Intelligence Agency, *Scientific Research Institute*, p. 3; Prados, *The Soviet Estimate*, p. 53; Harford, *Korolev*, p. 75; Michael Stoiko, *Soviet Rocketry: Past, Present, and Future* (New York: Holt, Reinhart and Winston, 1970), p. 72; Zaloga, *Target America*, p. 121.

11. G. Harry Stine, *ICBM: The Making of the Weapon That Changed the World* (New York: Orion, 1991), p. 132; Central Intelligence Agency, *Scientific Research Institute*, p. 4; Director of Central Intelligence, NIE 11-6-54, *Soviet Capabilities and Probable Programs in the Guided Missile Field*, October 5, 1954, p. 30.

12. Neufeld, *The Rocket and the Reich*, p. 272; Director of Central Intelligence, SNIE 11-10-57, *The Soviet ICBM Program*, December 17, 1957, p. 4; Stoiko, *Soviet Rocketry*, p. 73.

13. Stoiko, *Soviet Rocketry*, p. 73.

14. Stine, *ICBM*, p. 133; Alway, "The Soviet Legacy of the V-2"; Zaloga, *Target America*, p. 125.

15. Zaloga, *Target America*, p. 128.

16. Ibid., p. 129.

17. Ibid., p. 135; Timothy Varfolomeyev, "Soviet Rocketry That Conquered Space,

Part I: From the First ICBM to Sputnik Launches," *Spaceflight* 37, no. 8 (August 1995): 260–63.

18. Zaloga, *Target America,* pp. 140, 142; Arnold Horelick, *The Political Context of Soviet Defense Decisionmaking, 1953–1964,* pt. 1, *1953–1959* (Santa Monica, Calif.: RAND Corporation, 1975), p. 63.

19. Zaloga, *Target America,* pp. 143–44.

20. Ibid., p. 145; Ernest May, *History of the Strategic Arms Competition, 1945–1972,* pt. 1 (Washington, D.C.: Office of the Secretary of Defense Historical Office, 1981), p. 373; Director of Central Intelligence, SNIE 11-8-57, *Evaluation of Evidence Concerning Soviet ICBM Test Flights,* September 18, 1957, p. 1; Horelick, *The Political Context of Soviet Defense Decisionmaking,* p. 64.

21. Prados, *The Soviet Estimate,* p. 55; John Newhouse, *War and Peace in the Nuclear Age* (New York: Knopf, 1988), p. 121; Director of Central Intelligence, SNIE 11-9-56, *Sino-Soviet Intentions in the Suez Crisis,* November 6, 1956, p. 1.

22. Prados, *The Soviet Estimate,* p. 56.

23. Ibid., p. 57.

24. Director of Central Intelligence, NIE 11-6-54, *Soviet Capabilities and Probable Programs in the Guided Missile Field,* p. 1.

25. Ibid., p. 4.

26. Prados, *The Soviet Estimate,* pp. 57, 63; Ben Rich and Leo Janos, *SkunkWorks: A Personal Memoir of My Years at Lockheed* (Boston: Little, Brown, 1994), p. 124; Paul Lashmar, *Spy Flights of the Cold War* (Gloucestershire, UK: Sutton, 1996), pp. 76–83.

27. Prados, *The Soviet Estimate,* p. 63; Horelick, *The Political Context of Soviet Defense Decisionmaking,* p. 64; Director of Central Intelligence, SNIE 11-10-57, *The Soviet ICBM Program,* p. 8.

28. Director of Central Intelligence, NIE 11-4-57, *Main Trends in Soviet Capabilities and Policies 1957–1962,* November 12, 1957, pp. 6, 27.

29. Fred Kaplan, *The Wizards of Armageddon* (New York: Simon & Schuster, 1983), p. 135; Robert A. Divine, *The Sputnik Challenge: Eisenhower's Response to the Soviet Satellite* (New York: Oxford University Press, 1993), p. xiii.

30. Divine, *The Sputnik Challenge,* p. xv.

31. Stoiko, *Soviet Rocketry,* p. 83; May, *History of the Strategic Arms Competition, 1945–1972,* pt. 1, pp. 395–96.

32. Prados, *The Soviet Estimate,* p. 76; Director of Central Intelligence, SNIE 11-10-57, *The Soviet ICBM Program,* pp. 2–3.

33. Paul Bracken, *The Command and Control of Nuclear Weapons* (New Haven, Conn.: Yale University Press, 1983), p. 11; L. Wainstein, C. D. Cremeans, J. K. Moriarity, and J. Ponturo, *The Evolution of U.S. Strategic Command and Control and Warning, 1945–1972* (Arlington, Va.: Institute for Defense Analysis, 1975), pp. 203–4.

34. Bracken, *The Command and Control of Nuclear Weapons,* p. 15; L. Wainstein et al., *The Evolution of U.S. Strategic Command and Control and Warning, 1945–1972,* p. 217.

35. The existence of radiation outside the visible spectral region was discovered in 1800 by astronomer William Herschel, who had discovered Uranus, as a by-product of his search for a filtering technique to aid in his observations of the stars and planets. Further significant contributions on the properties of infrared radiation were made in 1847 by A.H.L. Fizeau and J.B.L. Foucault, who demonstrated that infrared radiation

has similar optical properties to visible light with respect to reflection, refraction, and interference. In 1881 Samuel Pierpont Langley discovered that the infrared region extended from 1.8 to 5.3 microns, an expanse far greater than previously believed. See E. Scott Barr, "The Infrared Pioneers: I. Sir William Herschel," *Infrared Physics* 1 (1961): 1–4; and "The Infrared Pioneers: III. Samuel Pierpont Langley," *Infrared Physics* 3 (1963): 195–206; James B. Campbell, *Introduction to Remote Sensing* (New York: Guilford Press, 1987), p. 193.

36. J. A. Curcio and J. A. Sanderson, *Further Investigations of the Radiation from Rocket Motor Flames*, Report No. N-3327 (Washington, D.C.: Naval Research Laboratory, 1948), p. iv; William W. Kellogg and Sidney Passman, *Infrared Techniques Applied to the Detection and Interception of Intercontinental Ballistic Missiles*, RM-1572 (Santa Monica, Calif.: RAND, 1955), p. 1.

37. Kellogg and Passman, *Infrared Techniques*, pp. 29–30, 46.

38. Ibid., p. 50.

39. Major Barkley G. Sprague, *Evolution of the Missile Defense Alarm System (MIDAS), 1955–1982* (Maxwell AFB, Ala.: Air Command and Staff College, Air University, 1985), p. 5; David N. Spires, *Beyond Horizons: A Half Century of Air Force Space Leadership* (Colorado Springs, Colo.: Air Force Space Command, 1997), p. 45.

40. Robert F. Piper, *The Space Systems Division: Background (October 1957–June 1962)* (Andrews AFB, Md.: Air Force Systems Command, 1963), p. 14.

41. Merton K. Davies and William R. Harris, *RAND's Role in the Evolution of Balloon and Satellite Observation Systems and Related U.S. Space Technology* (Santa Monica, Calif.: RAND, 1988), p. 101.

42. *Aerojet Magazine* 2, no. 2 (Summer 1992): 5–8; Iris Chang, *Thread of the Silkworm* (New York: Basic Books, 1995), p. 97.

2. MIDAS

1. Robert F. Piper, *The Space Systems Division: Background (October 1957–June 1962)* (Andrews AFB, Md.: Air Force Systems Command, 1963), p. 14; N. W. Watkins, "The MIDAS Project: Part I. Strategic and Technical Origins and Political Evolution, 1955–1963," *Journal of the British Interplanetary Society* 50, no. 6 (June 1997): 215–24.

2. Piper, *The Space Systems Division: Background (October 1957–June 1962)*, p. 14; ARPA Order No. 38-59, November 5, 1958, p. 1; ARPA Order No. 38-60, Amendment No. 7, August 26, 1959, p. 2; interview.

3. Interview with General Bernard Schriever, Washington, D.C., March 20, 1996; U.S. Congress, Senate Committee on Aeronautical and Space Sciences, *Missiles, Space, and Other Major Defense Matters* (Washington, D.C.: U.S. Government Printing Office, 1960), pp. 69–70.

4. Piper, *The Space Systems Division: Background (October 1957–June 1962)*, p. 14; DDR&E Ad Hoc Group, *Draft of a Report on MIDAS*, November 1, 1961, p. IV-1.

5. Letter from T. D. White to General Thomas S. Power, June 16, 1960, "SAMOS and MIDAS," Folder 4-J: Msls/Space/Nuclear, Box 36, General T. D. White Papers, Library of Congress.

6. Piper, *The Space Systems Division: Background (October 1957–June 1962)*, p. 14; DDR&E Ad Hoc Group, *Draft of a Report on MIDAS*, November 1, 1961, p. IV-1.

7. ARPA Order no. 38-60, Amendment no. 7, pp. 1–2, 5.

8. Ibid., p. 2.

9. President's Scientific Advisory Committee, *Report of the Early Warning Panel,* March 13, 1959, pp. 1–2.

10. Clarence L. Johnson, Vice President, Advanced Development Projects, Lockheed Aircraft Corporation, *ICBM Detection System,* Report no. SP-112, n.d.

11. President's Science Advisory Committee, *Report of the Early Warning Panel,* pp. 10–11.

12. Letter, General Thomas S. Power, CINCSAC, to General Thomas D. White, Air Force Chief of Staff, May 14, 1959.

13. Herbert F. York, *Military Space Projects, Report No. 10, March–April–May 1960* (Washington, D.C.: DDR&E, 1960), p. 35.

14. Ibid., p. 28.

15. Ibid., p. 29.

16. Ibid., p. 30.

17. "U.S. Fails with 'Spy' Satellite," *Miami Herald,* February 27, 1960, p. 3-A.

18. Jack Raymond, "Satellite to Spot Missile Firings Put into Orbit in Air Force Test; Allies Defend U.S. in U.N. Debate," *New York Times,* May 25, 1960, pp. 1, 4; "Midas Infrared Measurements Cut Short by Telemetry Failure," *Aviation Week,* June 6, 1960, p. 31.

19. "Midas Infrared Measurements Cut Short by Telemetry Failure"; Piper, *The Space Systems Division: Background (October 1957–June 1962),* pp. 14–15; Raymond, "Satellite to Spot Firings Put into Orbit in Air Force Test"; York, *Military Space Projects, Report No. 10,* p. 31.

20. Carl Berger, *The Air Force in Space Fiscal Year 1961* (Washington, D.C.: USAF Historical Division Liaison Office, 1966), p. 45.

21. Ibid.; Director of Central Intelligence, NIE 11-8-61, *Soviet Capabilities for Long Range Attack, Annex C,* June 7, 1961, p. 3; Director of Central Intelligence, NIE 11-8-60, *Soviet Capabilities for Long Range Attack through Mid-1965,* August 1, 1960, p. 3; "Soviet ICBM Tests Near Final Stages," *Air Intelligence Digest,* January 1959, pp. 4–5.

22. Berger, *The Air Force in Space Fiscal Year 1961,* p. 45.

23. Ibid.

24. Ibid., p. 46; Herbert York, *Military Space Projects, Report No. 11 June–July–August,* (Washington, D.C.: DDR&E, 1960), p. 17; Curtis Peebles, *Guardians: Strategic Reconnaissance Satellites* (Novato, Calif.: Presidio Press, 1987), pp. 308–9; "Midas Infrared Measurements Cut Short by Telemetry Failure."

25. Berger, *The Air Force in Space Fiscal Year 1961,* p. 46; "Ad Hoc MIDAS Panel—Agenda for Meeting at Lockheed, Sunnyvale, Calif., September 6–7, 1960." Interview with Dr. Sidney Drell, New York, N.Y., April 12, 1995. Drell's research, which he conducted with Professor M. A. Ruderman, resulted in the coauthored paper "Infrared Radiation from the Atmosphere Resulting from High Altitude Explosion X-Rays," *Infrared Physics* 2 (1962): 189–203.

26. DDR&E Ad Hoc Group, *Draft of a Report on MIDAS,* November 1, 1961, p. IV-3; Berger, *The Air Force in Space Fiscal Year 1961,* p. 46; telephone interview with Wolfgang Panofsky, August 6, 1992. Beyond the relatively straightforward question of technical feasibility, the panel also addressed a question that involved both Soviet capabilities and intentions—the vulnerability of any MIDAS system to Soviet counter-

measures. The possibility of the Soviets obtaining control over the satellites by sending them false commands was considered remote, as was the possibility of interfering with the data link from the satellite to the readout station. Soviet jamming of the command links was more likely—if they knew the location of the satellite.

One possibility for ensuring the survival of such satellites that was discussed was providing them with some stealth capability. It was thought that using the proper coatings could make it very difficult for the Soviets to detect or track the satellite with radar, although it would be difficult to make the satellite invisible to optical tracking systems. The advantage of such hidden satellites was that a very high degree of coverage might not be necessary. It was thought there could be quite large "holes" in the system if the Soviets were unable to determine when satellites were in a position to view missile launches. However, it was concluded that there was a fatal flaw in this approach: the satellites would have to avoid transmitting while over Soviet territory. The delays that would result would be sufficiently substantial so that much of the early warning advantage of MIDAS would be lost.

Ultimately, it was concluded that if the Soviets knew where the satellites were they could destroy them fairly easily, and that any countermeasures the United States might take would not change the conclusion. However, it was not considered clear that they would want to destroy the satellites, and several participants felt that their possible ability to do so was not a fatal flaw in a MIDAS system.

27. Berger, *The Air Force in Space Fiscal Year 1961*, pp. 46–47.

28. Ibid., p. 47.

29. U.S. Congress, Senate Armed Services Committee, *Military Procurement Authorization Fiscal Year 1962* (Washington, D.C.: U.S. Government Printing Office, 1961), p. 63; Philip J. Klass, *Secret Sentries in Space* (New York: Random House, 1971), p. 175. McNamara did, on January 16, 1961, settle one question that would be relevant if MIDAS achieved operational status—assigning operational command of the system to the Continental Air Command and operational control to NORAD. See Berger, *The Air Force in Space Fiscal Year 1961*, pp. 47–48.

30. Piper, *The Space Systems Division: Background (October 1957–June 1962)*, pp. 14–15; Berger, *The Air Force in Space Fiscal Year 1961*, p. 46; "Discoverer to Carry More Life Specimens," *Aviation Week*, January 2, 1961, p. 19; Chris Pocock, *Dragon Lady: The History of the U-2 Spyplane* (Shrewsbury, England: Airlife, 1989), pp. 126–29.

31. Piper, *The Space Systems Division: Background (October 1957–June 1962)*, p. 15; Peebles, *Guardians*, p. 309.

32. "Exchange of Notes between the Government of the United Kingdom of Great Britain and Northern Ireland and the Government of the United States of America on the Setting Up of a Missile Defense Alarm System Station in the United Kingdom, London, July 18, 1961," *Treaty Series No. 65 (1961)* (London: Her Majesty's Stationery Office, 1961), p. 2; Ian Clark, *Nuclear Diplomacy and the Special Relationship: Britain's Deterrent and America, 1957–1962* (Oxford: Clarendon Press, 1994), pp. 153–54.

33. Missile Defense Alarm System (Midas), Note by the Secretary of State for Air, Cabinet Defence Committee, 26th June 1961, CAB 131/26, XC6380, PRO; Clark, *Nuclear Diplomacy*, p. 154.

34. Carl Berger, *The Air Force in Space Fiscal Year 1962* (Washington, D.C.: USAF

Historical Division Liaison Office, 1966), p. 72; telephone interview with Harold Brown, August 2, 1996.

35. Berger, *The Air Force in Space Fiscal Year 1961*, p. 73.

36. Ibid.; Lawrence Freedman, *The Evolution of Nuclear Strategy* (New York: St. Martin's Press, 1981), p. 227.

37. Berger, *The Air Force in Space Fiscal Year 1962*, p. 73. A less acceptable argument to McNamara was using the warning from MIDAS as part of a launch on warning strategy. In at least one instance when that possibility was brought up, McNamara became furious. After an Air Force briefing, McNamara excused all but General Bernard Schriever, Jack Ruina of ARPA, and Harold Brown. McNamara asked Schriever what was MIDAS's value, assuming it would work. Somewhat taken aback, Schriever replied that it gave the United States a redundant warning capability. When McNamara then inquired about the value of the redundant warning capability, Schriever told him it would permit the United States to launch on warning. Ruina recalls that he had never seen McNamara as furious as he was then. The defense secretary, Ruina recalls, proceeded to tell Schriever that as long as he was Secretary of Defense and Jack Kennedy was President, the United States would never launch on warning, even if that required a force of 10,000 Minuteman ICBMs (telephone interview with Jack Ruina, March 19, 1992).

38. Berger, *The Air Force in Space Fiscal Year 1962*, p. 74.

39. Ibid.

40. Ibid., p. 75.

41. Ibid.; Lt. Gen. Robert M. Lee, Commander, Air Defense Command, Air Force Space Plan, October 4, 1961; Director of Central Intelligence, NIE 11-8.1-61, *Strength and Deployment of Soviet Long Range Ballistic Missile Forces*, September 21, 1961.

42. Berger, *The Air Force in Space Fiscal Year 1962*, p. 75.

43. Ibid., p. 73; Bill Becker, "Millions of Wires Orbited by U.S. in Radio Relay Test," *New York Times*, October 22, 1961, pp. 1, 44.

44. Berger, *The Air Force in Space Fiscal Year 1962*, p. 73; Air Force Missile Test Center, *Atlantic Missile Range Index of Missile Launchings, Supplement II, FY-62, July 1961–June 1962* (Patrick AFB, Fla.: Air Force Missile Test Center, 1962), p. 39; James Trainor, "Midas Faces Re-orientation," *Missiles and Rockets*, July 16, 1962, p. 12; private information.

45. Berger, *The Air Force in Space Fiscal Year 1962*, pp. 75–76.

46. DDR&E Ad Hoc Group, *Draft of a Report on MIDAS*, p. VI-1.

47. Ibid.

48. Berger, *The Air Force in Space Fiscal Year 1962*, p. 76.

49. Ibid.

50. Ibid., pp. 76–77.

51. Ibid., p. 77.

52. Ibid.

53. Ibid.

54. Ibid., pp. 77–78.

55. Ibid., p. 78.

56. Ibid., pp. 78–79.

57. Ibid., p. 79.

58. Ibid; Barry Miller, "USAF Explores Missile Plume Radiation," *Aviation Week and Space Technology*, April 2, 1962, pp. 63–65; Major Barkley G. Sprague, *Evolution of the Missile Defense Alarm System (MIDAS), 1955–1982* (Maxwell AFB, Ala.: Air Command and Staff College, Air University, 1985), p. 14.

59. Berger, *The Air Force in Space Fiscal Year 1962*, p. 80; Piper, *The Space Systems Division: Background (October 1957–June 1962)*, p. 10; Sprague, *Evolution of the Missile Defense Alarm System (MIDAS), 1955–1982*, p. 14.

60. Berger, *The Air Force in Space Fiscal Year 1962*, p. 80.

61. Ibid., p. 81.

62. Ibid.

63. Ibid. p. 82; Robert F. Piper and William D. Putnam, *History of the Space Systems Division, July–December 1962, Volume II* (Andrews AFB, Md.: Air Force Systems Command, 1963), p. 12; Ad Hoc Group on Ballistic Missile Warning, *Evaluation of Ballistic Missile Early Warning Systems*, July 10, 1962, pp. vi–vii.

64. Harold Brown, Memorandum for the Assistant Secretary of the Air Force (Research and Development), Subject: MIDAS System, June 25, 1962.

65. Ibid.

66. Ibid.

67. Piper and Putnam, *History of the Space Systems Division, July–December 1962, Volume II*, p. 11.

68. Ibid., p. 12; "Evaluation of Ballistic Missile Early Warning," pp. 8, 21, 25.

69. "Schriever Says Over-optimism Hurt Midas," *Aviation Week and Space Technology*, August 6, 1962, p. 28.

70. Piper and Putnam, *History of the Space Systems Division, July–December 1962, Volume II*, p. 13.

71. Berger, *The Air Force in Space Fiscal Year 1962*, p. 82.

72. B. A. Schriever, Commander, AFSC to Eugene M. Zuckert, DOD Program Change (4.4.040) on MIDAS (239A), 13 August 1962, Curtis LeMay Papers, Library of Congress, Box 141, "AFSC 1962"; Schriever interview.

73. B. A. Schriever, Commander, AFSC to Eugene M. Zuckert, DOD Program Change (4.4.040) on MIDAS (239A).

74. Ibid.

75. Ibid.

76. Ibid.

77. Ibid.

78. Ibid.

79. Robert S. McNamara, Memorandum for the Chairman, Joint Chiefs of Staff, Subject: Warning Against Ballistic Missiles, August 17, 1962; "Midas Delay Brings BMEWS Expansion," *Aviation Week and Space Technology*, August 6, 1962, pp. 26–27.

80. Piper and Putnam, *History of the Space Systems Division, July–December 1962, Volume II*, p. 14.

81. Ibid., p. 15; General B. A. Schriever, Commander, Air Force Systems Command, to Gen. Curtis LeMay, Hq USAF, "Warning Against Ballistic Missiles," September 19, 1962.

82. Piper and Putnam, *History of the Space Systems Division, July–December 1962, Volume II*, p. 16; Eugene M. Zuckert, Memorandum for the Chief of Staff, Subject: Tactical Warning Study, October 3, 1962, with Enclosure (Terms of Reference, Study

on Strategic Warning) in Library of Congress, Curtis LeMay Manuscript Collection, B-128 6—Air Force Office of the Secretary, 1962.

83. M. B. Carpenter, *Military Uses of Tactical Warning of an Attack on CONUS*, RM-3195-PR (Santa Monica, Calif.: RAND, 1962), pp. 27–46.

84. Piper and Putnam, *History of the Space Systems Division, July–December 1962, Volume II*, p. 16.

85. JCS 1899/743, Draft Memorandum for the Secretary of Defense, Subject: Warning Against Ballistic Missiles, September 29, 1962.

86. Piper and Putnam, *History of the Space Systems Division, July–December 1962, Volume II*, p. 16.

87. Ibid., p. 17.

88. Robert F. Piper and Joseph M. Rowe Jr., *History of the Space Systems Division, January–December 1963, Volume II* (Los Angeles AFS: SSD, 1965), p. 78; Roswell Gilpatric, Deputy Secretary of Defense, Memorandum for the Secretary of the Air Force, Subject: MIDAS, December 28, 1962.

89. Gilpatric, Memorandum for the Secretary of the Air Force, Subject: MIDAS.

90. Piper and Putnam, *History of the Space Systems Division, July–December 1962, Volume II*, pp. 16–17; interview.

3. Vindication

1. Department of Defense Directive S-5200.13, "Security Policy for Military Space Programs," November 15, 1962, NRO CORONA/ARGON/LANYARD Collection, 2/A/0078.

2. Robert F. Piper and Joseph M. Rowe, *History of the Space Systems Division, January–December 1963, Volume II* (Los Angeles AFS: SSD, 1965), p. 81.

3. Ibid., p. 83.

4. Ibid., p. 84.

5. Interview with Alexander Flax, Washington, D.C., July 29, 1992; telephone interview with Daniel Fink, August 13, 1992.

6. "Missile-Sensing Satellite Set for Mid-November Launch by Air Force," *Aerospace Daily*, October 16, 1970, pp. 257–59.

7. Fink interview.

8. Telephone interview with Harold Brown, August 2, 1996; "Half of Midas Spending Viewed as Waste," *Aviation Week & Technology*, June 17, 1963, p. 38; U.S. Congress, House Committee on Appropriations, *Department of Defense Approrpiations for 1964*, pt. 6 (Washington, D.C.: U.S. Government Printing Office, 1963), pp. 22–24.

9. "Half of MIDAS Spending Viewed as Waste"; U.S. Congress, House Committee on Appropriations, *Department of Defense Appropriations for 1964*, pt. 6, p. 26.

10. U.S. Congress, House Committee on Appropriations, *Department of Defense Appropriations for 1964*, pt. 6, p. 29.

11. Ibid., pp. 28–29; "Half of MIDAS Spending Seen as Waste."

12. Brown interview.

13. Adam Yarmolinsky, Memorandum for Timothy J. Rearden Jr., Special Assistant to the President, Subject: Supplement for the Weekly Report for the President,

May 31, 1963, p. 1; Gerald Cantwell, *The Air Force in Space Fiscal Year 1964* (Washington, D.C.: USAF Historical Division Liaison Office, 1967), p. 51; Piper and Rowe, *History of the Space Systems Division, January–December 1963, Volume II*, pp. 85, 87; Aerojet-General, "The W-37 Sensor: A Historical Perspective," n.d.; Marven R. Whipple, *Atlantic Missile Range Index of Missile Launchings, Fiscal Year 1963, July 1962–June 1963* (Patrick Air Force Base, Fla.: Air Force Missile Test Center, 1963), page numbers illegible.

14. Adam Yarmolinsky, Memorandum for Timothy J. Rearden Jr., Subject: Supplement for the Weekly Report for the President, May 31, 1963; Cantwell, *The Air Force in Space Fiscal Year 1964*, p. 51; Piper and Rowe, *History of the Space Systems Division, January–December 1963, Volume II*, pp. 85, 87.

15. Yarmolinsky, Memorandum for Timothy J. Rearden Jr., Subject: Supplement for the Weekly Report to the President, May 31, 1963; Adam Yarmolinksy, Memorandum for Timothy J. Rearden Jr., Special Assistant to the President, Subject: Supplement to the Weekly Report for the President, September 13, 1963; Cantwell, *The Air Force in Space Fiscal Year 1964*, p. 52; Major Barkley G. Sprague, *Evolution of the Missile Defense Alarm System (MIDAS), 1955–1982* (Maxwell AFB, Ala.: Air Command and Staff College, Air University, 1985), p. 16.

16. Telephone interview with Marvin Boatright, May 30, 1996.

17. Flax interview; Piper and Rowe, *History of the Space Systems Division, January–December 1963, Volume II*, p. 85.

18. Yarmolinsky, Memorandum for Mr. Timothy J. Rearden Jr., Special Assistant to the President, Subject: Supplement to the Weekly Report for the President, September 13, 1963; Cantwell, *The Air Force in Space Fiscal Year 1964*, p. 52; Piper and Rowe, *History of the Space Systems Division, January–December 1963, Volume II*, pp. 88, 91–92.

19. Cantwell, *The Air Force in Space Fiscal Year 1964*, p. 52; JCSM-865-63, Maxwell D. Taylor, Chairman, Joint Chiefs of Staff, Memorandum for the Secretary of Defense, Subject: Analysis and Evaluation of the Military Worth of a Satellite-Based Infrared Surveillance and Warning System, November 8, 1963, p. 1.

20. Cantwell, *The Air Force in Space Fiscal Year 1964*, p. 53.

21. Ibid.

22. Ibid.

23. Ibid., p. 54.

24. Ibid.

25. Ibid., pp. 54–55; Piper and Rowe, *History of the Space Systems Division, January–December 1963, Volume II*, pp. 93–94; private information.

26. "Analysis and Evaluation of the Military Worth of a Satellite-Based IR Surveillance System," Appendix to JCSM-865-63, Maxwell D. Taylor, Chairman, Joint Chiefs of Staff, Memorandum for the Secretary of Defense, Subject: Analysis and Evaluation of the Military Worth of a Satellite-Based Infrared Surveillance and Warning System, November 8, 1963.

27. Cantwell, *The Air Force in Space Fiscal Year 1964*, p. 55.

28. Ibid.

29. Ibid., pp. 55–56.

30. Ibid., p. 56.

31. Flax interview.

32. Air Force, SOR No. 209, Specific Operational Requirements for a Missile Defense Alarm System, January 28, 1964, pp. 1, 15; Warren E. Greene, *History of the Space Systems Division, January–December 1964, Volume II* (Los Angeles AFS: SSD, 1966), p. 40.

33. Cantwell, *The Air Force in Space Fiscal Year 1964,* p. 56.

34. Ibid., p. 57; Greene, *History of the Space Systems Division, January–December 1964, Volume II,* p. 41.

35. Cantwell, *The Air Force in Space Fiscal Year 1964,* p. 57.

36. Ibid.

37. Ibid., pp. 57–58.

38. Ibid., p. 58.

39. Ibid.; Greene, *History of the Space Systems Division, January–December 1964, Volume II,* p. 43.

40. Cantwell, *The Air Force in Space Fiscal Year 1964,* pp. 58–59; Robert E. McClellan, *History of the Space Systems Division, July–December 1965* (Los Angeles AFS: SAMSO, 1968), p. 36.

41. Cantwell, *The Air Force in Space Fiscal Year 1964,* p. 59.

42. Ibid.

43. Gerald Cantwell, *The Air Force in Space Fiscal Year 1965* (Washington, D.C.: USAF Historical Liaison Office, 1968), p. 35.

44. Ibid., pp. 36–37.

45. Ibid., pp. 37–38.

46. Ibid., p. 39; Robert E. McClellan, *History of the Space Systems Division, January–June 1965, Volume II* (Los Angeles: SSD, 1966), p. 29.

47. Cantwell, *The Air Force in Space Fiscal Year 1965,* pp. 39–40; Science Applications International Corp., *Fifty Year Commemorative History of Long Range Detection: The Creation, Development, and Operation of the United States Atomic Energy Detection System* (Patrick AFB, Fla.: Air Force Technical Applications Center, 1997), p. 124.

48. Harry C. Jordan and Robert E. McClellan, *History of the Space Systems Division, January–June 1966* (Los Angeles AFS: SAMSO, 1968), p. 19; Robert E. McClellan, *History of the Space Systems Division, July–December 1966, Volume I* (Los Angeles, AFS: SAMSO, 1968), p. 102.

49. David Baker, *The Shape of Wars to Come* (Briarcliff Manor, N.Y.: Stein & Day, 1982), p. 147; McClellan, *History of the Space Systems Division, July–Dec. 1966, Volume I,* p. 102.

50. *History of Space Systems Division, 1 January–30 June 1967* (Los Angeles AFS: SAMSO, 1969), p. 101; interview; "Series of SS-11 ICBM Tests to Pacific Impact Area Concluded," *EUCOM Intelligence Report,* September 7, 1966, p. 3; Office of the Director of Intelligence, *EUCOM Joint Intelligence Appraisal,* ca. November 1966, p. 10; Office of the Director of Intelligence, *EUCOM Joint Intelligence Appraisal,* ca. September 1966, p. 10; "Soviets Conduct First Long-Range Test of Solid-Propellant ICBM," *EUCOM Intelligence Report,* November 23, 1966, p. 1; Robert P. Berman and John Baker, *Soviet Strategic Forces: Requirements and Responses* (Washington, D.C.: Brookings Institution, 1982), pp. 104–105; Sprague, *Evolution of the Missile Defense Alarm System (MIDAS), 1955–1982,* pp. 21–22.

51. Interviews.

52. Interview; Robert S. Norris, Andrew S. Burrows, and Richard W. Fieldhouse,

Nuclear Weapons Databook, vol. 5, *British, French, and Chinese Nuclear Weapons* (Boulder, Colo.: Westview Press, 1994), pp. 407, 420.

53. Interview with Marvin Boatright, December 18, 1995.

54. McClellan, *History of the Space Systems Division, July–December 1966, Volume I,* p. 103.

55. Baker, *The Shape of Wars to Come,* p. 147; Jacob Neufeld, *The Air Force in Space 1970–1974* (Washington, D.C.: Office of Air Force History, 1976), p. 31; interview.

56. Director of Central Intelligence, NIE 11-8-66, *Soviet Capabilities for Strategic Attack,* October 20, 1966, p. 13; Asif A. Siddiqi, "Cold War in Space: A Look Back at the Soviet Union," *Spaceflight* 40, no. 2 (February 1998): 63–68.

57. Director of Central Intelligence, NIE 11-8-66, *Soviet Capabilities for Strategic Attack,* p. 13.

58. Ibid.

59. Robert Piper, *History of Space and Missile Systems Organization, 1 July 1967–30 June 1969, Volume 1* (Los Angeles AFS: SAMSO, 1970), p. 214.

60. Robert S. McNamara, Memorandum for the Chairman, Joint Chiefs of Staff, Subject: FOBS Threat, August 29, 1967.

61. "Soviet Missile and Space Activity Shows Increase with Total of 36 Operations in May," *EUCOM Intelligence Report,* June 30, 1967, p. 1.

62. Piper, *History of Space and Missile Systems Organization, 1 July 1967–30 June 1969, Volume 1,* p. 216; Siddiqi, "Cold War in Space."

4. Nurrungar and Buckley

1. Interview with Alexander Flax, Washington, D.C., July 29, 1992.

2. Ibid.

3. Correspondence from Ellis Lapin, undated.

4. Ibid.

5. Robert F. Piper, *History of Space and Missile Systems Organization, 1 July 1969–30 June 1970* (Los Angeles AFS: SAMSO, n.d.), p. 57; Major Barkely G. Sprague, *Evolution of the Missile Defense Alarm System (MIDAS), 1955–1982* (Maxwell AFB, Ala.: Air Command and Staff College, Air University, 1985), p. 24.

6. Private information.

7. *Weapons in the Wilderness: The Exploitation of the North-West of South Australia* (Adelaide: Anti-Bases Campaign, 1991), p. 23; Robert F. Piper, *History of Space and Missile Systems Organization, 1 July 1967–30 June 1969, Volume 1* (Los Angeles AFS: SAMSO, 1970), pp. 216–217; Peter Morton, *Fire Across the Desert: Woomera and the Anglo-Australian Joint Project 1946–1980* (Canberra: Australian Government Publishing Service, 1989), p. 535.

8. Morton, *Fire Across the Desert,* p. 535; Desmond Ball, *A Base for Debate: The U.S. Satellite Station at Nurrungar* (North Sydney: Allen & Unwin Australia, 1987), p. 39; Jacob Neufeld, *The Air Force in Space, 1970–1974* (Washington, D.C.: Office of Air Force History, 1976), p. 32; Robert F. Piper, *History of Space and Missile Systems Organization, 1 July 1967–30 June 1969, Volume I,* pp. 229, 231; *The History of the Aerospace Defense Command Fiscal Year 1969* (Peterson AFB, Colo.: ADCOM, n.d.), pp. 449–50.

9. Commonwealth of Australia, *Parliamentary Debates (Hansard), 1969,* House of

Representatives, Second Session of the Twenty-sixth Parliament (Third Period), from 25 February 1969 to 29 May 1969, Volume House of Representatives 62, from 25 February 1969 to 23 April 1969, p. 1400.

10. Piper, *History of Space and Missile Systems Organization, 1 July 1969–30 June 1970*, p. 57.

11. Commonwealth of Australia, *Parliamentary Debates (Hansard), 1969,* House of Representatives, Second Session of the Twenty-sixth Parliament (Third Period), Volume House of Representatives 63, from 24 April 1969 to 29 May 1969, pp. 1414–15; Commonwealth of Australia, *Parliamentary Debates (Hansard) 1969,* House of Representatives, Second Session of the Twenty-sixth Parliament (Third Period), from 25 February 1969 to 29 May 1969, Volume House of Representatives 63, from 24 April 1969 to 29 May 1969, p. 1458.

12. Commonwealth of Australia, *Parliamentary Debates (Hansard), 1969,* House of Representatives, Second Session of the Twenty-sixth Parliament (Third Period), from 25 February 1969 to 29 May 1969, Volume House of Representatives 63, from 24 April 1969 to 29 May 1969, pp. 1417–19.

13. Commonwealth of Australia, *Parliamentary Debates (Hansard), 1969,* House of Representatives, Second Session of the Twenty-sixth Parliament (Third Period), 25 February 1969 to 29 May 1969, Volume House of Representatives 63, from 24 April 1969 to 29 May 1969, p. 1431.

14. Ibid., pp. 1447, 1462.

15. "Total Secrecy," *Sydney Morning Herald,* April 25, 1969, p. 2.

16. Ibid.

17. Piper, *History of Space and Missile Systems Organization, 1 July 1969–30 June 1970*, pp. 57–58.

18. Ibid., p. 57.

19. Maximillian Walsh, "Joint Woomera Agreement Still Awaits the Top Signature," *Australian Financial Review,* October 22, 1969, p. 1; Ball, *A Base for Debate,* p. 39; Neufeld, *The Air Force in Space, 1970–1974,* p. 32; Piper, *History of Space and Missile Systems Organization, 1 July 1969–30 June 1970*, p. 58.

20. Ball, *A Base for Debate,* p. 39; Neufeld, *The Air Force in Space, 1970–1974,* p. 32; Piper, *History of Space and Missile Systems Organization, 1 July 1969–30 June 1970*, p. 58.

21. "Agreement between the Government of the Commonwealth of Australia and the Government of the United States of America Relating to the Establishment of a Joint Defence Space Communications Station in Australia," November 10, 1969.

22. Ball, *A Base for Debate,* p. 39; Neufeld, *The Air Force in Space, 1970–1974,* p. 32; Piper, *History of Space and Missile Systems Organization, 1 July 1969–30 June 1970*, p. 58.

23. Morton, *Fire across the Desert,* p. 535; Piper, *History of Space and Missile Systems Organization, 1 July 1969–30 June 1970,* p. 58, 58 n.; interviews; private information.

24. Ball, *A Base for Debate,* pp. 39, 42; Robert F. Piper, *History of the Space and Missile Systems Organization 1 July 1970–30 June 1971* (Los Angeles AFS: SAMSO, 1972), p. 13.

25. Piper, *History of Space and Missile Systems Organization, 1 July 1970–30 June 1971*, p. 14.

26. Ibid.

27. Ibid., pp. 14–15.

28. Ibid., pp. 15–16.

29. Ibid., p. 16.

30. Robert F. Piper, *History of Space and Missile Systems Organization, 1 July 1971–30 June 1972* (Los Angeles AFS: SAMSO, n.d.), p. 55.

31. Piper, *History of Space and Missile Systems Organization 1 July 1970–30 June 1971*, p. 16.

32. Ibid., pp. 16–17. Actually, it took several years before *Aviation Week* was able to sort out the functions of Nurrungar and Pine Gap. For example, see Philip J. Klass, "Australian Pressure on US Bases Eases," *Aviation Week & Space Technology*, April 30, 1973, pp. 67–68, which stated that Pine Gap "provides control signals and readout from a U.S. early warning satellite over the Indian Ocean," while Nurrungar "receives photos transmitted from reconnaissance satellites."

33. Piper, *History of Space and Missile Systems Organization, 1 July 1971–30 June 1972*, p. 17; "Missile Sensing Satellite Set for Mid-November Launch by Air Force," *Aerospace Daily*, October 16, 1970, pp. 257–59.

34. Neufeld, *The Air Force in Space, 1970–1974*, p. 33.

35. Piper, *History of Space and Missile Systems Organization, 1 July 1970–30 June 1971*, p. 17.

36. Telephone interviews with Daniel J. Brockway, May 28, 1996, and June 12, 1996.

37. Correspondence from Daniel J. Brockway, n.d.; Robert P. Berman and John C. Baker, *Soviet Strategic Forces: Requirements and Responses* (Washington, D.C.: Brookings Institution, 1982), pp. 106–107; Central Intelligence Agency, *The Developing Soviet Submarine Force*, September 13, 1968, p. 2.

38. Correspondence from Daniel J. Brockway; Mary Taylor, *A Case Study of the Implementation of the DSP: Summary of Initial Interviews* (Santa Monica, Calif.: RAND Corporation, n.d.).

39. Piper, *History of Space and Missile Systems Organization, 1 July 1967–30 June 1969, Volume 1*, pp. 229–30.

40. *The History of the Aerospace Defense Command Fiscal Year 1969*, pp. 449–50.

41. Neufeld, *The Air Force in Space, 1970–1974*, p. 32; Ball, *A Base for Debate*, p. 50; Dan Partner, "Buckley Will Help Track Super-Secret Sky Spies," *Denver Post*, April 27, 1969, pp. 1, 4.

42. Piper, *History of Space and Missile Systems Organization, 1 July 1969–30 June 1970*, pp. 59–60.

43. Ibid.; correspondence from Daniel J. Brockway, n.d.

44. Neufeld, *The Air Force in Space, 1970–1974*, p. 32 Piper, *History of the Space and Missile Systems Organization, 1 July 1969–30 June 1970*, p. 60; private information; William E. Burrows, *Deep Black: Space Espionage and National Security* (New York: Random House, 1986), pp. 223–24; Philip J. Klass, "NSA 'Jumpseat' Program Winds Down as Soviets Shift to Newer Satellites," *Aviation Week & Space Technology*, April 2, 1990, p. 46; Piper, *History of Space and Missile Systems Organization, 1 July 1970–30 June 1971*, p. 19.

45. Piper, *History of Space and Missile Systems Organization, 1 July 1970–30 June 1971*, p. 18.

46. Ibid., p. 19.

47. Interview.

48. Piper, *History of Space and Missile Systems Organization, 1 July 1970–30 June 1971*, p. 20.

49. Ibid., pp. 20–21; correspondence from Dan Brockway, n.d.

50. Piper, *History of Space and Missile Systems Organization, 1 July 1970–30 June 1971*, p. 21.

5. DSP's First Decade

1. Jacob Neufeld, *The Air Force in Space, 1970–1974* (Washington, D.C.: Office of Air Force History, 1976), p. 32 n.; *The History of the Aerospace Defense Command Fiscal Year 1972*, (Colorado Springs, Colo.: ADCOM, n.d.), p. 112; interview.

2. *Biography of Col. Frederick S. Porter Jr.* (Los Angeles AFS: Space and Missile Systems Organization, September 1971); Mary Taylor, *A Case Study of the Implementation of the DSP: Summary of Initial Interviews* (Santa Monica, Calif.: RAND Corporation, n.d.), pp. 2–3.

3. Director of Central Intelligence, NIE 11-8-70, *Soviet Forces for Operational Attack*, November 24, 1970, p. 18; John Prados, *The Soviet Estimate: U.S. Intelligence and Russian Military Strength* (New York: Dial Press, 1982), pp. 200–224; Russell Jack Smith, *The Unknown CIA: My Three Decades with the Agency* (New York: Berkeley, 1992), pp. 240–45.

4. Director of Central Intelligence, NIE 11-8-70, *Soviet Forces for Operational Attack*, p. 3; Robert P. Berman and John C. Baker, *Soviet Strategic Forces: Requirements and Responses* (Washington, D.C.: Brookings Institution, 1982), pp. 104–5.

5. Berman and Baker, *Soviet Strategic Forces*, pp. 104–5; *The History of the Aerospace Defense Command Fiscal Year 1972*, p. 111; Director of Central Intelligence, NIE 11-8-71, *Soviet Forces for Intercontinental Attack*, October 21, 1971, pp. 25–26.

6. William Beecher, "U.S. Lofts Satellite for Spotting Attack by Soviet or China," *New York Times*, November 7, 1970, pp. 1, 9; "U.S. Launches Spy Satellite to Watch for Rocket Attack," *International Herald Tribune*, November 7–8, 1970, p. 1.

7. Beecher, "U.S. Lofts Satellite for Spotting Attack."

8. Correspondence from Ellis Lapin, undated.

9. "A Well-Known Secret," *New York Times*, November 7, 1970, p. 9.

10. Beecher, "U.S. Lofts Satellite for Spotting Attack"; "U.S. Launches Spy Satellite to Watch for Rocket Attack."

11. Col. John Kidd and 1 Lt. Holly Caldwell, AIAA 92-1518, *Defense Support Program: Support to a Changing World*, AIAA Space Programs and Technologies Conference, March 24–27, 1992, Huntsville, Ala., chart no. 6; Robert F. Piper, *History of Space and Missile Systems Organization, 1 July 1970–30 June 1971* (Los Angeles AFS: SAMSO, 1972), p. 23; Desmond Ball, *A Base for Debate: The US Satellite Station at Nurrungar* (Sydney: Allen & Unwin, 1987), p. 21; Ellis Lapin, "Surveillance by Satellite," *Journal of Defense Research* 8, no. 2 (Summer 1976): 169–86.

12. Unclassified synthesis of SAMSO/FTD, "Preliminary Analysis of Project Hot Spot IR Signals," December 7, 1973; and Aerojet Electro Systems Company, "Applications of Infrared Tactical Surveillance," May 29, 1975, prepared by HQ Space and Missile Systems Center in response to an FOIA request by the author.

13. Timothy C. Hanley, Harry N. Waldron, and Elizabeth J. Levack, *History of Space*

Division, 1 October 1980–30 September 1981, Volume I, Narrative (Los Angeles AFS: Space Division History Office, n.d.), p. 300.

14. Robert F. Piper, *History of Space and Missile Systems Organization, 1 July 1967–30 June 1969, Volume I* (Los Angeles AFS: SAMSO, 1970), p. 232; Robert F. Piper, *History of Space and Missile Systems Organization, 1 July 1969–30 June 1970* (Los Angeles AFS: SAMSO), p. 60; Science Applications International Corp., *Fifty Year Commerative History of Long Range Detection: The Creation, Development, and Operation of the United States Atomic Energy Detection System* (Patrick AFB, Fla.: Air Force Technical Applications Center, 1997), p. 125.

15. Science Applications International Corp., *Fifty Year Commerative History of Long Range Detection*, pp. 124–25.

16. Ibid., p. 125.

17. Ibid., p. 126.

18. *The History of the Aerospace Defense Command Fiscal Year 1972*, p. 114; Senate Committee on Appropriations, *Department of Defense Appropriations for Fiscal Year 1974*, pt. 4 (Washington, D.C.: U.S. Government Printing Office, 1973), p. 717; Piper, *History of the Space and Missile Systems Organization, 1 July 1970–30 June 1971*, p. 23.

19. Piper, *History of Space and Missile Systems Organization, 1 July 1970–30 June 1971*, p. 18; Neufeld, *The Air Force in Space, 1970–1974*, pp. 32–33; *History of the Aerospace Defense Command Fiscal Year 1972*, p. 114.

20. Neufeld, *The Air Force in Space, 1970–1974*, pp. 32–33; *History of the Aerospace Defense Command Fiscal Year 1972*, p. 114.

21. Private information.

22. Correspondence from Ellis Lapin.

23. David Baker, *The Shape of Wars to Come* (New York: Stein & Day, 1982), p. 149; Piper, *History of Space and Missile Systems Organization, 1 July 1970–30 June 1971*, p. 29.

24. "Advanced Missile Warning Satellite Evolved from Smaller Spacecraft," *Aviation Week & Space Technology*, February 20, 1989, p. 45; Ball, *A Base for Debate*, p. 7.

25. Piper, *History of Space and Missile Systems Organization, 1 July 1970–30 June 1971*, p. 18; Neufeld, *The Air Force in Space, 1970–1974*, p. 34.

26. Philip J. Klass, "Australian Pressure on U.S. Bases Eases," *Aviation Week & Space Technology*, April 30, 1973, pp. 67–68; *Australian Labor Party Platform, Constitution and Rules as approved by the 29th Commonwealth Conference* (Adelaide: ALP Federal Secretariat, 1971), p. 34.

27. *Australian Labor Party Platform, Constitution and Rules*, p. 41; Klass, "Australian Pressure on U.S. Bases Eases."

28. Major Barkley G. Sprague, *Evolution of the Missile Defense Alarm System (MIDAS), 1955–1982* (Maxwell AFB, Ala.: Air Command and Staff College, Air University, 1985), p. 29.

29. Ball, *A Base for Debate*, p. 17; Baker, *The Shape of Wars to Come*, p. 151.

30. Lapin, "Surveillance by Satellite."

31. "Talon Shield Readied for DSP Operations," *Aviation Week & Space Technology*, April 4, 1994, p. 31.

32. James Bamford, *The Puzzle Palace: A Report on NSA, America's Most Secret*

Agency (Boston: Houghton Mifflin, 1982), pp. 190–91; Jeffrey T. Richelson, *The U.S. Intelligence Community*, 3d ed. (Boulder, Colo.: Westview Press, 1995), p. 28.

33. Thomas C. Snyder, Timothy C. Hanley, Robert J. Smith, Ronald Benesh, and Earl Goddard, *History of the Space and Missile Systems Organization, 1 July 1973–30 June 1975* (Los Angeles AFS: 6592 ABG, n.d.), p. 348.

34. Robert J. Smith, Timothy C. Hanley, Capt. Lothar P. de Temple, Robert F. Piper, and Dr. John T. Greenwood, *History of Space and Missile Systems Organization, 1 July 1972–30 June 1973* (Los Angeles AFS: Ca.: HQ 6592 Air Base Group History Office, n.d.), p. 127 n. 20; Office of the Director of Intelligence, *USEUCOM Joint Intelligence Appraisal*, January 17, 1973, p. 3.

35. Robert S. Norris, Andrew S. Burrows, and Richard W. Fieldhouse, *Nuclear Weapons Databook*, vol. 5, *British, French, and Chinese Nuclear Weapons* (Boulder, Colo.: Westview Press, 1994), pp. 409, 420–21; Sprague, *Evolution of the Missile Defense Alarm System (MIDAS), 1955–1982*, p. 29.

36. Lapin, "Surveillance by Satellite."

37. Interview with Ed Taylor, Vienna, Virginia, May 9, 1996.

38. Ibid.; unclassified synthesis of HQ SAMSO/SZ and HQ FTD, "Preliminary Analysis of Project Hot Spot IR Signals," December 7, 1973, and Final Report, Aerojet Electrosystems Company, Subj: Application of Infrared Tactical Surveillance, May 29, 1975, prepared by HQ Air Force Space and Missile Systems Center.

39. Peter Allen, *The Yom Kippur War* (New York: Scribner's, 1982), pp. 74, 142; correspondence from Ellis Lapin.

40. Unclassified synthesis of HQ SAMSO/SZ and HQ FTD, "Preliminary Analysis of Project Hot Spot IR Signals," December 7, 1973, and Final Report, Aerojet Electrosystems Company, Subj: Application of Infrared Tactical Surveillance, May 29, 1975; Snyder et al., *History of the Space and Missile Systems Organization, 1 July 1973–30 June 1975*, pp. 314–15, 315 n.

41. Neufeld, *The Air Force in Space, 1970–1974*, pp. 35, 39.

42. Director of Central Intelligence, NIE 11-8-72, *Soviet Forces for Intercontinental Attack*, October 26, 1972, p. 32; Robert L. Hewitt, Dr. John Aston, and Dr. John H. Milligan, *The Track Record in Strategic Estimating: An Evaluation of the Strategic National Intelligence Estimates, 1966–1975* (Washington, D.C.: Central Intelligence Agency, 1976), pp. 1, 12.

43. Hewitt, Aston, and Milligan, *The Track Record in Strategic Estimating*, p. 12; Berman and Baker, *Soviet Strategic Forces*, pp. 106–7; Capt. John M. Veilleux, "Probablility of ICBM and SLBM Reporting from Current Sensor Systems," *Technical Memorandum 75-3*, NORAD/CONAD/ADC, January 1975, pp. 5–6; Bruce G. Blair, *Strategic Command and Control: Redefining the Nuclear Threat* (Washington, D.C.: Brookings Institution, 1985), p. 140.

44. Neufeld, *The Air Force in Space, 1970–1974*, pp. 35, 39; private information.

45. Neufeld, *The Air Force in Space, 1970–1974*, pp. 35, 39; Kenneth Rush, Deputy Secretary of Defense, Memorandum for the Secretary of the Air Force, Subject: WWMCSS Warning Systems, December 19, 1972, with attachment, "WWMCSS Warning System Configuration (Alternative 4A)."

46. Correspondence from Dan Brockway, n.d.

47. Ibid.

48. William E. Burrows, *Deep Black: Space Espionage and National Security* (New

York: Random House, 1986), pp. 223–24; Philip J. Klass, "NSA 'Jumpseat' Program Winds Down as Soviets Shift to Newer Satellites," *Aviation Week & Space Technology,* April 2, 1990, pp. 46–47.

49. Interview; private information; Dr. Patrick E. Barry, "BSTS: First Line of Ballistic Missile Defense," *Grumman Horizons* 25, no. 2 (1989): 1–7. U.S. intelligence assessments in 1972 and 1973 suggested that the Soviets had not yet produced an ABM with the Sprint's characteristics. See Director of Central Intelligence, NIE 11-3-72, *Soviet Strategic Defenses,* November 2, 1972, pp. 16–18; Director of Central Intelligence, Memorandum to Holders NIE 11-3-72, *Soviet Strategic Defenses,* December 20, 1973, p. 9.

50. Interview; private information.

51. Barry Miller, "U.S. Moves to Upgrade Missile Warning," *Aviation Week & Space Technology,* December 2, 1974, pp. 16–18; Baker and Berman, *Soviet Strategic Forces,* pp. 104–5; Sprague, *Evolution of the Missile Defense Alarm System (MIDAS), 1955–1982,* p. 30.

52. Miller, "U.S. Moves to Upgrade Missile Warning."

53. Thomas S. Snyder, Timothy C. Hanley, Robert J. Smith, Ronald Benesh, and Earl Goddard, *History of the Space and Missile Systems Organization, 1 July 1973–30 June 1975* (Los Angeles AFS: HQ 6592 Air Base Group, n.d.), p. 312.

54. Ibid., p. 313; *History of Continental Air Defense Command and Aerospace Defense Command, 1 July 1974–30 June 1975* (Peterson AFB, Colo.: ADCOM, n.d.), p. 79; Msg, ADC to CSAF, 131200Z Mar 1974; Msg, AFSC to CSAF, 242205Z May 1974.

55. Interview with Marvin Boatright.

56. *History of Aerospace Defense Command, 1 July–31 December 1975,* p. 49; *History of Continental Air Defense Command and Aerospace Defense Command, 1 July 1974–30 June 1975,* pp. 80–81; Msg, ADC to JCS J-3, 142340Z February 1975; Msg, CSAF to ADC, 261846Z February 1975, p. 79; Msg, CSAF to ADC 181631Z December 1974; Msg, SAMSO to ADC, 232930Z December 1974; Msg, CSAF to ADC, 261846Z February 1975; Msg, ADC to CSAF, 142037Z April 1975, p. 80; Snyder et al., *History of the Space and Missile Systems Organization, 1 July 1973–30 June 1975,* p. 316 n.

57. Snyder et al. *History of the Space and Missile Systems Organization, 1 July 1973–30 June 1975,* p. 324; Smith et. al., *History of the Space and Missile Systems Organization, 1 July 1972–30 June 1973,* p. 132; Ball, *A Base for Debate,* p. 18; Kidd and Caldwell, "Defense Support Program: Support to a Changing World," chart no. 12.

58. ADCOM, *History of Aerospace Defense Command, 1 July–31 December 1975,* p. 51; Thomas Snyder, Timothy C. Hanley, Elizabeth J. Levack, and Earl Goddard, *History of Space and Missile Systems Organization, 1 July 1975–31 December 1976* (Los Angeles AFS: SAMSO, n.d.), p. 194; Dwayne A. Day, "Top Cover: Origins and Evolution of the Defense Support Program—Part 2," *Spaceflight,* February 1996, pp. 59–63.

59. Director of Central Intelligence, Memorandum to Holders, NIE 11-3-72, *Soviet Strategic Defenses,* p. 15.

60. Philip J. Klass, "Anti-Satellite Laser Use Suspected," *Aviation Week & Space Technology,* December 8, 1975, pp. 12–13.

61. Ball, *A Base for Debate,* p. 27.

62. Ibid.

63. Ibid.; "DOD Continues Satellite Blinding Investigation," *Aviation Week & Space Technology*, January 5, 1976, p. 18; John W. Finney, "Inflation and Growth Are Cited by Rumsfeld for Rise in Budget," *New York Times*, December 23, 1975, p. 22.

64. *History of the Aerospace Defense Command, 1 January 1976–31 December 1976* (Peterson AFB, Colo.: Aerospace Defense Command, n.d.), pp. 113–14; Ball, *A Base for Debate*, pp. 18–19; Timothy C. Hanley, Harry N. Waldron, and Elizabeth J. Levack, *History of Space Division, 1 October 1980–30 September 1981, Volume I, Narrative* (Los Angeles AFS: Space Division, n.d.), p. 300; ADCOM, "Flight 9 Justification Briefing," July 22, 1980.

65. *History of the Aerospace Defense Command, 1 January 1976–31 December 1976*, pp. 113–14; Hanley, Waldron, and Levack, *History of Space Division 1 October 1980–30 September 1981, Volume I, Narrative, p. 300; ADCOM, "Flight 9 Justification Briefing."*

66. Timothy C. Hanley, Harry N. Waldron, Thomas S. Snyder, and Elizabeth J. Levack, *History of Space and Missile Systems Organization, 1 January–31 December 1977* (Los Angeles AFS: SAMSO, n.d.), p. 235; *History of Aerospace Defense Command, 1 January–31 December 1976*, pp. 114–15.

67. ADCOM, "DSP Flight 3 Operations Test," August 12, 1977.

68. CINCAD, Subj: DSP Launch Initiation, February 15, 1979, p. 2. ADCOM, Defense Support Program (DSP) Improvement Status, April 24, 1979, p. 1; U.S. Congress, Senate Armed Services Committee, *Department of Defense Authorization for Appropriations Fiscal Year 1980*, pt. 5 (Washington, D.C.: U.S. Government Printing Office, 1979), p. 2660; ADCOM, "Request for IOT&E Extension," May 25, 1979, p. 2; Timothy C. Hanley, Harry N. Waldron, Julia M. Girard, and Elizabeth J. Levack, *History of Space and Missile Systems Organization, 1 January–30 September 1979, Volume I, Narrative* (Los Angeles AFS: Space Division, n.d.), pp. 196, 196 n.

69. Curtis Peebles, *High Frontier: The United States Air Force and the Military Space Program* (Washington, D.C.: U.S. Government Printing Office, 1997), pp. 41–42; Curtis Peebles, *Guardians: Strategic Reconnaissance Satellites* (Novato, Calif.: Presidio, 1987), p. 406; Desmond Ball, "The US Vela Nuclear Detection Satellite System and the Australian Connection," *Pacific Defence Reporter*, March 1982, pp. 15–16, 19.

70. Peebles, *High Frontier*, pp. 43–44; Ball, "The US Vela Nuclear Detection Satellite System and the Australian Connection."

71. David Albright and Corey Gay, "A Flash from the Past," *Bulletin of the Atomic Scientists*, November/December 1997, pp. 15–17; Ronald Walters, *The September 22, 1979, Mystery Flash: Did South Africa Deteriorate a Nuclear Bomb?* (Washington, D.C.: Washington Office for Africa Educational Fund, 1985), p. 6; Albright and Gay report that the uniqueness of the discrepancy has been challenged by one Los Alamos physicist who has been involved in the nuclear monitoring program since 1979.

72. Director of Central Intelligence, Interagency Intelligence Memorandum, *The 22 September 1979 Event*, December 1979, passim. The DIA belief is discussed on page 5.

73. See Jeffrey T. Richelson, "Double Flash: The Event of September 22, 1979," in preparation.

74. Harry G. Horak, *Vela Alert 747*, LA-8364-MS (Los Alamos, N.M.: Los Alamos Scientific Laboratory, 1980), pp. 2–3.

75. Official studies of the event include Horak, *Vela Event Alert 747;* Office of Sci-

ence and Technology Policy, *Ad Hoc Panel Report on the September 22 Event,* May 23, 1980; Dr. John E. Mansfield and Houston T. Hawkins, *The South Atlantic Mystery Flash: Nuclear or Not?* (Washington, D.C.: Defense Intelligence Agency, 1980).

76. Letter, Lt. Gen. James V. Hartinger, CINCAD, to Gerald Dineen, March 28, 1980.

77. HQ SAC to JCS, Subj: Analysis of DSP-E Outages, December 30, 1980.

78. CINCAD, "Termination of DSP-A Support," November 22, 1980.

79. ADCOM, "DSP Launch Decision Meeting—Space Division," December 17, 1980.

80. ADCOM, "Flight 9 Preparations," January 2, 1981; Col. Lloyd E. Thomas, USAF, Director of Missile Warning Systems to Lt. Col. Ellenberg, ADCOM, "Possible Requirement to Reposition Flight 6," January 30, 1981.

81. Sprague, *Evolution of the Missile Defense Alarm System (MIDAS), 1955–1982,* p. 31.

6. Surviving World War III

1. John Newhouse, *Cold Dawn: The Story of SALT* (New York: Holt, Reinhart, and Winston, 1973), pp. 165, 167; *Arms Control and Disarmament Agreements: Texts and Histories of Negotiations* (Washington, D.C.: U.S. Arms Control and Disarmanent Agency, 1980), pp. 148–49.

2. Gregg Herken, *Counsels of War* (New York: Knopf, 1985), p. 296.

3. PRM/NSC-10, "Comprehensive Net Assessment and Military Force Posture Review," February 18, 1977.

4. John Prados, *Keepers of the Keys: A History of the National Security Council from Truman to Bush* (New York: Morrow, 1991), p. 407.

5. Ibid., p. 408; Desmond Ball, "The Development of the SIOP, 1960–1983," in *Strategic Nuclear Targeting,* ed. Desmond Ball and Jeffrey Richelson (Ithaca, N.Y.: Cornell University Press, 1985), pp. 57–83; NSDM-242, "Planning the Employment of Nuclear Weapons," January 17, 1974, in William M. Arkin, Joshua M. Handler, Julia A. Morrissey, and Jacquelyn M. Walsh, *Encyclopedia of the U.S. Military* (New York: Harper and Row, 1990), pp. 487–89.

6. PD/NSC-18, "U.S. National Strategy," August 24, 1977.

7. Zbigniew Brzezinksi, *Power and Principle: The Memoirs of the National Security Adviser, 1977–1981* (New York: Farrar, Straus and Giroux, 1983), p. 456; Desmond Ball, "Developments in the U.S. Strategic Nuclear Policy under the Carter Administration," ACSI Working Paper no. 21, 1980.

8. Nuclear Targeting Policy Review Panel, *Nuclear Targeting Policy Review Panel, Phase I Report* (Washington, D.C: Department of Defense, 1978).

9. Ball, "The Development of the SIOP, 1960–1983."

10. Colin Gray, *Strategic Studies and Public Policy: The American Experience* (Lexington: University Press of Kentucky, 1982), p. 157; Jeffrey Richelson, "PD-59, NSDD-13 and the Reagan Strategic Modernization Program," *Journal of Strategic Studies* 6, no. 2 (June 1983): 125–46.

11. Gray, *Strategic Studies and Public Policy,* p. 157; Richelson, "PD-59, NSDD-13 and the Reagan Strategic Modernization Program."

12. U.S. Congress, House Appropriations Committee, *Department of Defense Appropriations for 1980*, pt. 3 (Washington, D.C.: U.S. Government Printing Office, 1979), p. 879.

13. Air Force, "Mission Element Need Statement (MENS) for Improved Missile Warning and Attack Assessment," November 19, 1979; correspondence from Dan Brockway, n.d.

14. U.S. Congress, House Armed Services Committee, *Department of Defense Authorization for Appropriations for FY 1981*, pt. 4, book 2 (Washington, D.C.: U.S. Government Printing Office, 1980), pp. 1911–12.

15. W. Graham Claytor Jr., Deputy Secretary of Defense, Memorandum for Secretary of the Air Force, Subject: DSARC I for Advanced Warning System, February 15, 1980; Timothy C. Hanley, Harry N. Waldron, Julia M. Girard, and Elizabeth J. Levack, *History of Space Division, 1 October 1979–30 September 1980, Volume I, Narrative* (Los Angeles AFS: Space Division, n.d.), p. 240; Robert J. Hermann, Assistant Secretary of the Air Force Research and Development and Acquisition, Memorandum for the Vice Chief of Staff, Subject: DSP DSARC I Implementation, March 3, 1980; Brockway correspondence.

16. DSARC I for Advanced Warning Systems; Hermann, DSP DSARC I Implementation; private information.

17. U.S. Congress, Senate Armed Services Committee, *Department of Defense Authorization for Appropriations for Fiscal Year 1978*, pt. 5 (Washington, D.C.: U.S. Government Printing Office, 1977), p. 3736; Barry Miller, "U.S. Moves to Upgrade Missile Warning," *Aviation Week & Space Technology*, December 2, 1974, p. 17; Desmond Ball, *A Base for Debate: The US Satellite Station at Nurrungar* (Sydney: Allen & Unwin, Australia, 1987), p. 59; Timothy C. Hanley, Harry N. Waldron, Thomas S. Snyder, and Elizabeth J. Levack, *History of Space and Missile Systems Organization, 1 January–31 December 1977* (Los Angeles AFS: SAMSO, n.d.), p. 243.

18. Hanley et al. *History of Space and Missile Systems Organization, 1 January–31 December 1977*, p. 242.

19. Timothy C. Hanley, Harry N. Waldron, Helena A. Behrens, and Elizabeth J. Levack, *History of Space and Missile Systems Organization, 1 January–31 December 1978, Volume I, Narrative* (Los Angeles AFS: SAMSO, n.d.), p. 228.

20. *History of Aerospace Defense Command, 1 January–31 December 1976* (Peterson AFB, Colo.: ADCOM, n.d.), p. 116.

21. Ibid.

22. ADCOM, Required Operational Capability (ROC) 3-77, *Required Operational Capability for a Simplified Processing Station (SPS)*, 1977, pp. 21, 23.

23. *History of Aerospace Defense Command, 1 January 1977–31 December 1978* (Peterson AFB, Colo.: ADCOM, n.d.), pp. 115–16.

24. Ibid, p. 116.

25. Timothy C. Hanley, Harry N. Waldron, Julia M. Girard, and Elizabeth J. Levack, *History of Space and Missile Systems Organization, 1 January–30 September 1979, Volume I, Narrative* (Los Angeles AFS: Space Division, n.d.), p. 207.

26. Department of the Air Force, *Supporting Data for Fiscal Year 1980 Budget Estimates. Descriptive Summaries: Research, Development, Test and Evaluation*, January 1979, p. 537.

27. ADCOM, "Simplified Processing Station Alternatives," January 12, 1979, p. 1.

28. ADCOM, "Talking Paper on Mobile Ground Terminal," December 8, 1980.

29. Hanley et al. *History of Space and Missile Systems Organization, 1 January–31 December 1978, Volume I, Narrative,* p. 231.

30. ADCOM, Defense Support Program (DSP) Improvement Status, April 24, 1979, p. 2.

31. ADCOM, Subj: Simplified Processing Station (SPS) Alternatives, January 12, 1979, p. 2; Ball, *A Base for Debate,* p. 60.

32. Ball, *A Base for Debate,* pp. 62–63.

33. Ibid.

34. Correspondence from Dan Brockway, n.d.

7. Slow Walkers, Fast Walkers, and Joggers

1. "Preparing for Nuclear War: President Reagan's Program," *Defense Monitor,* 10, no. 8, 1982.

2. Jeffrey Richelson, "PD-59, NSDD-13 and the Reagan Strategic Modernization Program," *Journal of Strategic Studies* 6, no. 2 (June 1983): 125–46.

3. Timothy C. Hanley, Harry N. Waldron, and Elizabeth J. Levack, *History of Space Division, 1 October 1980–30 September 1981, Volume I, Narrative* (Los Angeles AFS: Space Division History Office, n.d.), p. 303; ADCOM, "Talking Paper on DSP Flight 9 Handover," March 19, 1981.

4. ADCOM, "Flight 9 Deployment," March 2, 1981; HQ ADCOM, "Flight 9 Handover," March 30, 1981.

5. HQ ADCOM, "Flight 9 Handover," March 30, 1981.

6. HQ ADCOM, Flight 9 False Reports, April 21, 1981. If the noise is consistently higher than normal, the cell is a "soft ringer." If the return rate is excessively high and the amplitude is also high, it is a "hard ringer."

7. Ibid.

8. "Six Shots at a Nation's Heart," *Time,* April 13, 1981, pp. 24–38.

9. Ibid.

10. Patrick E. Tyler and Bob Woodward, "FBI Held War Code of Reagan," *Washington Post,* December 13, 1981, pp. 1, 27; Herbert L. Abrams, *"The President Has Been Shot": Confusion, Disability, and the 25th Amendment in the Attempted Assassination of Ronald Reagan* (New York: Norton, 1992), pp. 125–29.

11. ADCOM, Flight 9 Operational Status, April 14, 1981.

12. Secretary of State to US Mission NATO, "The Soviets and the False Missile Alert," November 28, 1979; Charles Perrow, *Normal Accidents: Living with High-Risk Technologies* (New York: Basic Books, 1984), p. 285; Robert Gates, *From the Shadows: The Ultimate Insider's Story of Five Presidents and How They Won the Cold War* (New York: Simon & Schuster, 1996), pp. 113–14.

13. U.S. Congress, Senate Armed Services Committee, *Report of Senator Gary Hart and Senator Barry Goldwater to the Committee on Armed Services United States Senate: Recent False Alerts from the Nation's Missile Attack Warning System* (Washington, D.C.: U.S. Government Printing Office, 1980), pp. 5–7. The Soviet KGB had a different view, informing its residencies that the "errors" were actually planned by the Defense

Department for training purposes. It also told its foreign outposts that the Soviet Union believed the false alerts were intended to give the impression that such errors were possible, and reduce Soviet concern over future alerts, providing possible cover for surprise attack. See Gates, *From the Shadows*, pp. 113–14.

14. U.S. Congress, Senate Armed Services Committee, *Report of Senator Gary Hart and Senator Barry Goldwater*, p. 8.

15. Perrow, *Normal Accidents*, p. 285. False alerts from other sensors would continue to be a problem in 1981. A September 9, 1981, memo from Lt. Col. Curl of NORAD, "Tactical Warning System Reliability," noted that "within the last two months there have been four occasions where data from PAVE PAWS [SLBM warning radars] could have generated false posturing actions. The first occurred on 10 Jul 81 when Beale [AFB, Calif.] reported a possible threat event and maintained site high confidence for 17 minutes." The memo also noted, "On 20 Aug 81, Otis [AFB, Mass.] processed a possible threat event and reported high confidence."

16. ADCOM, Flight 9 Operational Status, April 14, 1981.

17. SAC to ADCOM, Subj: DSP-W Flight 9 False Reports, 21 April 1981; ADCOM, Flight 9 Turnover, April 9, 1982, Attachment.

18. Hanley et al., *History of Space Division, 1 October 1980–30 September 1981, Volume I, Narrative*, p. 304.

19. HQ ADCOM, "DSP Flight 9 Handback," April 25, 1981; HQ ADCOM, "Flight 6 Deployment," April 28, 1981; CINCAD, "DSP Launch Alert," May 6, 1981.

20. HQ USAF to HQ AFSC, Defense Support Program (DSP) Launch Call, date illegible.

21. Timothy C. Hanley, Harry N. Waldron, and Elizabeth J. Levack, *History of Space Division, 1 October 1981–30 September 1982, Volume I, Narrative* (Los Angeles AFS: Space Division History Office, n.d.), p. 215; correspondence from Anthony Kenden.

22. ADCOM, DSP Flight 10 Handover Briefing, March 16, 1982; HQ ADCOM, Flight 10 Data Handling Procedures, March 19, 1982.

23. CINCAD, "Flight 10 Deployment," December 22, 1981; HQ ADCOM, "Initial Operating Capability—Flight 10," March 19, 1982, p. 2; Anthony Kenden, "US Military Satellites, 1983," *Journal of the British Interplanetary Society* 38, no. 2, February 1985, pp. 62–73.

24. HQ ADCOM, DSP-West Status, April 23, 1982, p. 1.

25. "Soviets Stage Integrated Test of Weapons," *Aviation Week & Space Technology*, June 28, 1982, pp. 21–22.

26. Ibid.

27. "Chinese Launch Ballistic Missile from Submarines," *Aviation Week & Space Technology*, October 25, 1982, p. 17.

28. Desmond Ball, *A Base for Debate: The US Satellite Station at Nurrungar* (Sydney: Allen & Unwin, Australia, 1987), pp. 24–25.

29. Ibid., p. 25; Robert P. Berman and John C. Baker, *Soviet Strategic Forces: Requirements and Responses* (Washington, D.C.: Brookings Institution, 1982), pp. 16–17.

30. U.S. Congress, Senate Armed Services Committee, *Department of Defense Authorization for Appropriations for Fiscal Year 1984*, pt. 6, *Sea Power and Force Projection* (Washington, D.C.: U.S. Government Printing Office, 1983), p. 2935.

31. Andrew Cockburn, *The Threat: Inside the Soviet Military Machine* (New York: Random House, 1983), p. 247.

32. Ibid., p. 250.

33. Director of Central Intelligence, NIE 11-15-82/D, *Soviet Naval Strategy and Programs through the 1990s*, March 1983, p. 11.

34. Ibid., p. 7.

35. William D. O'Neil, "Backfire: Long Shadow on the Sea-Lanes," *Proceedings of the U.S. Naval Institute*, March 1977, pp. 26–35.

36. U.S. Air Force, *A History of Strategic Arms Competition, Volume 3, A Handbook of Selected Soviet Weapon and Space Systems*, June 1976, pp. 12, 16; Director of Central Intelligence, NIE 11-6-79, *Soviet Strategic Forces for Peripheral Attack*, September 22, 1978, p. 1; O'Neil, "Backfire."

37. O'Neil, "Backfire"; Director of Central Intelligence, NIE 11-3/8-77, *Soviet Strategic Capabilities for Strategic Nuclear Conflict through the Late 1980s*, February 21, 1978, p. 14; Director of Central Intelligence, NIE 11-6-79, *Soviet Strategic Forces for Peripheral Attack*, 1978, p. 2; Paul H. Nitze, Leonard Sullivan Jr., and the Atlantic Council Working Group on Securing the Seas, *Securing the Sea: The Soviet Naval Challenge and Western Alliance Options* (Boulder, Colo.: Westview Press, 1979), p. 108; U.S. Congress, Senate Armed Services Committee, *Department of Defense Authorization for Appropriations for Fiscal Year 1982*, pt. 4, *Sea Power and Force Projection* (Washington, D.C.: U.S. Government Printing Office, 1981), p. 1634; U.S. Congress, Senate Armed Services Committee, *Department of Defense Authorization for Appropriations for Fiscal Year 1983*, pt. 6, *Sea Power and Force Projection* (Washington, D.C.: U.S. Government Printing Office, 1982), p. 3554.

38. U.S. Congress, House Committee on Armed Services, *Hearings on Military Posture and H.R. 6495, Department of Defense Authorization for Appropriations for Fiscal Year 1981*, pt. 3 (Washington, D.C.: U.S. Government Printing Office, 1980), p. 84. The range of each plane was not a fixed number but depended on the specifics of which alternative payload it carried, whether it was unrefueled or underwent prestrike refueling, and a variety of other factors. In addition, there was disagreement among the CIA, DIA, and military service intelligence units with respect to the best estimate of the Backfire's range. The DIA and military service agencies believed the Backfire had a greater capability than the CIA gave it credit for. See Director of Central Intelligence, *Soviet Strategic Capabilities for Strategic Nuclear Conflict Through the Late 1980s*, p. 15.

39. Director of Central Intelligence, NIE 11-15-82/D, *Soviet Naval Strategy and Programs Through the 1990s*, p. 20; U.S. Congress, Senate Armed Services Committee, *Department of Defense Authorization for Appropriations for Fiscal Year 1983*, pt. 6, *Sea Power and Force Projection*, p. 3567; Mark E. Miller, *Soviet Strategic Power and Doctrine: The Quest for Superiority* (Coral Gables, Fla.: Advanced International Studies Institute, 1982), p. 104; O'Neil, "Backfire"; Joel J. Sokolsky, "Soviet Naval Aviation and the Northern Flank: Its Military and Political Implications," *Naval War College Review*, January–February 1981, pp. 34–45.

40. Director of Central Intelligence, NIE 11-15-82/D, *Soviet Naval Strategy and Programs Through the 1990s*, p. 20; U.S. Air Force, *History of Strategic Arms Competition, 1945–1972*, p. 358; U.S. Congress, Senate Armed Services Committee, *Depart-*

ment of Defense Authorization for Appropriations for Fiscal Year 1982, pt. 4, *Sea Power and Force Projection*, p. 1634; Steven Zaloga, "The Tu-22M 'Backfire' Bomber—Part 2," *Jane's Intelligence Review*, August 1992, pp. 342–46; Norman Friedman, *The Naval Institute Guide to World Naval Weapons Systems 1991/92* (Annapolis, Md.: Naval Institute Press, 1991), p. 157.

41. Vita, Kenneth H. Horn, n.d.; correspondence from a former Aerospace DSP executive, April 1, 1998; private information.

42. Air Force Space Command, *History of Air Force Space Command/ADCOM, January–December 1984* (Peterson AFB, Colo.: AFSPACECOM/ADCOM, n.d.), p. 29; interviews.

43. HQ SPACECOM to HQ USAF, Subj: Navy Slow Walker Support, May 9, 1983; interview.

44. HQ SPACECOM to HQ USAF, Subj: Navy Slow Walker Support, May 9, 1983; HQ AFSPACECOM to HQ USAF, Subj: OGS Future Plans, January 10, 1986, pp. 9–10.

45. Lt. Paul J. Treutel, "Globetrotters!" *Space Tracks*, May–June 1994, p. 3; interview.

46. Treutel, "Globetrotters!" p. 3; HQ AFSPACECOM to HQ USAF, Subj: OGS Future Plans, January 10, 1986, pp. 9–10; Gary R. Wagner, "Shaping the Future of Tactical Space Ops," *Space Tracks*, May–June 1994, pp. 4–5.

47. Treutel, "Globetrotters!"; Wagner, "Shaping the Future of Tactical Space Ops." The detachment was originally to consist of forty sailors and an officer in charge but, according to Treutel, "for a number of reasons, not the least of which was the sheer shortage of available housing in the sleepy South Australian desert hamlet of Woomera, the detachment was scaled down to seven enlisted operators and a lieutenant."

48. Ronald S. Regehr, "Do Our Spy Satellites See UFOs?" unpublished memo, n.d.

49. Dennis Stacy and Patrick Huyghe, "Cosmic Conspiracy: Six Decades of Government UFO Cover-Ups, Part Five," *Omni*, August 1994, pp. 48ff.

50. AFSPACECOM Regulation 55-55, "Space Based Sensor (SBS) Large Processing Station (LPS) and European Ground Station (EGS) Tactical Requirements Doctrine (TRD)," September 30, 1992, p. 38; AFSPACECOM Regulation 55-27, "Space Based Sensor (SBS) Mobile Ground System (MGS) Tactical Requirements Doctrine (TRD)," October 15, 1992, p. 23.

51. Stacy and Huyghe, "Cosmic Conspiracy." A knowledgeable source provided the information about the nature of the FAST WALKER. An FOIA request for any report produced on the May 5, 1984, sighting produced a response from the Air Force Space Command that neither confirmed nor denied DSP interest in such events or the existence of a report.

52. Interview.

53. Gus W. Weiss, "The Life and Death of Cosmos 954," *Studies in Intelligence* 22 (Spring 1978): 1–7; Jack Manno, *Arming the Heavens: The Hidden Military Agenda for Space, 1945–1995* (New York: Dodd, Mead, 1984), p. 148; Correspondence from Ellis Lapin, n.d.; interview.

54. Paul B. Stares, *Space and National Security* (Washington, D.C.: Brookings Institution, 1987), pp. 85–87; interviews.

55. Richard Halloran, "U.S. Says Blast Hit Soviet Arms Base," *New York Times*,

June 23, 1984, p. 3; Rick Atkinson, "Soviet Arms Disaster Reported," *Washington Post,* June 22, 1984, pp. A1, A12.

56. "Soviet Naval Blast Called Crippling," *New York Times,* July 11, 1984, p. 6.

57. Mohammad Yousaf and Mark Adkin, *The Bear Trap: Afghanistan's Untold Story* (London: Leo Cooper, 1992), pp. 126, 150.

58. "Soviet Missile-Motor Plant Shut by Explosion, Pentagon Says," *Washington Post,* May 18, 1988, p. A29.

59. Amembassy London to Sec State, More on Explosion in Iraq, 7 September 1989; Joint Staff, Iraq Admits to "Oil Depot" Explosion, 081528Z Sept. 1989.

60. Don Phillips, "Fuselage Section of Jet Found," *Washington Post,* July 23, 1996, pp. A1, A8; David A. Fulghum, "ANG Eyewitnesses Reject Missile Theory," *Aviation Week & Space Technology,* July 29, 1996, p. 32; "Satellite Detected Flash Near Where Planes Vanished," *Washington Times,* September 17, 1997, p. A5; interview.

61. Barton Gellman, "Shift by Iran Fuels Debate over Sanctions," *Washington Post,* December 31, 1997, pp. A1, A16.

62. Interview with Brig. Gen. William King, Pasadena, Calif., December 20, 1995.

8. Evolutionary Developments and Suicidal Lasers

1. CINCAD to JCS et al., Subj: DSP Launch Alert, May 13, 1983; *History of Space Command/ADCOM, January–December 1983* (Peterson AFB, Colo.: AFSPACE-COM, n.d.), p. 28; Timothy C. Hanley, Harry N. Waldron, and Elizabeth J. Levack, *History of Space Division, 1 October 1982–30 September 1983, Volume I, Narrative* (Los Angeles AFS: Space Division, n.d.), p. 254.

2. CINCAD to JCS et al., Subj: DSP Launch Alert; *History of Space Command/ ADCOM, January–December 1983,* pp. 28–29; JCS, Memorandum for Chief of Staff, U.S. Air Force, Director, Defense Communications Agency, Subj: Launch Priority of Defense Support Program and Defense Satellite Communications Systems Satellite, November 18, 1983, SM-833-83.

3. *History of Space Command/ADCOM, January–December 1984* (Peterson AFB, Colo.: AFSPACECOM, n.d.), p. 55; Strobe Talbott, *Deadly Gambits: The Reagan Administration and the Stalemate in Nuclear Arms Control* (New York: Knopf, 1984), pp. 185–206; Ronald Reagan, National Security Decision Directive 15, "Theater Nuclear Forces (Intermediate-Range Nuclear Forces)," November 16, 1981; Ronald Reagan, National Security Decision Directive 104, "U.S. Approach to INF Negotations—II," September 21, 1983.

4. *History of Space Command/ADCOM, January–December 1984,* p. 55.

5. *History of the Space Division, 1 October 1983–30 September 1984* (Los Angeles AFS: SD History Office, n.d.), pp. 215–16.

6. *History of Space Command/ADCOM, January–December 1984,* pp. 55–57; *History of Space Division, 1 October 1983–30 September 1984,* p. 216; *History of 5th Defense Space Communications Squadron, 1 October–31 December 1984* (Woomera, Australia: 5th DSCS, 1985), p. 23.

7. *History of Space Division, October 1984–September 1985* (Los Angeles AFS: Space Division, n.d.), p. 299; *History of Space Division, 1 October 1983–30 September 1984,* p. 217; HQ USAF/XO message 231503Z July 1984; Space Division, Defense Support Systems, Mission Readiness Review, November 6, 1984.

8. Space Division, Defense Support Systems, Mission Readiness Review, n.p.; Col. John Kidd and 1 Lt. Holly Caldwell, *Defense Support Program: Support to a Changing World*, AIAA 92-1518 (Huntsville, Ala.: AIAA Space Programs and Technologies Conference, March 24–27, 1992), chart no. 12; Timothy C. Hanley, Harry N. Waldron, and Elizabeth J. Levack, *History of Space Division, 1 October 1981–30 September 1982, Volume I, Narrative*, p. 216; Timothy J. Hanley, Harry N. Waldron, Thomas S. Snyder, and Elizabeth J. Levack, *History of Space and Missile Systems Organization, 1 January–31 December 1977* (Los Angeles AFS: HQ 6592 Air Base Group History Office, n.d.), p. 239; Timothy C. Hanley, Harry N. Waldron, Julia M. Girard, and Elizabeth J. Levack, *History of Space Division, 1 October 1979–30 September 1980, Volume I, Narrative* (Los Angeles AFS: Space Division History Office, n.d.), p. 230.

9. Space Division, Defense Support Systems, Mission Readiness Review, November 6, 1984, n.p.; Kidd and Caldwell, *Defense Support Program*, chart no. 12; Timothy C. Hanley, Harry N. Waldron, and Elizabeth J. Levack, *History of Space Division, 1 October 1981–30 September 1982, Volume I, Narrative* (Los Angeles AFS: Space Division History Office, n.d.), p. 216; Hanley et al., *History of Space and Missile Systems Organization (U), 1 January–31 December 1977*, p. 239; interview.

10. Clyde R. Magill Jr., Deputy for Defense Support Systems, Space Division, to Lt. Gen. Forrest S. McCartney, Commander, Subject: Independent Readiness Review for Satellite 6R, June 29, 1984; Kidd and Caldwell, *Defense Support Program*, chart no. 12; *History of Air Force Space Command, January–December 1986* (Peterson AFB, Colo.: AFSPACECOM, n.d.), p. 42; Aerojet Electro Systems, *SED Familiarization Manual—DSP SED Sensor—DSP Sensor Evolutionary Development (SED)* (Azusa, Calif.: Aerojet, 1979), pp. 2–5.

11. Hanley et al., *History of Space Division, 1 October 1981–30 September 1982, Volume I, Narrative*, p. 216; Magill, Deputy for Defense Support Systems, Space Division, to Lt. Gen. Forrest S. McCartney, Commander, Subject: Independent Readiness Review for Satellite 6R; Hanley et al., *History of Space Division, 1 October 1979–30 September 1980, Volume I, Narrative*, p. 232; Hanley et al., *History of Space Division, 1 October 1980–30 September 1981, Volume I, Narrative*, p. 306.

12. *History of Space Division, October 1985–September 1986* (Los Angeles AFS: Space Division History Office, n.d.), pp. 292–97; Hanley et al., *History of Space Division, 1 October 1979–30 September 1980*, p. 232.

13. Hanley et al., *History of Space Division, 1 October 1981–30 September 1982, Volume I, Narrative*, p. 228; *History of Air Force Space Command, January–December 1986*, p. 42.

14. *History of USSPACECOM/ADCOM/AFSPACECOM, January–December 1985* (Peterson AFB, Colo.: USSPACECOM, n.d.), pp. 58–59; *History of Air Force Space Command, January–December 1986*, p. 36; *History of the 5th Defense Space Communications Squadron, January–March 1987* (Woomera, Australia: 5th DSCS, 1987), p. 10.

15. *History of USSPACECOM/ADCOM/AFSPACECOM, January–December 1985*, pp. 59–60; *History of the 5th Defense Space Communications Squadron, 1 April–30 June 1985* (Woomera, Australia: 5th DSCS, 1985), p. 13; "Los Alamos Energetic Particle Home Page," at huey.jpl.nasa.gov/~spravdo/neat.html.

16. *History of USSPACECOM/ADCOM/AFSPACECOM, January–December 1985*, pp. 59–60.

17. *History of the Space Division, October 1985–September 1986* (Los Angeles AFS:

Space Division, n.d.), p. 295; Jeffrey T. Richelson, *America's Secret Eyes in Space: The US KEYHOLE Spy Satellite Program* (New York: Harper & Row, 1990), pp. 206–8.

18. *History of Air Force Space Command, January–December 1986*, p. 39.

19. *History of Air Force Space Command, January–December 1987*, p. 41; Air Force Launches Satellite," *Washington Times*, November 30, 1987, p. A6; Edward H. Kolcum, "USAF Titan 34D Launches Missile Warning Satellite," *Aviation Week & Space Technology*, December 7, 1987, pp. 30–31.

20. *History of Air Force Space Command, January–December 1988* (Peterson AFB, Colo.: AFSPACECOM, n.d.), pp. 28–29.

21. Ibid., p. 29; private information.

22. *History of Air Force Space Command, January–December 1988*, p. 29.

23. Ibid., p. 31; *History of Air Force Space Command, January–December 1989* (Peterson AFB, Colo.: AFSPACECOM, n.d.), p. 48; "Los Alamos Energetic Particle Home Page."

24. *History of Air Force Space Command, January–December 1988*, pp. 29–30.

25. Ibid., p. 49.

26. *History of Air Force Space Command, January–December 1989*, p. 45.

27. *History of Space Division, October 1986–September 1987, Volume I, Narrative* (Los Angeles, AFB: Space Division, n.d.), p. 280; *History of the Air Force Space Command, January–December 1989*, pp. 46, 48; *History of Space Division, October 1988–September 1989, Volume I*, p. 418; William J. Broad, "Biggest U.S. Unmanned Rocket, Rival of Shuttle, Soars into Space," *New York Times*, June 15, 1989, pp. A1, B11; Kathy Sawyer, "New Titan IV Rocket Orbits Secret Satellite," *Washington Post*, June 15, 1989, p. A12.

28. *History of the Air Force Space Command, January–December 1989*, pp. 46, 48; *History of Space Division, October 1988–September 1989, Volume I*, p. 418.

29. Broad, "Biggest U.S. Unmanned Rocket, Rival of Shuttle, Soars into Space."

30. *History of Air Force Space Command, January–December 1989*, p. 48; "Los Alamos Energetic Particle Home Page."

31. Air Force Space Command, *History of Air Force Space Command, January–December 1989*, p. 47; Craig Covault, "New Missile-Warning Satellites to Be Launched on First Titan 4," *Aviation Week & Space Technology*, February 20, 1989, pp. 34–40; Hanley et al., *History of Space Division, 1 October 1979–30 September 1980, Volume I, Narrative*, pp. 230, 232; "U.S. IMEWS Space System and Creation of an Advanced Ballistic Missile Launch Detection System," *JPRS-UMA-95-011*, March 14, 1995, pp. 63–69. This last reference is a translation of an article by Lt. Col. A. Andronov and Capt. S. Garbuk, in *Zarubezhnoye Voyennoye Obozreniye* 12, 1994, pp. 34–40.

32. *History of Air Force Space Command, January–December 1989*, p. 47; Covault, "New Missile-Warning Satellites to Be Launched on First Titan 4," pp. 34–40; Edward H. Kolcum, "Titan 4, Delta 2 Launches Generate Confidence in Military Space Operations," *Aviation Week & Space Technology*, June 19, 1989, pp. 40–41; *History of Space Division, October 1987–September 1988, Volume I, Narrative* (Los Angeles AFB: Space Division, n.d.), p. 391; Timothy C. Hanley, Harry N. Waldron, and Elizabeth J. Levack, *History of Space Division, 1 October 1980–30 September 1981, Volume I, Narrative* (Los Angeles AFB: Space Division, n.d.), p. 311.

33. *History of Air Force Space Command, January–December 1989*, p. 47; Covault,

"New Missile-Warning Satellites to Be Launched on First Titan 4"; Kolcum, "Titan 4, Delta 2 Launches Generate Confidence in Military Space Operations."

34. *History of the Space Systems Division, October 1988–September 1989, Volume I* (Los Angeles AFB: SSD, n.d.), pp. 417, 421.

35. *History of Space Division, October 1988–September 1989, Volume I*, pp. 417, 421; Jeffrey M. Lenorovitz, "Joint Flight to Gather ICBM Tracking Data," *Aviation Week & Space Technology*, October 18, 1993, pp. 96–97; interview.

36. *History of Space Division, October 1988–September 1989, Volume I*, pp. 417, 421; HQ Air Force Space Command, "Defense Support Program (DSP) Experiments," September 10, 1990.

37. *History of the Space Systems Division, October 1988–September 1989, Volume I*, p. 441 n. 11.

38. Ibid.

39. *History of Space Division, October 1986–September 1987, Volume I, Narrative*, pp. 279–280; Adam Goodman, "Problems Plague McDonnell Laser," *St. Louis Post-Dispatch*, August 13, 1989, pp. 1, 9. On the subject of laser cross-links, see Geoffrey Hyde and Burton I. Edelson, "Laser Satellite Communications: Current Status and Directions," *Space Policy* 13, no. 1 (February 1997): 47–54.

40. *History of the Space Systems Division, October 1983–September 1984*, p. 226.

41. Ibid., p. 228.

42. Ibid.

43. Ibid.

44. U.S. Congress, Senate Appropriations Committee, *Department of Defense Appropriations for Fiscal Year 1986*, pt. 2 (Washington, D.C.: U.S. Government Printing Office, 1985), p. 345.

45. Neil Munro, "U.S. Air Force May Drop Satellite Laser-Link Program," *Defense News*, October 4–10, 1993, p. 46. *History of Space Division, October 1987–September 1988, Volume I, Narrative*, pp. 389–99; Goodman, "Problems Plague McDonnell Laser"; *History of the Space Systems Division, October 1989–September 1990, Volume I* (Los Angeles AFB: SSD, 1991), p. 554.

46. *History of Space Division, October 1989–September 1990, Volume I* (Los Angeles AFB: SSD, 1991), pp. 552–56; Goodman, "Problems Plague McDonnell Laser"; *History of Space Division, October 1987–September 1988, Volume I*, p. 414 n.31.

47. Goodman, "Problems Plague McDonnell Laser."

48. *History of Air Force Space Division, October 1988–September 1989, Volume I*, p. 426–27.

49. Ibid., p. 427; *History of the Space Systems Division, October 1989–September 1990, Volume I*, p. 554; interview.

50. *History of the Space Systems Division, October 1989–September 1990, Volume I*, p. 554.

51. Ibid.

52. Ibid., pp. 554–55.

53. Ibid., p. 555.

54. Ibid., p. 556.

55. "McDonnell-Douglas Delivers Second Production DSP Laser Crosslink," *Aerospace Daily*, January 19, 1993, p. 95; Vincent Kiernan and Daniel J. Marcus, "DSP Crosslink Trouble Hits McDonnell Financials," *Space News*, August 10–16, 1992, p. 4;

Munro, "U.S. Air Force May Drop Satellite Laser-Link Program." Such problems could have been avoided, according to several individuals involved in the DSP program, if radio cross-links had been chosen instead of laser cross-links. Such cross-links, they argue, would be sufficient for the volume of DSP data that would have to be transmitted. Further, the transmission frequency would have been in the middle of the oxygen absorption band, ensuring that the data could not be intercepted by ground stations of hostile nations. The laser cross-link option was heavily pushed by the "black world," which was also interested in cross-linking data from its imagery and signals intelligence satellites to the United States, eliminating the need for ground stations such as Pine Gap, Australia; Bad Aibling, Germany; and Menwith Hill, United Kingdom. Having the DSP community develop the cross-links would save the NRO and CIA a considerable sum of money.

56. Maj. Chesley, XRDD, AFSPACECOM, Point Paper on (DSP) System 1, March 26, 1992; Thomas Moorman, CINCAFSPACE to USCINCSPACE, DSP System 1 Status, May 3, 1990; interview.

9. Australia, Germany, and New Mexico

1. *Air Force Space Command Intelligence Threat Assessment: 5th Defense Space Communications Squadron, 1st Space Wing, Woomera, Australia* (Peterson AFB, Colo.: AFSPACECOM, 1991), pp. 1, 2.

2. David McKnight, *Australia's Spies and Their Secrets* (St. Leonards: Allen & Unwin, Australia, 1994), p. 292; John Pilger, *A Secret Country: The Hidden Australia* (New York: Knopf, 1991), p. 154.

3. National Security Study Memorandum 204, "U.S. Policy Toward Australia," July 1, 1974.

4. Pilger, *A Secret Country*, pp. 142–83; Central Intelligence Agency, "Australia Impasse Continues," *Weekly Review*, November 7, 1975, p. 22. The CIA analysis also referred to Fraser's handling of the challenge as "inept."

5. McKnight, *Australia's Spies and Their Secrets*, p. 294; Brian Toohey and William Pinwill, *OYSTER: The Story of the Australian Secret Intelligence Service* (Port Melbourne, Australia: William Heinemann, 1989), p. 145.

6. Desmond Ball, *Pine Gap: Australia and the US Geostationary Signals Intelligence Satellite Program* (Sydney: Allen & Unwin, Australia, 1988), pp. 88–89; Robert Lindsey, *The Falcoln and the Snowman: A True Story of Friendship and Espionage* (New York: Simon and Schuster, 1979), p. 58.

7. Desmond Ball, *A Suitable Piece of Real Estate: American Installations in Australia* (Sydney: Hale & Iremonger, 1980); Tom Ballantyne, "Australia's Nuclear Targets," *Sydney Morning Herald*, November 23, 1982; Andrew Mack, "Putting Mobile American Targets on Australian Highways," *National Times*, June 3–9, 1983, p. 4; William Pinwill, "Strike One!" *National Times*, April 27–May 3, 1984, pp. 12, 15; William Pinwill, "Nurrungar's Nuclear Attack Role Admitted," *National Times*, June 1–7, 1984.

8. Pinwill, "Nurrungar's Nuclear Attack Role Admitted."

9. Milton Cockburn, "US Bases Make Us 'Priority' Target: Hayden," *Sydney Morning Herald*, March 17, 1986, p. 1; Paul Kelly, "Star Wars: Best to Be Frank on Bases," *Australian*, February 15, 1985, p. 7.

10. Kelly, "Star Wars."

11. *History of Air Force Space Command, January–December 1987* (Peterson AFB, Colo.: AFSPACECOM, n.d.), p. 42.

12. Ibid.

13. Desmond Ball, *A Base for Debate: The US Satellite Station at Nurrungar* (Sydney: Allen & Unwin, Australia, 1987), pp. 67, 72.

14. Ibid., pp. 72–73.

15. Ibid., pp. 74, 75, citing Kelly, "Star Wars: Best to Be Frank on Bases," with respect to Hawke's statement.

16. Ball, *A Base for Debate*, pp. 75–76.

17. Ibid., p. 76, citing Clarence A. Robinson, "US Strategic Defense Options: Panel Urges Boost-Phase Intercepts," *Aviation Week & Space Technology*, December 5, 1983, p. 52; National Security Study Directive 6-83, "Study on Eliminating the Threat Threat Posed by Ballistic Missiles," April 18, 1983.

18. Ball, *A Base for Debate*, p. 80.

19. David Humphries, "Close US Base at Nurrungar, Says Professor," *The Age*, August 21, 1987, p. 3; Patrick Walters, "Shut US Base at Nurrungar, Says Defence Expert," *Sydney Morning Herald*, August 21, 1987, pp. 1, 13; Patrick Walters, "Australia Joins the Star Wars Program—in Secret," *Sydney Morning Herald*, August 21, 1987, p. 1.

20. *History of Air Force Space Command, January–December 1987*, pp. 42–43.

21. Defense Attache Office, Canberra, Subj: [Deleted], August 25, 1987.

22. *History of Air Force Space Command, January–December 1987*, pp. 42–43.

23. Ibid., p. 43.

24. Ibid.

25. Edmund Doogue, "Protesters Issue 'Eviction Notice' to US at Nurrungar," *The Age*, October 5, 1987, p. 23; *Unit Activity Report of the 5th Defense Space Communications Squadron, April–December 1987* (Woomera, Australia: 5th DSCS, 1988), n.p.

26. *Unit Activity Report of the 5th Defense Space Communications Squadron, April–December 1987*, n.p.

27. Gareth Evans, Minister for Foreign Affairs and Trade, to Mr. Laurence W. Lane Jr., Ambassador of the United States of America, Canberra ACT, 16 November 1988.

28. Ibid.

29. Embassy of the United States of America, Note No. 96, Canberra, November 16, 1988.

30. "Australia, U.S. Sign Accord on 2 Bases," *Washington Post*, November 23, 1988, p. A15.

31. *History of the Air Force Space Command, January–December 1989* (Peterson AFB, Colo.: AFSPACECOM, n.d.), p. 50; "Australia, U.S. Sign Accord on 2 Bases"; Air Force Space Command, *History of the Air Force Space Command, 1990* (Peterson AFB, Co.: AFSPACECOM, n.d.), p. 48; R. J. Hawke, "Joint Defence Facilities: Ministerial Statement," *Hansard* (House of Representatives), 22 November 1988, p. 2939. By late 1991 there were 186 Americans at Nurrungar (25 officers, 140 enlisted men, and 21 contractor personnel). There were between 177 and 182 Australians: 9 officers, 20 enlisted men, and 148–153 employees of Fairey Australasia Ltd. See Air Force Space Command, *Air Force Space Command Intelligence Threat Assessment, 5th Defense Space Communications Squadron, 1st Space Wing, Woomera, Australia*, 1 November 1991, p. 3.

32. E. C. Aldridge, Secretary of the Air Force, to Kim Beazley, Minister for Defence, November 28, 1988.

33. *History of the Air Force Space Command, 1990* (Peterson AFB, Colo.: AFSPACE-COM, n.d.), p. 48; Ltr., Col. Glenn Doss, DCS/Plans AFSPACECOM to AFSPACE-COM/CC et al., "Deputy Commander (Woomera) Controversy," 28 April 1990, with Background Paper on Australian Implementing Arrangement; *Implementing Arrangement between the United States Air Force and the Australian Department of Defence Concerning the Joint Defence Facility, Nurrungar.*

34. *History of the Air Force Space Command, 1990*, p. 48.

35. Ibid.

36. Ibid.

37. Ibid., p. 49.

38. Ibid.

39. Ibid.; AFSPACECOM to HQ USAF, "Nurrungar Implementing Arrangement," June 1990.

40. *History of the Air Force Space Command, 1990*, p. 49; AFSPACECOM, "Proposal for CC Trip to Australia," 25 June 1990.

41. *Air Force Space Command Intelligence Threat Assessment, 5th Defense Space Communications Squadron, 1st Space Wing*, Woomera, Australia, pp. 11, 17.

42. Ibid., p. 7.

43. Desmond Ball, "Sharing Nurrungar," *Asia-Pacific Defence Reporter*, November 1991, pp. 6–7, 10. The Soviet Union/Russia has relied on both elliptically orbiting and geosynchronous satellites to provide launch detection data. On the Soviet/Russian satellites, see Phillip S. Clark, "Russia's Geosynchronous Early Warning Satellite Programme," *Jane's Intelligence Review*, February 1994, pp. 61–64; Paul Podvig, "The Operational Status of the Russian Space-Based Early Warning System," *Science & Global Security* 4 (1994): 363–84; Phillip S. Clark, "Russian Early Warning Satellites," *Jane's Intelligence Review*, November 1994, pp. 493–95; Nicholas L. Johnson and David M. Rodvold, *Europe and Asia in Space, 1991–1992* (Colorado Springs, Colo.: Kaman Sciences Corporation, 1994), pp. 246–48; "Votintstev: Development of Missile Attack Warning System," *JPRS-UMA-94-008*, February 23, 1994, pp. 5–12.

44. Ball, "Sharing Nurrungar."

45. Ibid.

46. Derek Ballantine, "Anger at U.S. Base Invite," *Herald Sun*, November 4, 1991, p. 3.

47. Mark Metherell, "Evans Says US Did Not Ignore Opinion on Base," *The Age*, November 5, 1991, p. 13.

48. Ballantine, "Anger at U.S. Base Invite"; Metherell, "Evans Says US Did Not Ignore Opinion on Base."

49. Fm Amembassy Canberra to Sec State, "Opening Joint Facility in Australia to Soviets," November 4, 1991.

50. Fm: Amembassy Canberra to: Sec State, "U.S. Early Warning Initiative—Australian Request for Info Re Meeting with Soviets," November 21, 1991.

51. Fm: Sec State to: Amembassy Canberra, "U.S. Early Warning Initiative—Response on Request for Info Re Meeting with Soviets," November 23, 1991.

52. ADCOM, Subj: Simplified Processing Station (SPS) Overseas Siting, March 2, 1979, p. 1.

53. ADCOM, Subj: Simplified Processing Station (SPS) Overseas Siting, April 2, 1979, p. 2.

54. Ball, *A Base for Debate*, p. 61; interview.

55. HQ SPACECOM, SPS Relocation, September 29, 1982.

56. *History of Air Force Space Command, January–December 1986* (Peterson AFB, Colo.: AFSPACECOM, n.d.), p. 42.

57. *History of the Space Systems Division, October 1989–September 1990, Volume I* (Los Angeles AFB: Space Systems Division History Office, 1991), pp. 566–67.

58. Col. Fred H. Hirsch, Director of Missile Warning Systems, AFSPACECOM, to XPI, "European Ground Station (EGS) Contingency Planning," February 29, 1990.

59. John Schenck, AFSPACECOM, "European Ground Station (EGS), Contingency Planning," March 2, 1990.

60. Robert Gates, *From the Shadows: The Ultimate Insiders' Account of Five Presidents and How They Won the Cold War* (New York: Simon & Schuster, 1996), pp. 485–86.

61. *History of Air Force Space Command, 1990*, pp. 50–51; HQAFSPACECOM to HQ USAF, "Preliminary Site Evaluation for European Ground Station Relocation," July 26, 1990; HQ USAFE to HQ AFSPACECOM, "Release of Information on European Ground Station (EGS) to Germany," October 19, 1990.

62. *Space Division Chronology, 1984* (Los Angeles: Air Force Space Division, 1985), p. xxx; *Welcome to . . . Holloman Air Force Base* (San Diego: MARCOA Publishing, 1993), pp. 11–12.

63. Telephone interview with Jim Chambers, August 21, 1996.

64. Capt. Antonio V. Branco and Martin Whelan, *History of the 4 Satellite Communications Squadron, January–March 1987* (Holloman AFB, N.M.: 4SCS, 1988), p. 5.

65. *History of the 1025th Satellite Communications Squadron, January–March 1985* (Holloman AFB, N.M.: 1025th SCS, 1985), p. 2; *History of the Space Division, October 1983–September 1984* (Los Angeles AFS: Space Division History Office, n.d.), p. 239; *History of the Space Systems Division, October 1989–September 1990, Volume I*, p. 566.

66. *History of the 1025th Satellite Communications Squadron, January–March 1985*, p. 2; *History of the Space Division, October 1983–September 1984*, p. 239.

67. Daniel Ford, *The Button: The Pentagon's Command and Control System—Does it Work?* (New York: Simon & Schuster, 1985), p. 151; Bruce G. Blair, *Strategic Command and Control: Redefining the Nuclear Threat* (Washington, D.C.: Brookings Institution, 1985), pp. 108–11; Scott Sagan, *The Limits of Safety* (Princeton, N.J.: Princeton University Press, 1993), p. 229.

68. Ford, *The Button*, p. 149.

69. *History of the 1025th Satellite Communications Squadron, January–March 1985*, p. 12.

70. Private information.

71. Victor Semeyonov, "Through War and Peace: Russian Imaging Satellites," *Launchspace*, April 1998, pp. 46–49; Desmond Ball, "Soviet Signals Intelligence: Vehicular Systems and Operations," *Intelligence and National Security*, January 1989, pp. 5–27.

72. Space Division, *Defense Support Program (DSP) System Threat Assessment Report* (Los Angeles AFB: AFSC Space Division, May 31, 1989), p. x.

73. Major Burton A. Casteel Jr., *SPETsNAZ: A Soviet Sabotage Threat* (Maxwell AFB, Ala.: Air Command and Staff College, 1986), p. 12.

74. L. G. Gref, A. L. Latter, E. A. Martinelli, and H. P. Smith, *A Soviet Paramilitary Attack on U.S. Nuclear Forces: A Concept* (Santa Monica, Calif.: R&D Associates, 1973), p. 2-1.

75. *History of Air Force Space Command, January–December 1986* (Peterson AFB, Colo.: AFSPACECOM), p. 45.

76. Branco and Whelan, *History of the 4 Satellite Communications Squadron, January–March 1987*, p. 6; *Unit Activity Report of the 4th Satellite Communications Squadron, 1 January–30 June 1988*, n.d., n.p.

77. *History of Air Force Space Command, January–December 1989*, p. 52.

78. Ibid., pp. 52–53.

79. HQ AFSPACECOM, MGS Deployment Exercise ANCHOR READY, 25 May 1990; HQ AFSPACECOM, Anchor Ready Alert Order, 25 May 1990; HQ AF-SPACECOM, Warning Order, May 11, 1970.

80. *History of Air Force Space Command, 1 January–31 December 1991* (Peterson AFB, Colo.: AFSPACECOM, n.d.), p. 41.

81. Memo Col. James W. Knapp, Asst DCS/Requirements, AFSPACECOM to CC et al., "Mobile Ground Terminal-14 (MGT-14) Upgrade," August 6, 1990; *History of Air Force Space Command 1990* (Peterson AFB, Colo.: AFSPACECOM, n.d.), p. 12.

82. *History of Air Force Space Command, 1 January–31 December 1991*, p. 41.

83. Ibid.

84. MIN 171/92, "Australia and the United States Joint Exercise at Nurrungar," 17 July 1992; "USAF Tests Space Station Stand-In," *Jane's Defence Weekly*, August 1, 1992, p. 16. George W. Bradley III, Richard S. Eckert, Rick W. Sturdevant, Herbert M. Zolof, David L. Bullock, Freda X. Norris, and Karen V. Martin, *History of the Air Force Space Command, January 1992–December 1993* (Colorado Springs, Colo.: Air Force Space Command, n.d.), pp. 200, 201 n.

85. U.S. Space Command, Minutes of Third Nurrungar Review Conference, n.d., p. 2.

10. Desert Storm

1. U.S. News & World Report, *Triumph Without Victory: The Unreported History of the Persian Gulf War* (New York: Times Books, 1992), pp. 7–9; George Bush, National Security Directive 45, "U.S. Policy in Response to the Iraqi Invasion of Kuwait," August 20, 1990, p. 2.

2. Richard Hallion, *Storm over Iraq: Air Power and the Gulf War* (Washington, D.C.: Smithsonian Institution, 1992), p. 159.

3. U.S. News & World Report, *Triumph Without Victory*, p. 38.

4. Rick Atkinson, *Crusade: The Untold Story of the Persian Gulf War* (Boston: Houghton Mifflin, 1993), p. 141.

5. Duncan Lennox, "Inside the R-17 'Scud B' Missile," *Jane's Intelligence Review*, July 1991, pp. 302–5; "Improving on a Soviet Design," *Jane's Defence Weekly*, March 2, 1991, pp. 302–3; Thomas P. Christie, William J. Barlow, Donald L. Ockerman, Elliot S. Parkin, Frank Sevcik, Grant Sharp, Charles M. Waespy, and Howard C. Whetzel, *Desert Storm Scud Campaign* (Alexandria, Va.: Institute for Defense Analyses, 1992), p. I-7.

324   AMERICA'S SPACE SENTINELS

6. Norman Schwarzkopf, *It Doesn't Take a Hero* (New York: Bantam Books, 1993), p. 484.

7. Barbara L. Tuxbury, *TRAP Reporting of Iraqi Scud Missile Launches: Its Timeliness and Utility*, CRM 91-115 (Alexandria, Va.: Center for Naval Analyses, 1991), pp. A-2 to A-5; Lennox, "Inside the R-17 'Scud B' Missile"; Anthony H. Cordesman, *After the Storm: The Changing Balance in the Middle East* (Boulder, Colo.: Westview Press, 1993), p. 489; Theodore A. Postol, "Lessons of the Gulf War Experience with Patriot," *International Security* 16, no. 3 (Winter 1991–92): 119–71; General Accounting Office, *Patriot Missile Defense: Software Problem That Led to System Failure at Dhahran, Saudi Arabia* (Washington, D.C.: GAO, 1992), p. 6; Abdel Darwish and Gregory Alexander, *Unholy Babylon: The Secret History of Saddam's War* (New York: St. Martin's Press, 1991), p. 87; Department of Defense, *Conduct of the Persian Gulf War: Final Report to Congress* (Washington, D.C.: U.S. Government Printing Office, 1992), p. 16.

8. Atkinson, *Crusade: The Untold Story of the Persian Gulf War*, p. 96; Joseph S. Bermudez Jr., "Iraqi Missile Operations During 'Desert Storm,'" *Jane's Soviet Intelligence Review*, March 1991, p. 131.

9. Cordesman, *After the Storm*, pp. 484–85; W. Andrew Terrill, "The Gulf War and Ballistic Missile Proliferation," *Comparative Strategy* 11, no. 2 (1992): 163–76; Darwish and Alexander, *Unholy Babylon*, p. 61.

10. Cordesman, *After the Storm*, p. 485.

11. Dilip Hiro, *The Longest War: The Iran-Iraq Military Conflict* (New York: Routledge, 1991), pp. 134–35.

12. Craig Covault, "NORAD, Space Command Request System for Surveillance of Soviet Weapons," *Aviation Week & Space Technology*, April 6, 1987, pp. 73, 76; *Unit Activity Report of the 5th Defense Space Communications Squadron, January–June 1988* (Woomera AS, Australia: 5th DSCS, 1988), p. 2.

13. Mohammad Yousaf and Mark Adkin, *The Bear Trap: Agfhanistan's Untold Story* (London: Leo Cooper, 1992), pp. 215, 230; interview.

14. "Israel in Second Secret Test of Jericho IRBM," *Jane's Defence Weekly*, November 19, 1988, p. 1258; Richard M. Weinraub, "India Tests Mid-Range 'Agni' Missile," *Washington Post*, May 23, 1989, pp. A1, A21; Pushipindar Singh, "India's Agni Success Poses New Problems," *Jane's Defence Weekly*, June 3, 1989, pp. 1052–53; Barbara Crossette, "India Reports Successful Test of Mid-Range Missile," *New York Times*, May 23, 1989, p. A9.

15. Defense Intelligence Agency, "Special Assessment—South Africa: Missile Activity," July 5, 1989; Bill Gertz, "S. Africa on the Brink of Ballistic Missile Test," *Washington Times*, June 20, 1989, pp. A1, A9.

16. David B. Ottaway, "Israel Reported to Test Controversial Missile," *Washington Post*, September 16, 1989, p. A17.

17. Benjamin Beit-Hallahmi, *The Israeli Connection: Who Israel Arms and Why* (New York: Pantheon, 1987), pp. 108–74; David B. Ottaway and R. Jeffrey Smith, "U.S. Knew of Israel–South Africa Missile Deal," *Washington Post*, October 27, 1989, pp. A1, A27; Michael R. Gordon, "U.S. Says Data Suggest Israel Aids South Africa on Missile," *New York Times*, October 27, 1989, pp. A1, A7.

18. Planning Directive for: Appendix 28 to Annex C to USCINCSPACE OPORD 3401-90, DSP/MGS OPS, p. 1.

19. HQ USSPACECOM, Potential for Tactical Application of Ballistic Missile Launch Notification, June 20, 1990.

20. HQ AFSPACECOM, "Tactical Event Reporting System (TERS) Lessons Learned," September 13, 1990, n. 3; Msg. HQ USSPACECOM/SPJ3 to HQ AFSPACECOM/XR et al., "Potential Tactical Applications of Ballistic Missile Launch Notification," p. 1.

21. *History of the Space Systems Division, October 1989–September 1990, Volume I* (Los Angeles AFB: SSD, 1991), p. 544.

22. Ibid, p. 549.

23. Edward H. Kolcum, "Titan 4 Boosts Satellite to Monitor Persian Gulf," *Aviation Week & Space Technology*, November 19, 1990, p. 79; Vincent Kiernan, "Bush Report Reveals Payload Aboard Titan," *Space News*, October 5–11, 1992, p. 17.

24. TRADOC, *Certain Victory: The US Army in the Gulf War*, p. 18; Bill Gertz, "U.S. Detects Iraqi Missiles at Last Minute," *Washington Times*, December 10, 1990; Robert C. Toth, "Iraqi Missile Test Had U.S. Thinking War Had Started," *Los Angeles Times*, December 21, 1990, pp. A1, A10; "The Secret History of the War," *Newsweek*, March 18, 1991, pp. 28–39; Hallion, *Storm over Iraq*, p. 156; DIA Iraq Regional Intelligence Task Force, "Iraq Launches Multiple SRBMs Dec 2," December 3, 1990.

25. George Bush, National Security Directive 54, "Responding to Iraqi Aggression in the Gulf," January 15, 1991.

26. Doug Waller, *The Commandos: The Inside Story of America's Secret Soldiers* (New York: Simon & Schuster, 1994), p. 243; Michael R. Gordon and Bernard E. Trainor, *The General's War: The Inside Story of the Conflict in the Gulf* (Boston: Little, Brown, 1995), p. 212.

27. Joseph S. Bermudez Jr., "Iraqi Missile Operations During 'Desert Storm'—Update," *Jane's Soviet Intelligence Review*, May 1991, p. 225; U.S. News & World Report, *Certain Victory*, pp. 18–19; Robert W. Ward et al., *Desert Storm Reconstruction Report, Volume VIII: C³/Space and Electronic Warfare* (Alexandria, Va.: Center for Naval Analyses, 1992), p. I-1.

28. Department of Defense, *Conduct of the Persian Gulf War: Final Report to Congress* (Washington, D.C.: Department of Defense, 1992), p. 223.

29. Ibid., p. 240. The radars in question were the FPS-17 and FPS-79 detection and tracking radars operated by the Air Force Space Command from Diyarbakir AS, Pirinclik, Turkey. See *United States Space Command Operations Desert Shield and Desert Storm Assessment* (Peterson AFB, Colo.: USSPACECOM, 1992), p. 24.

30. Terrill, "The Gulf War and Ballistic Missile Proliferation"; Department of Defense, *Conduct of the Persian Gulf War, Final Report to Congress, Appendix T: Performance of Selected Weapon Systems* (Washington, D.C.: Department of Defense, 1992), p. T-152.

31. General Accounting Office, *Patriot Missile Defense*, p. 3; Department of Defense, *Conduct of the Persian Gulf War, Final Report to Congress, Appendix T: Performance of Selected Weapon Systems*, p. T-152.

32. General Accounting Office, *Patriot Missile Defense*, p. 4.

33. Ibid., pp. 4–5.

34. Vincent Kiernan, "Gulf War Led to Appreciation of Military Space," *Space News*, January 25–31, 1993, p. 8.

35. James W. Canan, "Space Support for Shooting Wars," *Air Force Magazine,* April 1993, pp. 30–34.

36. Ward et al., *Desert Storm Reconstruction Report, Volume VIII,* p. 8-26; Craig Covault, "USAF Missile Warning Satellites Providing 90-Sec. Scud Attack Alert," *Aviation Week & Space Technology,* January 21, 1991, pp. 60–61.

37. "DSPs Detected Fatal Scud Attack," *Aviation Week & Space Technology,* April 4, 1994, p. 32; U.S. Space Command, *United States Space Command Operations Desert Shield and Desert Storm Assessment* (Peterson AFB, Colo.: USSPACECOM, 1992), p. 24.

38. Ward et al., *Desert Storm Reconstruction Report, Volume VIII,* p. 6-20.

39. Canan, "Space Support for Shooting Wars"; Msg HQ USSPACECOM/SPJ3, "Potential for Tactical Applications of Ballistic Missile Launch Notification," June 29, 1990; Vincent Kiernan, "New Details on DSP War Usage," *Space News,* March 23–29, 1992, pp. 4, 29; U.S. Space Command, *United States Space Command Operations Desert Shield and Desert Storm Assessment,* p. 22.

40. Kiernan, "New Details on DSP War Usage"; "DSPs vs. SCUDs in Gulf War," *Orbiter,* March 11, 1992, pp. 1, 3; Covault, "USAF Missile Warning Satellites."

41. Randy E. Nees, "Targeting from Space," *Space Tracks,* March–April 1994, pp. 10–11; *History of Air Force Space Command, 1 January–31 December 1991* (Peterson AFB, Colo.: AFSPACECOM, 1993), p. 32; *United States Space Command Operations Desert Shield and Desert Storm Assessment,* pp. 16–18.

42. Sheldon B. Herskovitz, "High Marks for Military Radar: The Patriot Missile System," *Journal of Electronic Defense,* May 1991, pp. 55–59.

43. Ward et al., *Desert Storm Reconstruction Report, Volume VIII,* pp. 8-25 to 8-26, I-1 to I-3. The mean times are based on eighty-one reports. Time of receipt data was unavailable for five reports, while two launches had no TRAP reports.

44. Vincent Kiernan, "Cooper Lifts Veil of Secrecy to Applaud DSP," *Space News,* April 1–7, 1991, p. 6.

45. Barton Gellman, "New Study Cuts Patriot Missile Success Rate to 9 Percent," *Washington Post,* September 20, 1992, p. A4; General Accounting Office, *Operation Desert Storm: Data Does Not Exist to Conclusively Say How Well Patriot Performed* (Washington, D.C.: GAO, 1992), p. 3.

46. Tim Weiner, "Patriot Missile's Success a Myth, Israeli Aides Say," *New York Times,* November 21, 1993, p. 13.

47. Ibid. Some of the discrepancy between Army and Raytheon Corporation estimates of Patriot effectiveness and the less favorable views of others may be definitional, in that the Army definition of an intercept includes detonations that divert tumbling, disintegrating Scuds from their presumed targets. Whether it was possible to actually determine if such a deflection took place has been questioned. According to Peter D. Zimmerman, a physicist, formerly with the Center for Strategic and International Studies, "There is no way to be sure in a close encounter whether the Patriot got it or missed it. . . . The data are lousy."

It should also be noted that the Patriot version, the PAC-2, deployed to the Gulf was designed to counter a more advanced and stable tactical ballistic missile. When the Scuds, which began their dive into the atmosphere at 5,000 miles per hour, were slowed to 4,400 miles per hour by the increasingly dense air, they often began to break up. Intercepting a warhead that deviated from a stable course, amid debris from the missile, was a more difficult task than intercepting a more advanced missile that would have maintained its projected trajectory throughout the intercept stage—and necessitated further modifications to the computer software. See TRADOC, *Certain Victory*, pp. 19–20; Weiner, "Patriot's Missile's Success a Myth, Israeli Aides Say," p. 13; Seymour M. Hersh, "Missile Wars," *New Yorker*, September 26, 1994, pp. 86–99.

48. Ward et al., *Desert Storm Reconstruction Report, Volume VIII*, p. 8-28.

49. Gordon and Trainor, *The General's War*, p. 195; Moshe Arens, *Broken Covenant: American Foreign Policy and the Crisis between the U.S. and Israel* (New York: Simon & Schuster, 1995), p. 179.

50. Bermudez, "Iraqi Missile Operations During 'Desert Storm'"; Seymour Hersh, *The Samson Option: Israel's Nuclear Arsenal and American Foreign Policy* (New York: Random House, 1991), p. 130; Arens, *Broken Covenant*, p. 211.

51. Terrill, "The Gulf War and Ballistic Missile Proliferation"; Theresa Foley, "DSP Advocates, Foes Cited Dhahran Scud Attack," *Space News*, April 18–24, 1994, p. 4; Donatella Lorch, "Twisted Hulk of Warehouse Tells a Grim Story of Death," *New York Times*, February 27, 1991, p. A18; U.S. Congress, House Appropriations Committee, *Department of Defense Appropriations for 1992*, pt. 2 (Washington, D.C.: U.S. Goverment Printing Office, 1991), pp. 665–66.

52. *History of Air Force Space Command, 1 January–31 December 1991*, pp. 34–35; "DSPs Detected Fatal Scud Attack."

53. Space Systems Division, *DSP Desert Storm Summary Briefing* (Los Angeles AFB: Space Systems Division, June 1991), p. 48; "DSPs Detected Fatal Scud Attack."

54. Space Systems Division, *DSP Desert Storm Summary Briefing*, p. 48; "DSPs Detected Fatal Scud Attack."

55. "DSPs Detected Fatal Scud Attack"; Space Systems Divsion, *DSP Desert Storm Summary Briefing*, p. 48.

56. Schwarzkopf, *It Doesn't Take a Hero*, p. 484.

57. Hallion, *Storm over Iraq*, p. 179; U.S. Air Force, *Gulf War Air Power Survey, Draft Summary*, 15 April 1993, p. 3-28.

58. Hallion, *Storm over Iraq*, p. 181; Department of Defense, *Conduct of the Persian Gulf War: Final Report to Congress*, p. 226.

59. William Arkin, *Target Iraq: The Persian Gulf Air War and Its Implications*, forthcoming, chap. 7.

60. "Washington Whispers," *U.S. News & World Report*, January 28, 1991, p. 19; Department of Defense, *Conduct of the Persian Gulf War: Final Report to Congress*, p. 226.

61. Kiernan, "New Details on DSP War Usage"; "DSPs vs. SCUDs in Gulf War," *Orbiter*, March 11, 1992, pp. 1, 3; Covault, "USAF Missile Warning Satellites"; HQ USSPACECOM/SPJ3, "Potential for Tactical Applications of Ballistic Missile Notification"; Interview. According to an Aerospace Corporation study, "The benefit of improved tactical parameter accuracy from DSP stereo processing has been known for some time. . . . The tactical parameter improvements were considered a secondary benefit which was not of high interest. . . . The interest in improved accuracy has really been driven by the tactical missile threat. This threat drove the implementation of E-DUAL (dual processing at the EGS) in 1991 and O-DUAL (dual processing at

the OGS) in 1993." Guido W. Aru and Carl T. Lunde, *DSP-II Preserving the Air Force's Options* (El Segundo, Calif.: Aerospace Corporation, 1993), p. B-28.

62. Defense Intelligence Agency, *Mobile Short-Range Ballistic Missile Targeting in Operation DESERT STORM,* OGA-1040-23-91, November 1991, p. iii; Arkin, *Target Iraq,* chap. 7, pp. 16, 29; U.S. Air Force, *Gulf War Air Power Survey, Draft Summary,* pp. 3-28 to 3-30; "Scud Launch Procedures May Hold Key to Defeat of Mobile Launchers," *Aerospace Daily,* January 28, 1991, pp. 149–50; Gordon and Trainor, *The General's War,* p. 230; Hallion, *Storm over Iraq,* p. 181; communication with William Arkin, June 6, 1996. An IDA study states that there were "five major groupings of launch areas." See Christie et al., *Desert Storm Scud Campaign,* p. I-17.

63. Arkin, *Target Iraq,* pp. 16, 29; U.S. Air Force, *Gulf War Air Power Survey, Draft Summary,* pp. 3-28 to 3-30; "Scud Launch Procedures May Hold Key to Defeat of Mobile Launchers," U.S. Congress, House Committee on Armed Services, *Intelligence Successes and Failures in Operations Desert Shield/Desert Storm* (Washington, D.C.: HCAS, August 1993), p. 18; Gordon and Trainor, *The General's War,* p. 498; Christie et al., *Desert Storm Scud Campaign,* pp. 12, D-3; Department of Defense, *Conduct of the Persian Gulf War: Final Report to Congress,* p. 223.

64. Communication from William Arkin, February 13, 1995; Craig Covault, "Recon Satellites Lead Allied Intelligence Effort," *Aviation Week & Space Technology,* February 4, 1991, pp. 25–26.

65. Communication from William Arkin.

66. Hallion, *Storm over Iraq,* p. 181.

67. Dick Cheney, Secretary of Defense, *Annual Report to the President and the Congress* (Washington, D.C.: U.S. Government Printing Office, 1992), p. 85.

68. Australian Department of Defence, Press Release No. 136/91, "The Role of the Joint Defence Facility Nurrungar in the Gulf War," November 5, 1991.

69. Letter from Newton Mas to Ellis Lapin, July 22, 1991.

11. False Starts

1. *History of the Space and Missile Systems Center, October 1991–September 1992, Volume I* (Los Angeles AFB: Air Force Space and Missile Systems Center, n.d.), pp. 332–34; *History of the Space Systems Division, October 1990–September 1991* (Los Angeles AFB: Air Force Space Systems Division, n.d.), p. 337.

2. *United States Space Command, January to December 1992, Narrative* (Peterson AFB, Colo.: U.S. Space Command Office of History, 1996), p. 159.

3. Edward H. Kolcum, "NASA Overcomes Series of Problems to Launch Atlantis with DSP Payload," *Aviation Week & Space Technology,* December 2, 1991, pp. 24–25; Mark C. Cleary, *The Cape: Military Space Operations, 1971–1992* (Patrick AFB, Fla.: 45th Space Wing History Office, 1994), p. 133; National Aeronautics and Space Administration, *Post-Mission Report, Flight: STS-44,* December 11, 1991, p. 1; Warren Wright, "Space Shuttle Deploys DSP Satellite," *Space Trace,* January 1992, p. 8.

4. Kolcum, "NASA Overcomes Series of Problems"; Craig Covault, "Shuttle Deploys DSP Satellite, Crew Performs Reconnaissance Tests," *Aviation Week & Space Technology,* December 2, 1991, pp. 23–24; Cleary, *The Cape,* p. 133.

5. NASA Press Release 91-176, Defense Satellite Deploy, Observation Highlights,

STS-44, p. 1; Craig Covault, "Premature Atlantis Landing Reduces Reconnaissance, Medical Test Results," *Aviation Week & Space Technology*, December 9, 1991, pp. 70–71; National Aeronautics and Space Administration, *STS-44 Space Shuttle Mission Report* (Houston: Lyndon B. Johnson Space Center, 1992), p. 21.

6. Vincent Kiernan, "Equipment Trouble, Hazy View, Cut into Atlantis Mission," *Space News*, December 9–14, 1991, p. 10.

7. *History of the Space and Missile Systems Center, October 1991–September 1992, Volume I*, p. 334; "Los Alamos Energetic Profile Home Page," @ www.

8. *U.S. Space Command, January–December 1992* (Peterson AFB, Colo.: U.S. Space Command History Office, 1996), p. 197.

9. Vincent Kiernan, "Defense Satellites Carry Secret Science Instruments," *Space News*, July 19–25, 1993, p. 25; D. J. McComas, S. J. Bame, B. L. Barraclough, J. R. Donart, R. C. Elphic, J. T. Gosling, M. B. Moldwin, K. R. Moore, and M. F. Thomsen, "Magnetospheric Plasma Analyzer: Initial Three-Spacecraft Observations from Geosynchronous Orbit," *Journal of Geophysical Research* 98, no. A8 (August 1993): 13, 453–13, 465.

10. McComas et al., "Magnetospheric Plasma Analyzer"; Kiernan, "Defense Satellites Carry Secret Science Instruments."

11. Kiernan, "Defense Satellites Carry Secret Science Instruments."

12. Ibid.

13. Ibid.

14. Ibid.

15. Duane Andrews, "Report to the Congress on the Follow-On Early Warning System," April 15, 1991, p. 1; HQ USAF, Program Management Directive for Advanced Warning System, May 1, 1981; Department of the Air Force, Supporting Data for Fiscal Year 1984, Budget Estimates, Submitted to Congress, January 31, 1983, Descriptive Summaries, Research, Development, Test, and Evaluation, p. 229.

16. Paul Stares, *The Militarization of Space: U.S. Policy, 1945–1984* (Ithaca, N.Y.: Cornell University Press, 1985), pp. 224–25; National Security Decision Directive 85, "Eliminating the Threat from Ballistic Missiles," March 25, 1983.

17. NSSD 6-83, "Study on Eliminating the Threat Posed by Ballistic Missiles," April 18, 1983; "Studies in Support of the President's Defensive Weapons Initiative," attachment to Joint Chiefs of Staff, "Unclassified Summary of NSSD-6-83," May 6, 1983.

18. Andrews, "Report to the Congress on the Follow-On Early Warning System."

19. Ibid.

20. Private information. According to one individual involved in the DSP program, alternative combinations of high-altitude detection sensors and low-altitude weapons sensors were not considered.

21. David J. Lynch, "Star Wars Satellite Put on Hold for Now," *Defense News*, November 12, 1985, pp. 1, 11.

22. Ibid.

23. Notes, Press Briefing by Dr. Mel Brashears, October 18, 1988.

24. Ibid.

25. Ibid.

26. Ibid.

27. *History of the Space Systems Division, October 1990–September 1991*, p. 363.

28. Ibid., pp. 363–64.

29. James B. Schultz, "New Infrared Sensors Critical to SDI Surveillance Satellites," *Armed Forces Journal International*, May 1989, pp. 78–79; "Dual Debates," *Aviation Week & Space Technology*, March 13, 1989, p. 15.

30. *History of the Space Systems Division, October 1989–September 1990, Volume I* (Los Angeles AFB: SSD, 1991), p. 590.

31. Andrew Lawler, "Pentagon Revamping BSTS; Project Moving to Air Force," *Space News*, May 14–20, 1990, pp. 1, 20; HQ USSPACECOM, BSTS Strategy Meeting, April 26, 1990.

32. Lawler, "Pentagon Revamping BSTS."

33. Vincent Kiernan, "Senate Says No to BSTS Funding," *Space News*, May 14–20, 1990, pp. 1, 20.

34. Ibid.

35. Ibid.

36. Ibid.

37. Ibid.

38. Vincent Kiernan, "Air Force Readies Competition for Advanced Warning Satellites," *Space News*, November 19–25, 1990, p. 6.

39. House Appropriations Committee, *Department of Defense Appropriations for 1994*, pt. 5 (Washington, D.C.: U.S. Government Printing Office, 1993), p. 455.

40. Vincent Kiernan, "Air Force Puts BSTS Development Plans on Back Burner," *Space News*, September 3–9, 1990, p. 30; Kiernan, "Air Force Readies Competition for Advanced Warning Satellites"; *History of Space Systems Division, October 1989–September 1990, Volume I*, p. 601; *History of Space Systems Division, October 1990–September 1991*, p. 367.

41. *History of Space Systems Division, October 1989–September 1990, Volume I*, pp. 603–4.

42. Robert Boczkiewicz, "New Satellite System to Spot Mobile Attacks," *Washington Times*, April 12, 1991, p. A5.

43. Ibid.

44. U.S. Congress, House Appropriations Committee, *Department of Defense Appropriations for 1992*, pt. 6 (Washington, D.C.: U.S. Government Printing Office, 1991), p. 464; "FEWS Won't Use SDI Technology," *Military Space*, January 13, 1992, pp. 3–4.

12. *The Unwanted Option*

1. "Navy, Army May Fight with Air Force over Funds for Intelligence Data Processor," *Inside the Navy*, May 11, 1992, pp. 1, 11; *United States Space Command January–December 1992, Narrative* (Peterson AFB, Colo.: USSPACECOM, 1996), p. 193.

2. Naval Space Command, NAVAL TAGS OVERVIEW, May 1992, chart 2.

3. Memorandum of Agreement between the U.S. Army Strategic Defense Command and the Space and Naval Warfare Systems Command, Subject: Joint Development and Procurement of Prototype for User Test and Evaluation of Surveillance and Reporting of Tactical Infrared Events, April 24, 1992; "Navy, Army Sign Agreement to Develop Theater Level Data Processor," *Inside the Navy*, May 11, 1992, pp. 1, 11.

4. Neil Munro, "Services Plan to Develop Missile Detection Satellites," *Defense News*, July 19–25, 1993, p. 44; USSPACECOM, Point Paper on Tactical Initiatives, June 15, 1992; USSPACECOM, Background Paper on RADIANT IVORY, August 25, 1992; *United States Space Command, January–December 1992, Narrative*, p. 193.

5. USSPACECOM, DSP Tactical Support Issues—Read Ahead Package, February 27, 1991; USSPACECOM, "Tactical Exploitation of DSP," February 22, 1991.

6. USSPACECOM, DSP Tactical Support, March 1991.

7. Ibid.

8. Ibid.; Munro, "Services Plan to Develop Missile Detection Satellites."

9. Dan J. Brockway, "World Peace through Vigilance," briefing given at Space Congress, Cocoa Beach, Fla., April 1995.

10. George Hayward, "TALON," *Space Trace*, March 1993, pp. 10–11.

11. Duane Andrews, "Report to the Congress on the Follow-on Early Warning System," April 15, 1991, p. 4.

12. "FEWS Won't Use SDI Technology," *Military Space*, January 13, 1992, pp. 3–4.

13. Philip Finnegan and Neil Munro, "Pentagon Seeks Next-Generation Missile-Warning Satellites," *Defense News*, May 20, 1991, p. 8.

14. Ibid. It is probably fair to assume that the industry official was not associated with TRW or Aerojet-General. A former Aerojet executive remarked that "we weren't lucky," and argued that the ability of DSP to respond so well to a situation that it was not designed for was a sign of its robustness.

15. C3I Systems Committee, Integrated Program Assessment for the Follow-On Early Warning System (FEWS), 1991, p. 3. The study in question is Air Force Space Command, *Cost and Operational Effectiveness Analysis (COEA) for the Advanced Space-Based Tactical Warning and Attack Assessment System* (Peterson AFB, Colo.: AFSPACE-COM, 1991). The COEA was originally to have been completed by July 1991, but the Office of the Secretary of Defense decided that it should consider two additional options: one inclined DSP satellite and one with FEWS with less onboard data processing. See *United States Space Command Command History, January 1990–December 1991, Narrative* (Peterson AFB, Colo.: USSPACECOM, 1995), p. 94.

16. C3I Systems Committee, Integrated Program Assessment for the Follow-On Early Warning System (FEWS), p. 4.

17. *United States Space Command, Command History, January 1990–December 1991, Narrative*, p. 94–96; David E. Jeremiah, VCJCS, JROCM-057-91, Memorandum for the Under Secretary of Defense for Acquisition, Subject: Follow-on Early Warning System, 18 October 1991.

18. General Accounting Office, *Early Warning Satellites: Funding for Follow-On System Is Premature* (Washington, D.C.: GAO, 1991), p. 3.

19. "USAF Told to Look Again at FEWS," *Jane's Defence Weekly*, January 11, 1992, p. 40; "Air Force to Restructure Warning Satellite Program," *Defense Week*, January 6, 1992, p. 3; *History of the Space and Missile Systems Center, October 1991–September 1992, Volume I* (Los Angeles AFB: Space and Missile Systems Center, n.d.), p. 364.

20. "FEWS Must See 'Stealth' Missiles," *Military Space*, May 4, 1992, pp. 3–4.

21. "Air Force Touts Brilliant Eyes Intel Value," *Military Space*, June 1, 1992, pp. 1, 8.

22. U.S. Congress, House Appropriations Committee, *Department of Defense Appropriations for 1993*, pt. 5 (Washington, D.C.: U.S. Government Printing Office, 1992), p. 425.

23. Major General Garry Schnelzer, *Air Force Space Sensor Study*, AFPEO/SP, April 12, 1993, pp. 4, 6.

24. Ibid., p. 11.

25. Ibid., p. 12.

26. Ibid., pp. 11–13.

27. Ibid., p. 13.

28. Ibid., p. 14.

29. Ibid., p. 26.

30. Ibid., pp. 39, 42. A study conducted by the Aerojet Corporation for the Air Force Foreign Technology Division concluded that ICBM modifications that would surpress radiation, and result in dim targets, were impractical.

31. U.S. Congress, House Appropriations Committee, *Department of Defense Appropriations for 1994*, pt. 1 (Washington, D.C.: U.S. Government Printing Office, 1993), pp. 351, 391–92.

32. Ibid., p. 392.

33. Ibid., p. 393.

34. Statement of Guido William Aru before the House of Representatives, Committee on Government Operations, February 2, 1994, p. iii. The statement appears in U.S. Congress, House Committee on Government Operations, *Strategic Satellite Systems in a Post–Cold War Environment* (Washington, D.C.: U.S. Government Printing Office, 1994), pp. 51–87.

35. Guido W. Aru and Carl T. Lunde, *DSP-II: "Preserving the Air Force's Options"* (El Segundo, Calif.: Aerospace Corporation, 1993), p. iii; interview.

36. Aru and Lunde, *"DSP-II: Preserving the Air Force's Options,"* p. 4.

37. Ibid., Executive Summary, p. 8.

38. Ibid., Executive Summary, p. 10.

39. Ibid., Executive Summary p. 12.

40. Ibid., pp. 58–59; Executive Summary, pp. 10, 12.

41. Letter, Charles A. Horner, to Edward C. Aldridge, May 24, 1993.

42. The quote is from a copy of the letter in the author's possession.

43. Letter from Edward C. Aldridge, President, the Aerospace Corporation to General Charles A. Horner, Commander, Air Force Space Command, May 27, 1993.

44. Col. Joseph Bailey, "Point Paper on DSP-II TOR" (Los Angeles AFB: Space-Based Early Warning Systems Progam Office, Air Force Space and Missile Systems Center, May 24, 1993), p. 1.

45. Ben Iannotta and Neil Munro, "Report: Brilliant Eyes Constellation Could Be Halved," *Space News*, June 14–20, 1993, p. 10.

46. Briefing Slides, "Follow-On Early Warning System (FEWS)," July 13, 1993, charts 4, 6, 7, 9.

47. Ibid., charts 11, 12, 15.

48. Ibid., chart 14.

49. George R. Schneiter, Director, Strategic and Space Systems, Office of the Under Secretary of Defense, Memorandum for the Secretaries of the Military Departments, "Study on Space-Based Tactical Warning/Attack Assessment Alternatives,"

June 22, 1993 w/Terms of Reference, Task Force on Ballistic Missile Defense Assessment of Tactical Warning/Attack Assessment Alternatives.

50. Neil Munro, "Horner Proposes Shifting DSP Funds to New System," *Space News*, June 21–27, 1993, p. 10.

51. David A. Fulghum, "Half-Size FEWS to Detect Aircraft, Small Missiles," *Aviation Week & Space Technology*, August 9, 1993, pp. 24–25.

52. Ibid.

53. Ibid.

54. Ibid.

55. William J. Lynn, Director, Program Analysis and Evaluation, Office of the Secretary of Defense, Memorandum for the Under Secretary of Defense (Acquisition and Technology), Subject: Possible Low-Cost Alternative to FEWS, June 8, 1993.

56. Colonel Edward R. Dietz, Memorandum for: Colonel Bailey, Subject: "I am Concerned," September 7, 1993.

57. Roger L. Hall, "Personal for Col. Q.," October 12, 1993.

58. Briefing Slides, Report of the Space-Based IR Sensors Technical Support Group, October 1993.

59. Ibid.

60. Ibid.; interview.

61. Briefing slides, Report of the Space-Based IR Sensors Technical Support Group.

62. Private information.

63. Private information.

64. Briefing slides, Report of the Space-Based IR Sensors Technical Support Group.

65. Ibid.

66. U.S. Congress, House Committee on Government Operations, *Strategic Satellite Systems in a Post–Cold War World*, pp. 352–53; Untitled, Undated IDA Briefing Slides on IDA Review of FEWS Program.

67. U.S. Congress, House Committee on Government Operations, *Strategic Satellite Systems in a Post–Cold War World*, p. 351; Untitled, Undated IDA Briefing Slides on IDA Review of FEWS Program.

68. "Think Tank Report to Pentagon Stresses Modest Satellite Upgrades," *Defense Week*, November 15, 1993, p. 5.

69. U.S. Congress, House Committee on Government Operations, *Strategic Satellite Systems in a Post–Cold War Environment*, pp. 37–38.

70. Ibid., pp. 74, 97.

71. Ibid., p. 365.

72. Statement of Guido William Aru, pp. 5, 7.

73. Ibid., pp. 13, 27.

74. Ibid., p. 14.

75. Ibid., p. 19.

76. Ibid., p. 24.

77. U.S. Congress, House Committee on Appropriations, *Strategic Satellite Systems in a Post–Cold War Environment*, p. 378.

78. Ibid., pp. 184, 230.

79. Ibid., pp. 230, 374.

80. Ibid., pp. 374–75.

81. Transcript, *60 Minutes*, May 22, 1994.

13. High Now, Low Later

1. Jeffrey M. Lenorovitz, "U.S. Copes with Aging DSP Warning Network," *Aviation Week & Space Technology*, March 28, 1994, pp. 20–21; USCINCSPACE to CJCS, Subj: DSP Launch Call, July 6, 1992.

2. Lenorovitz, "U.S. Copes with Aging DSP Warning Network."

3. Ibid.

4. Ibid.

5. Larry Zamos, *TALON SHIELD/ALERT Overview Briefing* (Azusa, Calif.: The Aerospace Corporation, 1994), p. 2; U.S. Congress, House Committee on Appropriations, *Department of Defense Appropriations for 1995*, pt. 4 (Washington, D.C.: U.S. Government Printing Office, 1994), pp. 6, 210; U.S. Congress, Senate Armed Services Committee, *Department of Defense Authorization for Appropriations for Fiscal Year 1995 and the Future Years Defense Program*, pt. 7 (Washington, D.C.: U.S. Government Printing Office, 1994), p. 569; "House to Cut DSP, RDT&E Will Delay Ground Station Consolidation: DOD," *Aerospace Daily*, September 14, 1994, p. 415.

6. Zamos, *TALON SHIELD/ALERT Overview Briefing*, pp. 5–6, 21.

7. Ibid., pp. 18–19.

8. Barbara Starr, "N. Korea Casts a Longer Shadow with TD-2," *Jane's Defence Weekly*, March 12, 1994, p. 1.

9. Jeffrey M. Lenorovitz, "Mobile DSP Station to Improve Detection of Korean Missiles," *Aviation Week & Space Technology*, April 4, 1994, p. 31; U.S. Congress, House Committee on Appropriations, *Department of Defense Appropriations for 1995*, pt. 4, p. 209.

10. U.S. Congress, House Committee on Appropriations, *Department of Defense Appropriations for 1994*, pt. 4, p. 148; Ben Iannotta, "DoD ALARM Briefing Draws Mixed Reviews," *Space News*, March 7–13, 1994, pp. 1, 28.

11. Iannotta, "DoD ALARM Briefing Draws Mixed Reviews."

12. Ibid.

13. U.S. Congress, House Committee on Appropriations, *Department of Defense Appropriations for 1995*, pt. 4, pp. 114–15.

14. U.S. Congress, House Committee on Appropriations, *Department of Defense Appropriations for 1995, Part 1* (Washington, D.C.: U.S. Government Printing Office, 1994), pp. 113, 560; U.S. Congress, House Committee on Appropriations, *Department of Defense Appropriations for 1995*, pt. 4, p. 135.

15. U.S. Congress, House Committee on Appropriations, *Department of Defense Appropriations for 1995*, pt. 4, pp. 114, 116, 121. According to a former Aerojet executive, what Moorman described as overheating came from a misunderstanding of the design of the satellite/focal plane, which very slowly increased in temperature over a very long period of time. Often the temperature increase is a determinant of satellite lifetime, and maintenance of high detectability levels could be attained by replacing the satellite.

16. Ibid., p. 140.

17. U.S. Congress, House Committee on Appropriations, *Department of Defense Appropriations for 1995*, pt. 3 (Washington, D.C.: U.S. Government Printing Office, 1994), p. 749.

18. U.S. Congress, House Committee on Appropriations, *Department of Defense Ap-*

propriations for 1995, pt. 4, pp. 117–18; Ben Iannotta, "ALARM's Economic Value Questioned," *Space News*, April 18–24, 1994, pp. 3, 21.

19. U.S. Congress, House Committee on Appropriations, *Department of Defense Appropriations for 1995*, pt. 4, p. 120.

20. Ibid., p. 149.

21. Ibid., p. 136.

22. Iannotta, "ALARM's Economic Value Questioned."

23. Ben Iannotta, "Pentagon Reopens Old Missile Warning Debate," *Space News*, June 27–July 3, 1994, p. 3; General Accounting Office, *Risk and Funding Implications for the Space-Based Infrared Low Component* (Washington, D.C.: GAO, 1997), p. 12.

24. "House-Senate Authorizers Accelerate ALARM but Terminate Dem/Val," *Aerospace Daily*, August 11, 1994, p. 229.

25. Ibid.; Andrew Lawler, "Missile-Spotting System Splits U.S. Congress," *Space News*, August 15–28, 1994, p. 4.

26. Lawler, "Missile-Spotting System Splits U.S. Congress."

27. "Appropriators Fund $111 Million to Adapt Sensor for ALARM," *Aerospace Daily*, September 28, 1994, p. 491; *Congressional Record—House*, September 26, 1994, p. H9648.

28. Steve Weber, "Panel Probes Infrared Possibilities," *Space News*, September 5–11, 1994, p. 10; Steve Weber, "Plans for Heritage Sensor Endorsed," *Space News*, October 3–9, 1994, pp. 4, 21.

29. U.S. Congress, House Committee on Appropriations, *Department of Defense Appropriations for 1995*, pt. 3, p. 797; Lenorovitz, "U.S. Copes with Aging DSP Warning Network"; Ben Iannotta, "Who Will Finance the Sensor?" *Space News*, January 3–9, 1994, p. 3.

30. Ben Iannotta, "MSTI Team Struggles for Support to Continue Research," *Space News*, June 20–26, 1994, p. 15.

31. U.S. Congress, Senate Armed Services Committee, *Department of Defense Authorization for Appropriations for Fiscal Year 1995 and the Future Years Defense Program*, pt. 1 (Washington, D.C.: U.S. Government Printing Office, 1994), p. 1023; Ben Iannotta and Neil Munro, "Radiant Agate May Carry Secret Sensor to Boost Support," *Space News*, December 13–19, 1993, p. 22; Weber, "Panel Probes Infrared Possibilities." A decision to piggyback the infrared sensors on, for example, geosynchronous SIGINT satellites would raise a number of questions, including how to position the satellites in a crisis given different optimal positions for the SIGINT function versus the launch detection function, and what to do about satellite replacement given that the SIGINT sensor will generally have a greater lifetime than the infrared sensor.

32. Steve Weber, "Deutch Approves SBIR Missile Warning Plan," *Space News*, November 7–13, 1994, p. 3.

33. Ibid.; Weber, "Plans for Heritage Sensor Endorsed"; Steve Weber, "Competition Heating Up for Air Force's ALARM Program," *Space News*, October 24–30, 1994, p. 8; "EW Plan: Consolidate ALARM, DSP; Repackage Brilliant Eyes as a Prototype," *Inside the Pentagon*, October 27, 1994, pp. 1, 6.

34. "Air Force Scraps ALARM, Announces New Infrared Competition," *Aerospace Daily*, November 29, 1994, p. 291.

35. U.S. Air Force, *Space Based Infrared Systems (SBIRS) Pre-Solicitation Conference*, January 13, 1995, p. 11.

36. Ibid., p. 13; Maj. Gen. Robert S. Dickman in U.S. Congress, Senate Committee on Armed Services, *Department of Defense Authorization for Appropriations for Fiscal Year 1996 and the Future Years Defense Program*, pt. 7 (Washington, D.C.: U.S. Government Printing Office, 1996), p. 139.

37. "TRW Wins Brilliant Eyes," *Aviation Week & Space Technology*, May 8, 1995, p. 26; Cheri Privor, "Rockwell Retains Role in Warning System," *Space News*, May 29–June 4, 1995, p. 6; Warren Ferster, "Changes, Technical Troubles Blamed for SBIRS Program Delays, Cost Hike," *Space News*, June 9–15, 1997, pp. 3, 19.

38. Office of the Under Secretary of Defense for Acquisition and Technology, *Report of the Defense Science Board Task Force on Space and Missile Tracking System*, August 1996, pp. 1–2.

39. U.S. Air Force, *Space Based Infrared Systems (SBIRS) Pre-Solicitation Conference*, p. 13; "Air Force Unveils New Plans for Early Warning System," *Aerospace Daily*, January 19, 1995, p. 85A.

40. "Senate Appropriators Boost Funding to Accelerate LEO," *Aerospace Daily*, July 31, 1995, p. 147A; U.S. Congress, Senate Armed Services Committee, *National Defense Authorization Act for Fiscal Year 1996 Report* (Washington, D.C.: U.S. Government Printing Office, 1996), pp. 100–101; U.S. Congress, House of Representatives, House Report 104-406, *National Defense Authorization Act for Fiscal Year 1996* (Washington, D.C.: U.S. Government Printing Office, 1995), pp. 715–16; John Donnelly, "Air Force Needs Billions More to Pay for Tracking Satellites," *Defense Week*, January 22, 1996, p. 3; "SBIRS Acceleration Said Underfunded by about $2 Billion," *Aerospace Daily*, April 24, 1996, pp. 143B–144; Jennifer Heronema, "Speeding SMTS to Cost $2 Billion," *Space News*, April 29–May 5, 1996, pp. 1, 29; Senate Armed Services Committee, "Advance Questions for Lieutenant General Howell M. Estes III, n.d. [ca. June 1996], p. 10; John Donnelly, "Missile-Tracking Satellites Set for Quicker Deployment," *Defense Week*, September 9, 1996, pp. 1, 13; "DSB Says SMTS Acceleration Possible," *Aerospace Daily*, October 3, 1996, p. 18; "Space Tracking System May Be Accelerated," *Jane's Defence Weekly*, October 9, 1996, p. 6; Office of the Under Secretary of Defense for Acquisition and Technology, *Report of the Defense Science Board on Space and Missile Tracking System*, pp. 5–6; Paul Kaminski, Deputy Under Secretary of Defense for Acquisition and Technology, Memorandum for Secretaries of the Military Departments et al., Subject: Defense Science Board Task Force on the Space and Missile Tracking System (SMTS), September 3, 1996; General Accounting Office, *Risk and Funding Implications for the Space-Based Infrared Low Component*, pp. 13–14.

41. "Army, Air Force Consider JTAGS Follow-On," *Aerospace Daily*, April 11, 1996, pp. 65–67.

42. Dwayne A. Day, "Capturing the High Ground: The U.S. Military in Space 1987–1995, Part I," *Countdown*, January/February 1995, pp. 30–41.

43. U.S. Air Force, *Space-Based Infrared Systems (SBIRS) Pre-Solicitation Conference*, p. 15; "Military Loads," *Aviation Week & Space Technology*, June 23, 1997, p. 13; "First Flight of Titan 4B Marks Major USAF Advance," *Aviation Week & Space Technology*, March 3, 1997, p. 29.

44. Michael A. Dornheim, "In-orbit Calibration Sharpens DSP," *Aviation Week & Space Technology*, February 10, 1997, p. 22.

45. Headquarters, Air Force Space Command, Special Order GD-040, September 27, 1994; Elton Price, "Theater Missile Warning Unit Activates," *Guardian*, No-

vember 1994, p. 12; "Alert Replaces Grease Pencil, String for Theater Warning Squadron," *Space News*, April 3–9, 1995, advertising section, p. A4; William B. Scott, "Air Force 'Alert' System Speeds Missile Warnings," *Aviation Week & Space Technology*, May 1, 1995, pp. 56–57. The unit reached initial operational capability on March 10, 1995. See John Kennedy, "Theater Missile Warning Unit Reaches Operations Milestone," *Guardian*, April 1995, p. 13.

46. "Alert Replaces Grease Pencil, String for Theater Warning Squadron."

47. Brian Orban, "The Next Generation," *Guardian*, February 1995, pp. 8–9.

48. Scott, "Air Force 'Alert' System Speeds Missile Warnings"; "Tactical Surveillance Detachments Move," *Space Tracks*, spring 1995, p. 20; USCINCSPACE, UI 10-17, "Tactical Event System," July 17, 1995; Naval Space Command, *1992–1993 Naval Space Command History*, March 29, 1994, p. 10; "Lt. Thompson Heads JTAGS Detachment," *Space Tracks*, March/April 1998, p. 21.

49. Diane Leite, "Space Links Enhance Missile Defense," *Space Tracks*, spring 1995, pp. 7–8; Ballistic Missile Defense Organization, *1994 Report to Congress on Ballistic Missile Defense* (Washington, D.C.: BMDO, 1994), pp. 2–14; "Army, Air Force Consider JTAGs Follow-On," pp. 65–67; "Command of Korea Detachment Transfers," *Space Tracks*, winter 1997, p. 15.

50. Warren Ferster, "U.S. Army Seeks Greater Influence in Satellite Design," *Space News*, May 5–11, 1997, p. 6; David Hughes, "JTAGS to Expedite Space-Based Warnings," *Aviation Week & Space Technology*, March 3, 1997, p. 61.

14. Future Missions

1. Joseph C. Anselmo, "New Sensors Drive Winning SBIRS Bid," *Aviation Week & Space Technology*, November 18, 1996, p. 23.

2. Ibid.

3. "MSX Satellite Passes Rocket Tracking Test," *Space News*, April 24–30, 1997, p. 10; "BMDO Needs More Money to Keep MSX Going," *Aerospace Daily*, April 1, 1997, pp. 1, 2; Kristi Marten, "MSX Turns Attention to Earth-Orbiting Objects," *APL News*, April 1998, p. 3.

4. "MSTI-3 Finishes Mission to Collect Infrared Imagery," *Space News*, June 2–8, 1997, p. 11.

5. John L. Morris, Principal Deputy Director, CMO, *MASINT: Progress and Impact, Brief to NMIA*, November 19, 1996, slide 14; U.S. Congress, House Committee on National Security, *National Defense Authorization Act for Fiscal Year 1996*, Report *104-131* (Washington, D.C.: U.S. Government Printing Office, 1995), p. 122; Department of Defense, *Department of Defense Space Program: An Executive Overview for Fiscal Year 1998–2003* (Washington, D.C.: Office of the Undersecretary of Defense [Acquisition and Technology], 1997), p. 28; interview.

6. "USAF Lists Force Structure, Realignment Changes," *Aerospace Daily*, July 6, 1995, p. 14; Ian Olgeirson, "Details Emerge on State's Role as Key Spy Hub," *Denver Business Journal*, April 5–11, 1996, pp. 1A, 27A–28A.

7. *Nurrungar News*, March 1993, p. 1; interview with Patrick Carroll, Counselor, Defense Policy, Australian Embassy, Washington, D.C., June 11, 1996; Office of the Minister of Defence, "Australian Cooperation with US Missile Early Warning Program to Continue," July 26, 1996; Nancy Andreas, "Space-Based Infrared System

(SBIRS) of Systems," *IEEE Aerospace Applications Conference Proceedings* 4 (1997): 429–40.

8. Andreas, "Space-Based Infrared System (SBIRS) System of Systems"; William B. Scott, "First Phase of DSP-SBIRS Transition Underway," *Aviation Week & Space Technology*, September 29, 1997, pp. 57–59.

9. Andreas, "Space-Based Infrared System (SBIRS) System of Systems"; Scott, "First Phase of DSP-SBIRS Transition Underway."

10. Andreas, "Space-Based Infrared System (SBIRS) System of Systems."

11. Ibid.; Boeing, "Low-Altitude Demonstration System," n.d. (ca. 1997); Lockheed-Martin/Boeing, "Team SBIRS-Low," n.d. (ca. 1997); "Raytheon Finishes Building SBIRS-Low Sensor," *Space News*, April 13–19, 1998, p. 9; "RFP in the Works," *Aviation Week & Space Technology*, May 18, 1998, p. 21.

12. Warren Ferster, "Firm Tempted by SBIRS," *Space News*, April 13–19, 1998, pp. 1, 19; "Spectrum Astro to Bid for SBIRS-Low," *Aviation Week & Space Technology*, April 13, 1998, pp. 57–58; "Spectrum Astro Commits to SBIRS-Low," *Aerospace Daily*, May 7, 1998, pp. 212–13.

13. Warren Ferster, "Missile Tracking Tops Military Space Plan," *Space News*, September 15–21, 1997, p. 8; William B. Scott, "Funding Fracas Imperils 'Nudet' Sensor," *Aviation Week & Space Technology*, January 5, 1998, pp. 29–30.

14. "Russian Typhoon Sub in First Launch of Missile from North Pole," *Aerospace Daily*, August 30, 1995, p. 326; "Modified SS-25 Launched in Second Test from Silo," *Aerospace Daily*, September 7, 1995, p. 367; "Russia Test Fires Fourth Topol-M Missile," *Aerospace Daily*, July 10, 1997, p. 51; Nikolai Novichkov, "Russia Will Field Topol-M ICBM by End of This Year," *Jane's Defence Weekly*, July 16, 1997, p. 3; David A. Fulghum, "Russian Missile Tests Yield Mixed Results," *Aviation Week & Space Technology*, January 19, 1998, p. 30; "Russians Deploy First of Missiles," *Washington Times*, December 25, 1997, p. A9; "Russia Inaugurates First Topol-M ICBM in Refurbished Silo," *Aerospace Daily*, January 7, 1998, pp. 25–26; "Upgraded SS-N-20 Fails in Test Launch as Sergeyev Watches," *Aerospace Daily*, November 25, 1997, p. 304; General Eugene E. Habiger, Commander-in-Chief, U.S. Strategic Command, "Strategic Deterrence in the 21st Century," Speech to Air Force Association, April 2, 1998, http://www.stratcom.af.mil/testimony; "Russia Holds Strategic Command Exercise," *Jane's Intelligence Review*, November 1997, p. 2; Steven J. Zaloga, "RVSN Revival Rests with the Topol-M," *Jane's Intelligence Review*, April 1998, pp. 5–7.

15. Michael R. Gordon, "U.S. to China: Be More Open on Arms Plan," *New York Times*, October 19, 1994, p. A14; Duncan Lennox, "Ballistic Missiles Hit New Heights," *Jane's Defence Weekly*, April 30, 1994, pp. 24–28; Joseph C. Anselmo, "U.S. Eyes China Missile Threat," *Aviation Week & Space Technology*, October 21, 1996, p. 23; Wyn Bowen and Stanley Shepard, "Living under the Red Missile Threat," *Jane's Intelligence Review*, December 1996, pp. 560–64; Bill Gertz, "China's Nukes Could Reach Most of U.S.," *Washington Times*, April 1, 1998, pp. A1, A18.

16. Steven Mufson, "Chinese Ships, Planes Maneuver Near Taiwan," *Washington Post*, March 13, 1996, p. A18; Seth Faison, "Tensions Seen as Receding as China Ends War Games," *New York Times*, March 26, 1996, p. A8; Patrick E. Tyler, "War Games off Taiwan to Expand, Beijing Says," *New York Times*, March 10, 1996, p. 12; Office of Naval Intelligence, *Chinese Exercise Strait 961: 8–25 March 1996* (Suitland, Md.: ONI, 1996), pp. 1, 7.

17. On the Condor II see William E. Burrows and Robert Windrem, *Critical Mass: The Dangerous Race for Superweapons in a Fragmenting World* (New York: Simon & Schuster, 1994), pp. 466–80.

18. Barbara Starr, "N. Korea Casts a Long Shadow with TD-2," *Jane's Defence Weekly,* March 12, 1994, p. 1; Commander Jonathan Sears, "The Northeast Asia Nuclear Threat," *Naval Institute Proceedings,* July 1995, pp. 43–46; Bill Gertz, "N. Korean Missile Could Reach U.S., Intelligence Warns," *Washington Times,* September 29, 1995, p. A3; Office of the Secretary of Defense, *Proliferation: Threat and Response* (Washington, D.C.: U.S. Government Printing Office, 1997), pp. 7–8; National Air Intelligence Center, *Ballistic and Cruise Missile Threat* (Wright-Patterson AFB, Ohio: NAIC, 1998), p. 9.

19. "Indian Lawmakers Back Missile Program," *Washington Times,* May 3, 1997, p. A8; Ranjan Roy, "Defiant India Eyes New Missiles," *Washington Times,* August 1, 1997, p. A14; Pakistan Missile Firing Triggers Indian Protest," *Aviation Week & Space Technology,* July 21, 1997, p. 35; Ben Sheppard, "Too Close for Comfort: Ballistic Ambitions in South Asia," *Jane's Intelligence Review,* January 1998, pp. 32–35; Manoj Joshi, "Operation Defreeze," *India Today International,* August 11, 1997, p. 68; Tim Weiner, "U.S. Says North Korea Helped Develop New Pakistani Missile," *New York Times,* April 11, 1998, p. A3; Kathy Gannon, "Pakistan Test Fires Missile, Irking India," *Washington Times,* April 7, 1998, p. A13; Rahul Bedi, "India Warns It Won't Back Off in Missile Race with Pakistan," *Washington Times,* April 8, 1998, p. A13; Andrew Koch and Waheguru Pal Singh Sidhu, "Subcontinental Missiles," *Bulletin of the Atomic Scientists,* July/August 1998, pp. 44–49; Joseph Bermudez, "A Silent Partner," *Jane's Defence Weekly,* May 20, 1998, pp. 16–17; Commission to Assess the Ballistic Missile Threat to the United States, *Executive Summary of the Report of the Commission to Assess the Ballistic Missile Threat to the United States* (Washington, D.C.: U.S. Congress, 1998), pp. 15–17.

20. Lennox, "Ballistic Missiles Hit New Heights"; Alan George, "Iran Puts Together Scud-B Missiles," *Washington Times,* December 5, 1994, p. A14; Stewart Stogel, "Missile Plans by Iraq May Aim at Europe," *Washington Times,* February 16, 1996, pp. A1, A19; Kenneth J. Cooper, "India Halts Development of Medium-Range Missile," *Washington Post,* December 6, 1996, p. A46; "Going Ballistic," *Aviation Week & Space Technology,* January 12, 1998, p. 387; Bill Gertz, "Russia, China Aid Iran's Missile Program," *Washington Times,* September 10, 1997, pp. A1, A11; Bill Gertz, "Pentagon Confirms Details on Iranian Missiles," *Washington Times,* March 27, 1998, p. A10; Michael R. Gordon, "U.S. Is Pressing Moscow on Iran and Missile Aid," *New York Times,* March 9, 1998, pp. A1, A8; "A Wake-Up Call from Iran," *Aviation Week & Space Technology,* March 9, 1998, p. 21; Bill Gertz and Martin Sieff, "Iran's Missile Test Alarms Clinton," *Washington Times,* July 24, 1998, pp. A1, A13; Tim Weiner, "Iran Said to Test Missile Able to Hit Israel and Saudis," *New York Times,* July 23, 1998, pp. A1, A6; Duncan Lennox, "Iran's Ballistic Missile Projects: Uncovering the Evidence," *Jane's Intelligence Review,* June 1998, pp. 24–28.

21. System Planning Corporation, *Ballistic Missile Proliferation: An Emerging Threat 1992* (Arlington, Va.: System Planning Corporation, 1992), p. 32.

22. "Alert Replaces Grease Pencil, String for Theater Warning Squadron," *Space News,* April 3–9, 1995, advertising section, p. A4; Roger Cohen, "Serbs Attack Muslim Town with Antiaircraft Missiles," *New York Times,* November 5, 1994, p. 5; Joel

Brand, "U.N. Denounces Serb Firing of 4 Missiles," *Washington Post,* November 5, 1994, p. A17.

23. "Iran Fires Missiles at Rebel Base in Iraq," *New York Times,* November 7, 1994, p. A6; "Newsbreaks," *Aviation Week & Space Technology,* November 14, 1994, p. 14.

24. Defense Department efforts in the field of missile defense were severely criticized by an outside review group in early 1998. See Bradley Graham, "Panel Fires at Antimissile Programs," *Washington Post,* March 22, 1998, pp. A1, A24; Paul Mann, "Missile Defense Riddled with Diverse Failures," *Aviation Week & Space Technology,* March 30, 1998, pp. 22–23.

25. David Hughes, "Patriot PAC-3 Upgrade Aimed at Multiple Threats," *Aviation Week & Space Technology,* February 24, 1997, pp. 59–61; Inspector General, Department of Defense, *The Patriot Advanced Capability—3 Program,* Report No. 97-018 (Arlington, Va.: IG, DoD, 1996), p. i; Statement of Lt. Gen. Lester L. Lyles, Director, BMDO, before U.S. Congress, Senate Armed Services Committee, March 24, 1998, p. 6.

26. Ballistic Missile Defense Organization, *1994 Report to Congress on Ballistic Missile Defense* (Washington, D.C.: BMDO, 1994), p. 2-14; Naval Space Command, *1992–1993 Naval Space Command History,* March 29, 1994, p. 10; Barbara Starr, "USA Fields 'Scuds' to Test Theatre Missile Defense," *Jane's Defence Weekly,* May 7, 1997, p. 3.

27. Michael A. Dornheim, "Thaad Program Future Tied to Test Results," *Aviation Week & Space Technology,* March 3, 1997, pp. 64–65; Bradley Graham, "Low-Tech Flaws Stall High-Altitude Defense," *Washington Post,* July 27, 1998, pp. A1, A12.

28. U.S. Congress, House Committee on Appropriations, *Department of Defense Appropriations for 1995,* pt. 4 (Washington, D.C.: U.S. Government Printing Office, 1994), pp. 227–28.

29. Diane Leite, "Space Links Enhance Missile Defense," *Space Tracks,* Spring 1995, pp. 7–8; John M. Pollin, "Fielding a Theater Ballistic Missile Defense," *Naval Institute Proceedings,* July 1996, pp. 63–64; David Hughes, "Navy Readies Fleet for Anti-Scud Warfare," *Aviation Week & Space Technology,* February 24, 1997, pp. 61–63; U.S. Congress, House Committee on National Security, *Proliferation Threats and Missile Defense Responses* (Washington, D.C.: U.S. Government Printing Office, 1997), pp. 34–35.

30. Michael A. Dornheim, "'Theater Wide' Missile Defense Appealing, Controversial, Difficult," *Aviation Week & Space Technology,* March 3, 1997, pp. 62–63.

31. Stanley W. Kandebo, "NMD System Integrates New and Updated Components," *Aviation Week & Space Technology,* March 3, 1997, pp. 47–51; Ballistic Missile Defense Organization, *1997 Report to the Congress on Ballistic Missile Defense* (Washington, D.C.: BMDO, 1997), pp. 3-6 to 3-12.

32. Kandebo, "NMD System Integrates New and Updated Components."

33. Joseph C. Anselmo, "Outgoing BMDO Chief Seeks Better Threat Assessment," *Aviation Week & Space Technology,* June 3, 1996, p. 32; John F. Harris, "Clinton Warns GOP on Missile Defense," *Washington Post,* May 23, 1996, pp. A1, A12. In May 1996 President Clinton charged that a Republican plan to move immediately to establish a national defense system would be premature and would waste billions. The administration has defended that policy on the basis of the 1995 National Intelligence Estimate 95-19, *Emerging Missile Threats to North America.* The estimate characterizes the

prospect of an unauthorized Russian or Chinese ICBM launch as a result of a political crisis as low. It also estimates that the range of the North Korean Taepo Dong 2 will be between 2,480 and 3,720 miles. The higher figure would allow North Korea to strike portions of Alaska and the far western portion of the Hawaiian Island chain. The estimate also concludes that "North Korea is unlikely to obtain the technological capability to develop a longer range operational ICBM," that no evidence is in hand that suggests Iran wants to develop an ICBM, and if "Tehran wanted to . . . it would not be able to do so before 2010." In addition, "Iraq's ability to develop an ICBM is severely constrained by international sanctions and the intrusive U.N. inspection and monitoring regime." The estimate was criticized by a number of prominent former national security officials, including former DCI R. James Woolsey and just-retired BMDO director Lieutenant General Malcolm O'Neill. Critics have charged that the estimate gave insufficient attention to the clandestine acquisition of technology and missile development, ignored potential intelligence gaps, failed to consider worst-case scenarios, and implicitly treated portions of Alaska and Hawaii as expendable. Some Republicans suggested the estimate had been politicized in an attempt to justify administration policy. An independent panel chaired by former DCI Robert Gates concluded that the study was not politicized but was politically naive and not as useful as it could have been. A General Accounting Office report also criticized the estimate for a number of weaknesses, including overstating the certainty that there would be no threat to the contiguous forty-eight states for the next fifteen years. See Bill Gertz, "General Faults Forecast of Missile Threat," *Washington Times*, June 14, 1996, p. A8; Bill Gertz, "Intelligence Report Warns of Missile Launches against U.S.," *Washington Times*, May 14, 1996, p. A3; "DCI National Estimate President's Summary, Emerging Missile Threats to North America during the Next 15 Years, NIE 95-19, November 1995," *Washington Times*, May 14, 1996, p. A15; Bill Gertz, "Probe Urged of Missile-Threat Report," *Washington Times*, January 29, 1996, p. A12; Rowan Scarborough and Bill Gertz, "Missile-Threat Report 'Politicized,' GOP Says," *Washington Times*, January 30, 1996, pp. A1, A14; statement of R. James Woolsey, House Committee on National Security, March 14, 1996, pp. 4, 5; Bill Gertz, "Independent Panel Faults Estimate of Missile Threat," *Washington Times*, December 5, 1996, p. A5; General Accounting Office, *Analytic Soundness of Certain National Intelligence Estimates*, GAO/ NSIAD-96-225 (Washington, D.C.: GAO, August 1996), pp. 3–4; Central Intelligence Agency, *NIE 95-19: Independent Panel Review of "Emerging Missile Threats to North America During the Next 15 Years,"* December 23, 1996.

34. David E. Mosher, "The Grand Plans," *IEEE Spectrum*, September 1997, pp. 28–39.

35. Naoki Usui, "U.S., Japan Discuss Sharing Missile Warnings," *Space News*, January 23–29, 1995, p. 6; Steve Rodan, "Arrow 2 Blows Up Scud-like Target," *Washington Times*, August 21, 1996, p. A12.

36. Jeffrey M. Lenorovitz, "U.S., Russia to Share Missile Warning Data," *Aviation Week & Space Technology*, April 11, 1994, pp. 24–25.

37. Philip Shenon, "U.S. to Aid Israel with Technology Against Attacks," *New York Times*, April 29, 1996, pp. A1, A10; David A. Fulghum and Bruce D. Nordwall, "Israel's Missile Defenses Rate U.S. High-Tech Boost," *Aviation Week & Space Technology*, May 6, 1996, pp. 22–23; "Newsbreaks," *Aviation Week & Space Technology*, July 15, 1996, p. 16.

38. Naoki Usui, "Pentagon to Supply Early Warning Data to JDA," *Space News,* June 10–16, 1996, p. 28; Clifford Krauss, "Japan Hesitant About U.S. Antimissile Project," *New York Times,* February 15, 1997, p. 3; Norman Friedman, "Japan Rejects Missile Defense," *Naval Institute Proceedings,* April 1997, pp. 107–8.

39. John Donnelly, "U.S. Giving NATO Nations, Israel Access to Missile Warning," *Defense Week,* December 23, 1996, pp. 1, 13.

40. "U.S. Offers Allies Access to SBIRS Warning Data," *Space News,* Janaury 6–12, 1997, p. 2.

41. William B. Scott, "Russian Pitches Common Warning Network," *Aviation Week & Space Technology,* January 9, 1995, pp. 46–47.

42. Cheri Privor, "Joint Efforts Could Strengthen Defense Capabilities," *Space News,* August 14–27, 1995, p. 11; Ballistic Missile Defense Organization, Subject: Russian–American Observation Satellite (RAMOS) Program, n.d.

43. Craig Covault, "U.S., Russia Teamed on Tracking Debris," *Aviation Week & Space Technology,* December 18/25, 1995, p. 14.

44. "RAMOS Dead without Russian Support," *Aerospace Daily,* August 22, 1996, p. 278; "U.S., Russia Move ahead on Scaled-Down RAMOS," *Aerospace Daily,* January 21, 1997, p. 91.

45. William J. Broad, "For Killer Asteroids, Respect at Last," *New York Times,* May 14, 1996, pp. C1, C7; Leon Jaroff, "A Shot Across Earth's Bow," *Time,* June 3, 1996, p. 62.

46. J. Kelly Beatty, "'Secret' Impacts Revealed," *Sky & Telescope,* February 1994, pp. 26–27; William J. Broad, "Meteoroids Hit Atmosphere in Atomic-Size Blasts," *New York Times,* January 26, 1994, p. C1.

47. Broad, "Meteoroids Hit Atmosphere in Atomic-Size Blasts"; "It Was Only a Meteor, Mr. President," *New Scientist,* April 2, 1994, p. 4.

48. Beatty, "'Secret' Impacts Revealed."

49. "Satellites Detect Record Meteor," *Sky & Telescope,* June 1994, p. 11; Thomas B. McCord, John Morris, David Persing, Edward Tagliaferri, Cliff Jacobs, Richard Spalding, LouAnn Grady, and Ronald Schmidt, "Detection of a Meteoroid Entry into the Earth's Atmosphere on February 1, 1994," *Journal of Geophysical Research* 100, E2, (February 25, 1995): 3245–49.

50. Broad, "Meteoroids Hit Atmosphere in Atomic-Size Blasts."

51. Ibid; Beatty, "'Secret' Impacts Revealed."

52. Director of Central Intelligence, "A Description of Procedures and Findings Related to the Report of the U.S. Environmental Task Force," January 21, 1994, p. 6; "Disaster Tracking Discussed," *Space News,* February 3–9, 1997, p. 17; Department of Defense, "[Deleted] Satellite Support to National Fire Detection, Global Volcano Monitoring," n.d.; Memorandum of Agreement between the Deputy Under Secretary of Defense, Space and the U.S. Geological Survey, U.S. Department of the Interior for Cooperation in Wildland Fire Detection, Volcanic Activity Monitoring, and Volcanic Ash Cloud Tracking, April 30, 1997.

Notes to Epilogue

1. Warren Ferster, "Poor Directions Led to USAF Satellite Loss," *Space News*, September 6, 1999, p. 8; Warren Ferster, "U.S. Air Force Strives to Recover from Losses," *Space News*, September 20, 1999, p. 8; "USA-142 Information," http://heavens-above.com, accessed January 29, 2012.

2. Justin Ray, "U.S. Air Force Satellite to Be Checked for Possible Damage," *Spaceflight Now*, January 6, 2000, www.spaceflightnow.com; Associated Press, "Titan Rocket and Satellite Launched by Air Force," May 9, 2000; Stew Magnuson, "Air Force Titan 4B Launches DSP Missile-Warning Satellite," *Space News*, May 22, 2000, p. 18. DSP-Flight 20 was actually the twenty-first DSP produced (i.e., DSP-21), since DSP-18 was scheduled to be DSP-Flight 22. See *History of Space and Missile Systems Center, 1 October 1998–30 September 2001* (El Segundo, Calif.: Air Force Space and Missile Systems Center, n.d.), 1:194; Craig Covault, "Titan IV Flight Validates Quality Control Reforms," *Aviation Week & Space Technology*, May 15, 2000, pp. 24–25. The conclusion that Flight 20 is the Pacific satellite is based on data appearing in the "Geostationary Satellite Log" of *Jonathan's Space Report*, www.planet4589.org/space/jsr/jsr.html, accessed January 28, 2012.

3. Annette Wells, "'Beautiful!': Titan IVB Rocket Takes DSP-21 Satellite into Orbit," *Astro News*, August 10, 2001, p. 1; "Defective IUS Part Delays Titan 4 Launch," *Space News*, July 2, 2001, p. 2. Identification of Flight 21 as the Eastern Hemisphere satellite is based on data appearing in the "Geostationary Satellite Log" of *Jonathan's Space Report*, www.planet4589.org/space/jsr/jsr.html, accessed January 28, 2012.

4. *History of Space and Missile Systems Center*, 1:197; Information provided by Air Force Space and Missile Systems Center, January 27, 2012.

5. Problems included both a gap in radar coverage as well as a limited satellite launch detection system. See Steven Mufson and Bradley Graham, "U.S. Offers Aid to Russia on Radar Site," *Washington Post*, October 17, 1999, pp. A1, A26; Michael R. Gordon, "U.S. Asks Russia to Alter Treaty for Help on Radar," *New York Times*, October 17, 1999, pp. 1, 14; David Hoffman, "Russia 'Blind' to Attack by U.S. Missiles," *Washington Post*, June 1, 2000, pp. A1, A19; Geoffrey Forden, "Reducing a Common Danger: Improving Russia's Early Warning System," *Policy Analysis*, no. 399, May 3, 2001; Geoffrey Forden, Pavel Podvig, and Theodore A. Postol, "False Alarm, Nuclear Danger," *IEEE Spectrum*, March 2000, pp. 31–39.

6. Bob Brewin and Daniel Verton, "Russians Ask for Help with Y2K Nukes," *Federal Computer Week*, September 6, 1999, http://athena.fcw.com; "Intel Inside," *Aviation Week & Space Technology*, July 26, 1999, p. 27.

7. Lisa Hoffman, "Russian Nuclear Staffers to Hold 2000 Watch in U.S.," *Washington Times*, September 11, 1999, p. A4; Stephen Barr, "U.S., Russia Agree to Establish Y2K Center," *Washington Post*, September 11, 1999, p. A9; Elizabeth Becker, "Russia to Join U.S. in Battle to Ward Off Y2K Debacle," *New York Times*, October 28, 1999, p. A14; Tom Kenworthy, "Former Nuclear Arms Adversaries Are Now Comrades Against Y2K," *Washington Post*, December 31, 1999, pp. A29; John D. Steinbrunner, *The Significance of Joint Missile Surveillance*, American Academy of Arts and Sciences Committee on International Security Studies Occasional Paper, July 2001, p. 5.

8. Tim Weiner, "Iran Said to Test Missile Able to Hit Israel and Saudis," *New York Times*, July 23, 1998, pp. A1, A6; Steven Erlanger, "Washington Casts Wary Eye at Missile Test," *New York Times*, July 24, 1998, p. A6; Sheryl WuDunn, "North Korea

Fires Missile over Japanese Territory," *New York Times*, September 1, 1998, p. A6; Rowan Scarborough and Bill Gertz, "North Korea Fires New Missile over Japanese Airspace," *Washington Times*, September 1, 1998, pp. A1, A8; Nicholas D. Kristof, "North Koreans Declare They Launched a Satellite, Not a Missile," *New York Times*, September 5, 1998, p. A5; Dana Priest, "N. Korea May Have Launched Satellite," *Washington Post*, September 5, 1998, p. A21; Joseph C. Anselmo and Robert Wall, "Missile Test Extends North Korea's Reach," *Aviation Week & Space Technology*, September 7, 1998, pp. 56–57; David A. Fulghum, "North Korean Space Attempt Verified," *Aviation Week & Space Technology*, September 21, 1998, pp. 30–31.

9. Bill Gertz, "China Prepared to Test ICBM with Enough Range to Hit U.S.," *Washington Times*, November 12, 1998, pp. A1, A6; Patrick Worship, "Russians Test-Fire New Ballistic Missile," *Washington Times*, December 10, 1998, p. A19.

10. Arthur Max, "India Rejects Appeals, Test-Fires Missiles," *Washington Times*, April 12, 1999, p. A13; Barry Bearak, "India Test Missile Able to Hit Deep into Neighbor Lands," *New York Times*, April 12, 1999, p. A3; Michael Mecham, "India Test-Fires Agni Missile; Pakistan Responds with Ghauri," *Aviation Week & Space Technology*, April 19, 1999, p. 37; Bill Gertz, "China Tests New Long-Range Missile," *Washington Times*, August 3, 1999, pp. A1, A10; Shirley A. Kan, Congressional Research Service, *China: Ballistic and Cruise Missiles*, August 10, 2000, p. 14.

11. Reuters, "Russia Launches Third Ballistic Missile in Two Days," *Russia Today*, October 4, 1999, www.russiatoday.com; Kenneth Silber, "Did U.S. Satellites See Attack on Chechnaya?," October 25, 1999, www.space.com; Bill Gertz and Rowan Scarborough, Inside the Ring, *Washington Times*, October 29, 1999, p. A10; Bill Gertz and Rowan Scarborough, "Inside the Ring," *Washington Times*, November 12, 1999, p. A9; "Russia Tests Missile Again," *New York Times*, November 18, 1999, p. A11; "World News Roundup," *Aviation Week & Space Technology*, December 20/27, 1999, p. 8.

12. Bill Gertz, "Iran Missile Test Fails After Takeoff," *Washington Post*, September 22, 2000, p. A5; Reuters, "Russia Tests New Nuclear-Capable Missile," *Russia Today*, September 29, 2000, www.russiatoday.com; Bill Gertz, "China Runs 2nd Test of Long-Range Missile," *Washington Times*, December 12, 2000, pp. A1, A14; Bill Gertz, "Pentagon Confirms Chinese Test-Fired Long-Range Missile," *Washington Times*, December 13, 2000, p. A8; Bill Gertz, and Rowan Scarborough, Inside the Ring, *Washington Times*, December 22, 2000, p. A8.

13. "India Tests Enhanced Version of Missile," *Washington Post*, January 18, 2001, p. A17; Susan B. Glasser, "Russia Tests Strategic Missiles," *Washington Post*, February 17, 2001, p. A25; Bill Gertz and Rowan Scarborough, Inside the Ring, *Washington Times*, November 2, 2001, p. A10; Bill Gertz and Rowan Scarborough, "Inside the Ring," *Washington Times*, December 21, 2001, p. A8.

14. Private information.

15. Sharon LaFraniere and Lee Hockstader, "Russian Plane Explodes over Black Sea," *Washington Post*, October 5, 2001, pp. A1, A32; John J. Lumpkin, Associated Press, "U.S. Intelligence Believes Ukrainian Surface-to-Air Missile Brought Down Airliner," October 5, 2001, www.globalsecurity.org; Sabrina Tavernise and Michael Wines, "Accident Suspected in Black Sea Crash," *New York Times*, October 6, 1991, p. A5; Sharon LaFraniere, "Ukranian Won't Deny Rocket Hit Russian Jet," *Washington Post*, October 6, 2001, p. A25; Sabrina Tavernise, "Ukrainians Shift Stance on

the Cause of Air Crash," *New York Times*, October 7, 2001, p. A7; Michael A. Dornheim, "Ukranian Missile Exercises Likely Cause of Downed Tu-154," *Aviation Week & Space Technology*, October 15, 2001, pp. 62–63.

16. William J. Broad, "Military Warning System Also Tracks Bomb-Size Meteors," *New York Times*, May 29, 2001, p. D4. In addition, the trajectory of a meteorite that fell over a frozen lake in Canada in January 2000 was "closely tracked" by "American military satellites" according to one account. See John Noble Wilford, "Clues to Life May Come from Meteorite," *New York Times*, October 13, 2000, p. A25.

17. Dee W. Pack, Carl J. Rice, Barbara J. Tressel, Carolyn J. Lee-Wagner, and Edgar M. Oshika, "Civilian Uses of Surveillance Satellites," *Crosslink*, January 2000, pp. 2–8; Civil Applications Committee, *1998–99 Activity Report, Civil Applications Committee*, n.d., p. 6; Civil Applications Committee, *Civil Applications Committee 2000–2001 Activity Report*, April 3, 2002, p. 7.

18. Air Force Space Command, *SBIRS Overview Brief: Combat Air Force Commander's Conference, 16–17 November 1998*, n.d., p. 3.

19. Ibid., n.p. Since some pages have been removed because they contained classified material, the pages numbers of the remaining pages cannot be determined unless they are numbered.

20. Ibid. Several years later, it was reported that DSP satellites were capable of detecting tanks and other ground vehicles, and that SBIRS would improve upon that capability. See Jeremy Singer, "SBIRS Expected to Improve Estimates of Missile's Ultimate Target," *Space News*, January 17, 2005, p. 16.

21. Lt. Bruce Dickey, "Space-Based Infrared System to Provide Critical Battlespace Awareness," *Space Tracks*, October 2000, pp. 6–7; Richard J. Newman, "Space Watch, High and Low," *Air Force Magazine*, July 2001, pp. 35, 37–38.

22. *History of Space and Missile Systems Center*, 1:208; "Tests Begin on Troubled SBIRS-High Software," *Space News*, July 17, 2000, p. 2; Jeremy Singer, "U.S. Air Force Proceeds with Modified SBIRS High Design," *Space News*, December 4, 2000, pp. 4, 28. In December 2000, it was reported that SBIRS had corrected its software problems. See Bryan Bender, "USAF Satellite Effort Clears Software Hurdle," *Jane's Defence Weekly*, December 6, 2000, p. 10; "House Funding Panel Clobbers U.S. Military Space Programs," *Space News*, October 29, 2001, p. 4.

23. Jeremy Singer, "U.S. Air Force's SBIRS High Price Tag Continues to Rise," *Space News*, November 13, 2000, p. 15; Warren Ferster and Jeremy Singer, "Pentagon Proposes Staggering Launches of SBIRS Satellites," *Space News*, December 11, 2000, pp. 1, 9; "First SBIRS High Launch Moved to Late 2004," *Space News*, March 5, 2001, p. 2; Jeremy Singer, "SBIRS High Could Face More Cost Growth, Schedule Delays," *Space News*, November 5, 2001, p. 6; Anne Marie Squeo, "Crucial Lockheed Satellite System for Use in Missile Defense Is over Cost and Late," *Wall Street Journal*, November 16, 2001, p. B2.

24. Jeremy Singer, "Problems Plague SBIRS High Program," *Space News*, December 17, 2001, pp. 1, 3.

25. Warren Ferster, "Flight-Test Debate Hinders SBIRS Low Design Awards," *Space News*, July 5, 1999, pp. 1, 20; General Accounting Office, *Space-Based Infrared System-Low at Risk of Missing Initial Deployment Date*, GAO-01-6 (Washington, D.C.: GAO, 2001), pp. 7–8.

26. Robert Wall, "Spectrum Astro, TRW To Design SBIRS-Low," *Aviation Week &*

Space Technology, August 23, 1999, pp. 37, 40–41; Warren Ferster, "Spectrum Astro Secures Its Biggest Deal Ever," *Space News*, September 6, 1999, p. 3.

27. "SBIRS Low Constellation to Cost $10.6 Billion," *Space News*, May 8, 2000, p. 2; Gopal Ratnam and Jeremy Singer, "Support Falters for SBIRS Low," *Space News*, September 18, 2000, pp. 1, 36; "Pentagon Defers SBIRS Decision," *Space News*, October 23, 2000, p. 18; "Air Force Plans to Change Design of SBIRS Low Craft," *Space News*, November 27, 2000, p. 2; Carla Anne Robbins, "One Troubled System Shows Hurdles Facing Missile-Defense Plans," *Wall Street Journal*, June 15, 2001, pp. A1, A6.

28. General Accounting Office, *Space-Based Infrared System-Low at Risk*, pp. 3–4, 16.

29. Jeremy Singer, "Pentagon to Stretch Out SBIRS," *Space News*, March 5, 2001, p. 4; Jeremy Singer, "Air Force Official Slams GAO Report About SBIRS Low," *Space News*, March 12, 2001, pp. 3, 20; *History of Space and Missile Systems Center*, 1:218–222; Arthur L. Money, to Louis J. Rodriguez, December 14, 2000, in General Accounting Office, *Space-Based Infrared System-Low at Risk*, p. 30; see also pp. 32–33 of the report.

30. Warren Ferster and Jeremy Singer, "Pentagon Moves to Reduce Risk of SBIRS Low," *Space News*, August 27, 2001, p. 4; Jeremy Singer, "Congressional Moves Threaten SBIRS Low Risk Reduction Plan," *Space News*, October 8, 2001, pp. 3, 26; "Slicing and Dicing," *Aviation Week & Space Technology*, October 29, 2001, p. 31; Jeremy Singer, "Hopes Rest with U.S. Senate to Save SBIRS Low," *Space News*, October 29, 2001, p. 6.

31. Singer, "Hopes Rest with U.S. Senate to Save SBIRS Low"; "Key U.S. Lawmaker Considers Keeping SBIRS Low on Track," *Space News*, December 3, 2001, p. 4.

32. Bradley Graham, "Missile Defense Plan Scores a Direct Hit," *Washington Post*, October 3, 1999, p. A22; Robert Wall, "Intercept Boosts NMD Design," *Aviation Week & Space Technology*, October 11, 1999, p. 4.

33. Bradley Graham, "Panel Faults Antimissile Program on Many Fronts," *Washington Post*, November 14, 1999, pp. A1, A18; Paul Mann, "Missile Defense Still 'Troubled,'" *Aviation Week & Space Technology*, November 22, 1999, p. 31; Elizabeth Becker, "Missile Is Unable to Hit Its Target in Pentagon Test," *New York Times*, January 19, 2000, pp. A1, A14; Roberto Suro, "Missile Sensor Failed in Test's Final Seconds, Data Indicate," *Washington Post*, January 20, 2000, p. A4; James Glanz, "Experts Play Down ABM Test Failure," *New York Times*, July 15, 2000, p. A4; Panel on Reducing Risk in BMD Flight Test Programs, *National Missile Defense Review*, 1999.

34. William J. Broad, "Missile Contractor Doctored Tests, Ex-Employee Charges," *New York Times*, March 7, 2000, pp. A1, A19; William J. Broad, "Antimissile Testing Is Rigged to Hide a Flaw," *New York Times*, June 9, 2000, pp. A1, A22; General Accounting Office, *Status of the National Missile Defense Program*, GAO/NSIAD-00-131 (Washington, D.C.: GAO, 2000), p. 3; Roberto Suro and Thomas E. Ricks, "More Doubts Are Raised on Missile Shield," *Washington Post*, June 18, 2000, pp. A1, A16; Robert Wall, "More NMD Testing, Decoy Work Urged," *Aviation Week & Space Technology*, June 26, 2000, pp. 30–31; Eric Schmitt, "Pentagon Likely to Delay New Test for Missile Shield," *New York Times*, September 1, 2000, pp. A1, A5; James Glanz, "Other Systems Might Provide a U.S. Missile Shield," *New York Times*, September 4,

2000, pp. A1, A6; William J. Broad, "Defense Came in Several Packages, All Flawed," *New York Times*, September 2, 2000, p. A7; National Missile Defense Independent Review Team, *Executive Summary*, June 13, 2000.

35. General Accounting Office, *Schedule for Navy Theater Wide Program Should Be Revised to Reduce Risk*, GAO/NSIAD-00-121 (Washington, D.C.: GAO, 2000), pp. 3, 5.

36. Glanz, "Other Systems Might Provide a U.S. Missile Shield."

37. Lockheed Martin, "Air Force Accepts New Missile Warning Control Station from Lockheed Martin," January 7, 2002, www.lockheedmartin.com; information provided by Air Force Space and Missile Systems Center, January 27, 2012.

38. U.S. Strategic Command, *Theater Event System (TES) Architecture and Operations*, SD 523-2, May 2, 2004, pp. 6–7, 38; "NNSOC: Spanning the Information Technology Domain," *Domain*, Winter 2003, pp. 16–18. See also Eric Schmitt and Philip Shenon, "New Mobile Radar System Looking Out for Iraqi Missile Launchings in Gulf," *New York Times*, December 26, 2002, p. A14; "New Commander for JTaGS Pacific," *Domain*, Spring 2005, p. 7; Jeremy Singer, "U.S. Army Space Support Teams Practice for Possible Korea Fight," *Space News*, September 5, 2005, p. 4.

39. Craig Covault, "Eagle Eye on Threats," *Aviation Week & Space Technology*, February 23, 2004, pp. 99–100.

40. Covault, "Eagle Eye on Threats"; "Northrop Grumman Ships Last DSP Craft to Cape," *Space News*, May 16, 2005, p. 8; "DSP Launch on Heavy-Lift Delta 4 Delayed to 2007," *Space News*, May 22, 2006, p. 3; Jeremy Singer, "U.S. Air Force Unshaken by Delta 4 Heavy Performance," *Space News*, January 10, 2005, p. 6.

41. Bill Gertz and Rowan Scarborough," "Inside the Ring," *Washington Times*, January 4, 2002, p. A9; "Asia-Pacific," *Aviation Week & Space Technology*, February 11, 2002, p. 21; Bill Gertz, "China Tests Arms Designed to Fool Defense Systems," *Washington Times*, July 23, 2002, pp. A1, A13; Bill Gertz and Rowan Scarborough, "Inside the Ring," *Washington Times*, August 30, 2002, p. A10; "Russia Continues Missile Testing," *Newsday*, December 5, 2003, www.missilethreat.com.

42. "Iran Fails To Develop Shihab-3 Missile Engine," www.menewsline.com, accessed February 13, 2002; 32nd Army Air and Missile Defense Command, *Operation Iraqi Freedom*, n.d., n.p.; 32nd Army Air and Missile Defense Command, *Active Defense: The Iraqi Missile Fight*, n.d, n.p.

43. "DSP Role Changes from First Gulf War to Recent War in Iraq," *Space News*, October 13, 2003, p. 17; "Washington Outlook," *Aviation Week & Space Technology*, December 22, 2003, p. 21; U.S. Central Air Forces, OPERATION IRAQI FREEDOM—*By the Numbers*, April 30, 2003, p. 9; 32nd Army Air and Missile Defense Command, *Active Defense: The Iraqi Missile Fight*, n.d, n.p. *Active Defense* states that early warning was provided by "DSP-A, SBIRS." OPERATION IRAQI FREEDOM states that twenty-six missile launches were detected from space, without explicitly identifying DSP as the space system in question.

44. David A. Fulghum, "Offensive Gathers Speed," *Aviation Week & Space Technology*, March 24, 2003, pp. 22–23; U.S. Central Air Forces, OPERATION IRAQI FREEDOM— *By the Numbers*, p. 9; "William B. Scott, "Milspace Technology Coups," *Aviation Week & Space Technology*, January 19, 2004, pp. 417–18.

45. Peter Baker, "Missile Self-Destructs During Russian Military Exercises," *Washington Post*, February 19, 2004, p. A20; "Pakistan Tests Longest-Range Missile Yet,

Military Reports," March 9, 2004, www.usatoday.com; Bill Gertz, "China Flexes Muscle on Eve of Taiwan Vote," *Washington Times*, March 18, 2004, p. A3; Paul Hughes, Reuters, "Iran Tests Missile Capable of Hitting Israel," August 11, 2004; "Russia Tests Intercontinental Ballistic Missile, August 12, 2004, www.smh.com.au.

46. "Japan Reports Missile Test By North Korea," *Washington Post*, May 2, 2005, p. A11; Steven Erlanger, "Syria Test-Fires 3 Scud Missiles, Israelis Say," *New York Times*, June 3, 2005, p. A10; Bill Gertz, "Russian Warhead Alters Course Midflight in Test," November 21, 2005, www.washingtontimes.com; Craig Covault, "Iran's Sputnik," *Aviation Week & Space Technology*, November 29, 2004, pp. 36–37.

47. MSNBC, "500 Dead After Fiery Blast in Nigerian Capital," January 28, 2002, www.msnbc.com.

48. "Flash Point," *Aviation Week & Space Technology*, October 14, 2002, p. 17; Robert Roy Britt, "DoD Satellite Tracked Siberian Fireball That Might Have Hit Earth," October 14, 2002, www.space.com. See also Philip Ball, "Satellites Spy on Meteorite Explosions," November 21, 2002, www.nature.com; P. Brown, R. E. Spalding, D. O. Revelle, E. Tagliaferri, and S. P. Worden, "The Flux of Small Near-Earth Objects Colliding with the Earth," *Nature*, November 21, 2002, pp. 294–96.

49. MSNBC, "Space Shuttle Crashes on Re-Entry," February 1, 2003, www.msnbc.com; Bill Gertz, "Orbiting Supercameras Aid Search for Shuttle Debris," *Washington Times*, February 5, 2003, p. A11; "In Orbit," *Aviation Week & Space Technology*, March 8, 2004, p. 19.

50. K. S. Jayaraman, "Fatal Blast at Rocket Plant Will Not Affect ISRO Programs," *Space News*, March 1, 2004, p. 20; Anthony Faiola, "Huge Blast Reported in N. Korea," *Washington Post*, April 23, 2004, pp. A1, A20; James Brooke, "'3,000 Casualties Reported in North Korean Rail Blast," *New York Times*, April 23, 2004, p. A3; Bill Gertz and Rowan Scarborough, "Inside the Ring," *Washington Times*, September 24, 2004, p. A6; Christopher Torchia, "Powell Says N. Korea Blast Not Nuclear," September 12, 2004, www.latimes.com; Christopher Torchia, "N. Korea Says Blast Was from Excavation," *Washington Times*, September 14, 2004, p. A14.

51. Singer, "U.S. Army Space Support Teams Practice for Possible Korea Fight"; Civil Applications Committee, *Civil Applications Committee 2004 Activity Report*, n.d., p. 13; Civil Applications Committee, *Civil Applications Committee 2006 Activity Report*, n.d., p. 9.

52. Fred Simmons and Jim Creswell, "DSP: The Defense Support Program," *Crosslink*, Summer 2000, pp. 18–25; U.S. Geological Survey, "About the Volcano Hazards Program," http://volcanoes.usgs.gov/vhp, accessed July 21, 2011; Pack et al., "Civilian Uses of Surveillance Satellites"; Civil Applications Committee, *Civil Applications Committee 2003 Activity Report*, n.d., p. 16.

53. "NRO Exploring Alternatives For Space-Based Missile Warning," *Space News*, February 11, 2002, p. 4; "Veteran Rocket Scientist Takes On New Mission," *Space News*, April 8, 2002, p. 46.

54. Jeremy Singer, "In Shaky Times, Pentagon Eyes Future," *Space News*, April 8, 2002, p. 6; Michael Mecham and Robert Wall, "It's High Noon for SBIRS-High," *Aviation Week & Space Technology*, April 8, 2002, pp. 85–87. Independent Review Team conclusions are also reported in General Accounting Office, *Military Space Operations: Common Problems and Their Effects on Satellite and Related Acquisitions* (Washington, D.C.: GAO, 2003), p. 18; General Accounting Office, *Despite Restructuring,*

SBIRS High Program Remains at Risk of Cost and Schedule Overruns, GAO-04-48 (Washington, D.C.: GAO, 2009), p. 7.

55. Department of Defense, "Nunn-McCurdy Certification—Space Based Infrared System (SBIRS) High Program," May 2, 2002; General Accounting Office, *Despite Restructuring, SBIRS High Program Remains at Risk*, p. 8; "News Transcript: Under Secretary of Defense Aldridge Roundtable," May 2, 2002, www.defenselink-mil; "SBIRS High Work to Continue but Could Still Be Cancelled," *Space News*, May 6, 2002, p. 4; Anne Marie Squeo, "Officials Say Space Programs Facing Delays, Are 'in Trouble,'" *Wall Street Journal*, December 4, 2002, pp. A1, A13.

56. Jeremy Singer, "Rogue Radio Waves Delay Delivery of First SBIRS High Sensor," *Space News*, February 17, 2003, p. 8; General Accounting Office, *Despite Restructuring, SBIRS High Program Remains at Risk*, pp. 2–3; Defense Science Board/Air Force Scientific Advisory Board Joint Task Force, Office of the Under Secretary of Defense for Acquisition, Technology, and Logistics, *Acquisition of National Security Space Programs*, May 2003, pp. 30–31.

57. General Accounting Office, *Despite Restructuring, SBIRS High Program Remains at Risk*, pp. 2–3.

58. Ibid., pp. 1, 3, 9, 14, 20.

59. Ibid., pp. 18–20.

60. Jeremy Singer, "Air Force Says New SBIRS High Problems Are Manageable," *Space News*, October 20, 2003, p. 8; Warren Ferster, "Availability of Initial SBIRS Sensor Delayed Until Mid-2005," *Space News*, November 17, 2003, p. 21. Late November also brought a critical report by the Defense Department's Inspector General concerning software testing for the SBIRS program. See Inspector General, Department of Defense, *Development Testing of Space-Based Infrared System Mission Critical Software*, November 24, 2003.

61. "SBIRS Payload Passes Key Testing Milestone," *Space News*, January 23, 2004, p. 14; "U.S. Air Force Studies Alternatives to Last 3 SBIRS Satellites," *Space News*, April 5, 2004, p. 9.

62. Jeremy Singer, "U.S. Air Force Needs Additional Money for SBIRS High Program," *Space News*, March 29, 2004, pp. 1, 3; Jeremy Singer, "Air Force Options Are Limited for Troubled SBIRS Program," *Space News*, June 28, 2004, p. 9.

63. "Lockheed Martin Delivers First SBIRS Sensor to U.S. Air Force," *Space News*, August 9, 2004, p. 4; Robert Wall, "Demanding Attention," *Aviation Week & Space Technology*, September 6, 2004, pp. 36–37; "SBIRS High Launches May Slip To Autumn 2008," *Space News*, October 4, 2004, p. 4; "SBIRS Software Clears Performance Review," *Space News*, November 22, 2004, p. 8.

64. Jeremy Singer, "SBIRS Report To Include Update on Health of Defense Support Program," *Space News*, November 29, 2004, p. 8; Office of the Secretary of Defense, *Report to the Defense and Intelligence Committees of the Congress of the United States on the Status of the Space Based Infrared System Program*, March 2005, p. i.

65. Office of the Secretary of Defense, *Report to the Defense and Intelligence Committees of the Congress of the United States on the Status of the Space Based Infrared System Program*, pp. 28–38.

66. Letter from Peter B. Teets, Acting Secretary of the Air Force, to Hon. John W. Warner, March 10, 2005; Jeremy Singer, "DoD Notifies Congress of Higher SBIRS Cost," *Space News*, March 14, 2005, p. 10; Amy Butler, "A Billion Here, A Billion

There," *Aviation Week & Space Technology*, March 21, 2005, pp. 22–23; Peter B. Teets, *Statement by Under Secretary of the Air Force, Director, National Reconnaissance Office, Department of Defense Executive Agent for Space, The Honorable Peter B. Teets*, March 16, 2005, p. 9.

67. "SBIRS High Slip," *Aviation Week & Space Technology*, March 14, 2005, p. 17.

68. Jeremy Singer, "Teets: Pentagon Must Launch SBIRS by 2015," *Space News*, March 28, 2005, p. 16; "Taming Space Acquisition," *C4ISR Journal*, May 2005, pp. 46–47.

69. "Price of SBIRS Program Goes Up Another $1 Billion," *Space News*, April 26, 2004, p. 20; "Groundhog Day," *Aviation Week & Space Technology*, July 18, 2005, p. 21; "Missile Warning System Costs Again Rise over 25 Percent," *Space News*, August 1, 2005, p. 3; "Initial SBIRS Launch Could Slip Yet Again," *Space News*, August 8, 2005, p. 3.

70. "Lockheed Martin Ships Second SBIRS Sensor," *Space News*, October 3, 2005, p. 8; "Northrop Integrates Sensor For First SBIRS Satellite," *Space News*, October 10, 2005, p. 9.

71. Letter from Kenneth Krieg, Under Secretary of Defense for Acquisition, Technology, and Logistics, to Hon. John Warner, December 12, 2005; Jeremy Singer, "Pentagon Scales Back SBIRS Program," *Space News*, December 19, 2005, pp. 1, 4.

72. Letter from Kenneth Krieg, Under Secretary of Defense for Acquisition, Technology, and Logistics, to Hon. John Warner.

73. "Profile: Gary E. Payton, Deputy Undersecretary for Space Programs, U.S. Air Force," *Space News*, February 20, 2006, p. 22; Jeremy Singer, "Air Force Budget, New Development Approach Reflect Congressional Scrutiny of Cost Growth," *Space News*, February 13, 2006, p. 13.

74. Government Accountability Office, *Assessments of Selected Major Weapon Programs*, GAO-06-391 (Washington, D.C.: GAO, 2006), p. 101; Government Accountability Office, *Improvements Needed in Space Systems Acquisitions and Keys to Achieving Them*, GAO-06-626T (Washington, D.C.: GAO, 2006), p. 15.

75. Jeremy Singer, "New Technology Offers Low-Risk Approach to SBIRS Replacement," *Space News*, April 17, 2006, p. 12.

76. Col. Roger Teague, Command, Space Group, Space Based Infrared Systems Wing, Space and Missile Systems Center (SMC), *SBIRS Transformational Capability*, November 30, 2006, p. 3.

77. Ibid., pp. 4, 9.

78. Jeremy Singer, "SBIRS Low Competitors Could Become a Team," *Space News*, February 25, 2002, p. 6.

79. Ed Taylor, "Gilbert, Ariz.-Based Aerospace Firm Loses Satellite Contract to TRW," *East Valley Tribune*, April 19, 2002, from www.globalsecurity.org; "Weak Signals," *Aviation Week & Space Technology*, April 22, 2002, p. 21; Jeremy Singer, "Pentagon Names TRW Prime Contractor in SBIRS Low," *Space News*, April 22, 2002, p. 11; Jeremy Singer, "Pentagon to Launch Smaller SBIRS Low Constellation," *Space News*, August 26, 2002, p. 3; Ann Roosevelt, "TRW Moves Out on SBIRS-Low," *Defense Week*, September 3, 2002, p. 6; Jeremy Singer, "SBIRS Low Demonstration to Rely On Old Hardware," *Space News*, September 9, 2002, p. 3.

80. "Spectrum Astro Expects Role in STSS Satellites," *Space News*, March 1, 2004, p. 24; "Northrop Grumman Completes Inspection of STSS Hardware," *Space News*,

April 5, 2004, p. 9; Amy Butler, "Space Tracking for MDA," *Aviation Week & Space Technology*, October 3, 2005, p. 51; Jeremy Singer, "MDA Seeks Design Funds for Full Missile Tracking System," *Space News*, May 23, 2005, p. 14.

81. Bradley Graham, "Rise and Fall of a Navy Missile," *Washington Post*, March 28, 2002, p. A3.

82. Steven A. Hildreth (coordinator), Congressional Research Service, *Missile Defense: The Current Debate*, March 23, 2005, p. CRS-4; General Accounting Office, *Additional Knowledge Needed in Developing System for Intercepting Long-Range Missiles* (Washington, D.C.: GAO, 2003), p. 5. The Bush administration's statement of its ballistic missile defense policy was laid out in George W. Bush, National Security Presidential Directive 23, Subject: National Policy on Ballistic Missile Defense, December 16, 2002, available at www.fas.org.

83. Hildreth, , *Missile Defense*, pp. CRS-28–CRS-35.

84. Ibid., pp. CRS-35–CRS-36; Bradley Graham, "Interceptor System Set, but Doubts Remain," *Washington Post*, September 29, 2004, pp. A1, A16; Robert Wall, "Ready or Not . . . ," *Aviation Week & Space Technology*, June 28, 2004, pp. 46, 48.

85. Congressional Research Service, *Missile Defense*, pp. CRS-38–CRS-39.

86. Ibid., pp. CRS-41–CRS-45. The T in THAAD was altered from Theater to Terminal in February 2004.

87. Bill Gertz, "S. Korea Receives Advanced Patriots," *Washington Times*, November 20, 2003, p. A2; Robert Wall, "Changing Defenses," *Aviation Week & Space Technology*, April 19, 2004, pp. 73–75; Anthony Faiola, "U.S. to Deploy Patriot Missiles in Japan to Counter North Korea," *Washington Post*, June 27, 2006, p. A15.

88. Bradley Graham, "U.S. Missile Defense Test Fails," *Washington Post*, December 16, 2004, p. A5; Randy Barrett, "Patriot System Performance Again Raises Questions," *Space News*, April 21, 2003, p. 7; Bradley Graham, "Missile Defense Testing May Be Inadequate," *Washington Post*, January 22, 2004, p. A4; Bradley Graham, "Missile Defense Director Moves to End Test Glitches," *Washington Post*, March 10, 2005, p. A19; Bradley Graham, "Panel Faults Tactics in Rush to Install Antimissile System," *Washington Post*, June 10, 2005, p. A7.

89. E-mail from Colleen [Shipman], U.S. Air Force, to William A. Ofelein, Johnson Space Center, Subject: Re: FW: Crew trip to Europe, January 17, 2008; "U.S. Air Force Delays Last DSP Launch until Summer," *Space News*, March 19, 2007, p. 3; "DSP Launch Delay Will Affect Other Satellites," *Space News*, March 26, 2007, p. 3.

90. David Dobrydney, "Sharks Prepare Final DSP," *Air Force Print News Today*, July 26, 2007, www.patrick.af.mil; "Last DSP Missile-Warning Satellite Launches Atop Delta 4 Heavy," *Space News*, November 19, 2007, p. 9; Justin Ray, "Largest U.S. Unmanned Rocket Launches Surveillance Satellite," *Spaceflight Now*, November 12, 2007, www.spaceflightnow.com; Stephen Clark, "Last-of-Its-Kind Surveillance Satellite Reaches Orbit," *Spaceflight Now*, November 11, 2007, www.spaceflightnow.com; Jeremy Singer, "DSP Poses Operational Test for Delta 4 Heavy," *Space News*, November 5, 2007, p. 10. The conclusion that Flight 23 was to serve as the European satellite is based on data appearing in the "Geostationary Satellite Log" of *Jonathan's Space Report*, www.planet4589.org/space/jsr/jsr.html, accessed January 28, 2012.

91. Northrop Grumman, "Northrop Grumman-Built Defense Support Program Flight Satellite Successfully Launched," November 11, 2007; Craig Covault, "Fire

and Brimstone," *Aviation Week & Space Technology*, November 5, 2007, p. 31; "U.S. Air Force Buoyed by DSP-23 Launch," *Space News*, November 26, 2007, p. 17; "Nothing to Fear," *AeroSpace Briefing*, November 20, 2007, p. 2.

92. "DSP Decommissioned by the U.S. Air Force," *Space News*, August 4, 2008, p. 3; Northrop Grumman, "DSP Support Program Satellite Decommissioned," July 31, 2008; Turner Brinton, "DSP Constellation Health Concerns Prompt Plan for Gap-Filler Satellite," *Space News*, November 24, 2008, pp. 1, 4; Andrea Shalal-Esa, Associated Press, "U.S. Missile-Warning Satellite Fails," November 24, 2008; Peter B. de Selding, "U.S. DSP-23 Satellite Drifts near Vicinity of Other Craft," *Space News*, December 8, 2008, p. 4.

93. Craig Covault, "Secret Inspection Satellites Boost Space Intelligence Ops," *Spaceflight Now*, January 14, 2009, www.spaceflightnow.com.

94. Ibid.; Andrea Shalal-Esa, "U.S. Still Probing Security Satellite Failure," January 6, 2009, www.reuters.com. See also Brian Weeden, "The Ongoing Saga of DSP Flight 23," *The Space Review*, January 19, 2009, www.thespacereview.com.

95. Amy Butler, "Turning Heads," *Aviation Week & Space Technology*, April 16, 2007, pp. 35–36; Amy Butler, "In Recovery," *Aviation Week & Space Technology*, June 4, 2007, pp. 24–26; Lockheed Martin, *SBIRS HEO*, n.d.

96. Government Accountability Office, *Assessments of Selected Weapons Programs* (Washington, D.C.: GAO, 2007), p. 124.

97. Jeremy Singer, "Air Force Plans August Award for AIRSS Flight Demo," *Space News*, March 5, 2007, p. 16; Amy Butler, "Refocusing," *Aviation Week & Space Technology*, June 4, 2007, pp. 26–27; "USAF Decides to Buy Third SBIRS High from Lockheed," *Space News*, June 25, 2007, p. 3; "Airing Differences," *AeroSpace Briefing*, June 8, 2007, p. 2.

98. Ben Iannotta, "SBIRS Returns," *C4ISR Journal*, July 2007, pp. 26–27.

99. "Sbirs Delivery," *Aviation Week & Space Technology*, August 13, 2007, p. 15; "Lockheed Martin Mates SBIRS Spacecraft, Sensor," *Space News*, September 17, 2007, p. 8; "Makeover," *Aviation Week & Space Technology*, September 24, 2007, p. 31.

100. Michael W. Wynne, Secretary of the Air Force, Memorandum for Under Secretary of Defense for Acquisition, Technology, and Logistics, September 26, 2007.

101. "How Much More for SBIRS?" *AeroSpace Briefing*, October 5, 2007, p. 2; Jeremy Singer, "DoD Mulls Missile Warning Options amid Latest Travails with SBIRS," *Space News*, October 8, 2007, pp. 1, 18.

102. "Same Old Story on SBIRS," *Space News*, October 8, 2007, p. 32.

103. Jeremy Singer, "More Tests Ordered on 1st SBIRS Craft," *Space News*, December 3, 2007, p. 17.

104. Government Accountability Office, *Assessments of Selected Weapon Programs*, GAO-08-467SP (Washington, D.C.: GAO, 2008), pp. 153–54.

105. "Third SBIRS Satellite Ordered From Lockheed," *Space News*, March 17, 2008, p. 8; "Moving On," *Aviation Week & Space Technology*, April 14, 2008, p. 23; Lockheed Martin, "Lockheed Martin Achieves Key Milestone on New Missile Warning Satellite," April 30, 2008, www.lockheedmartin.com; "USAF Finishes Checkout of 2nd SBIRS Sensor in Space," *Space News*, June 23, 2008, p. 3; "Launch Warning," *Aviation Week & Space Technology*, June 30, 2008, p. 19; "1st On-Orbit SBIRS Sensor Moves to New Test Phase," *Space News*, September 1, 2008, p. 16.

106. Lockheed Martin, "Lockheed Martin SBIRS Team Completes On-Orbit Handover of First HEO Payload to U.S. Air Force," August 5, 2008, www.lock heedmartin.com; "U.S. Air Force Assumes Control of SBIRS Payload," *Space News*, August 11, 2008, p. 8; Lockheed Martin, "U.S. Air Force/Lockheed Martin Team Achieves Major Operational Milestone on First SBIRS HEO System," August 28, 2008, www.lockheedmartin.com; "1st On-Orbit SBIRS Sensor Moves to New Test Phase," *Space News*, September 1, 2008, p. 16.

107. Government Accountability Office, *DOD's Goal for Resolving Space Based Infrared System Software Problems Are Ambitious*, GAO-08-1073 (Washington, D.C.: GAO, 2008); "What GAO Found" (Washington, D.C.: GAO, 2008); Turner Brinton, "Nagging SBIRS Software Issue Tempers Optimism on Program," *Space News*, October 6, 2008, pp. 1, 12.

108. "U.S. Air Force to Order Fourth SBIRS Satellite," *Space News*, September 15, 2008, p. 8; Amy Butler, "Shaken, Not Stirred," *Aviation Week & Space Technology*, October 27, 2008, p. 44.

109. "Raytheon Demonstrates Infrared AIRSS Sensor," *Space News*, March 17, 2008, p. 3; "SAIC Delivers Prototype Missile Warning Sensor," *Space News*, March 31, 2008, p. 8; Amy Butler, "Innovative Infrared," *Aviation Week & Space Technology*, August 4, 2008, pp. 31–32.

110. Turner Brinton, "Pentagon Space Procurement Agenda Marked by Uncertainty," *Space News*, November 24, 2008, p. 12; John J. Young Jr., Under Secretary of Defense for Acquisition, Technology, and Logistics, Memorandum, Subject: Infrared Augmentation Satellite [IRAS]), Acquisition Decision Memorandum, December 1, 2008; "Air Force May Sole-Source Missile Warning Satellite," *Space News*, December 15, 2008, p. 3; "Pentagon Kills Gap-Filler Missile Warning Satellite," *Space News*, April 30, 2009, p. 8; Turner Brinton, "ORS Office Eyes Missile Warning Mission," *Space News*, May 4, 2009, p. 4.

111. "U.S. Air Force Adds $102M To Lockheed SBIRS Deal," *Space News*, January 5, 2009, p. 3; "New SBIRS Flight Software Passes Performance Test," *Space News*, February 2, 2009, p. 8; "SBIRS Payload Completes Thermal Vacuum Testing," *Space News*, March 9, 2009, p. 9; "Clearing a Hurdle," *Aviation Week & Space Technology*, December 7, 2009, p. 19; "First SBIRS Satellite Passes Thermal Vacuum Testing," *Space News*, December 7, 2009, p. 8.

112. Warren Ferster, "Pentagon Seeks New Missile Warning Satellites," *Space News*, April 6, 2009, p. 7; "Northrop Delivers SBIRS Payload to Lockheed Martin," *Space News*, April 27, 2009, p. 8; "Infrared Sensor Installed on Second SBIRS Satellite," *Space News*, June 15, 2009, p. 8; "$1.5-Billion Sbirs Contract," *Aviation Week & Space Technology*, June 8, 2009, p. 16;. "SBIRS Payload Enters On-Orbit Testing Phase," *Space News*, July 6, 2009, p. 16.

113. Turner Brinton, "Software Fix Adds $750M to SBIRS Price Tag," *Space News*, July 27, 2009, p. A3; Colin Clark, "SBIRS Problems Persist: New Sats Needed," September 14, 2009, www.dodbuzz.com; Stephen Clark, "First SBIRS Early Warning Satellite Delayed Until 2011," *Spaceflight Now*, December 10, 2009, www.space flightnow.com.

114. "Step Forward for Sbirs," *Aviation Week & Space Technology*, January 18, 2010, p. 15; "2nd SBIRS Craft Completes Integrated Testing Milestone," *Space News*,

February 22, 2010, p. 9; "2nd SBIRS Hosted Payload Ready to Gather Tech Intel," *Space News*, May 17, 2010, p. 8; Amy Butler, "Missile Warning Alert," *Aviation Week & Space Technology*, April 19, 2010, pp. 35–36.

115. Turner Brinton, "Defense Spending Act Pushes Smarter Investments in Space," *Space News*, January 10, 2010, p. 4; "Big Changes in Store for Missile Warning Tech Effort," *Space News*, January 25, 2010, p. 3; Turner Brinton, "U.S. Air Force Scales Back Missile Warning Technology Program," *Space News*, February 8, 2010, p. 12; Amy Butler and Michael Bruno, "Finding in a Vacuum," *Aviation Week & Space Technology*, February 8, 2010, pp. 27–28.

116. Debra Werner, "Long-Delayed 1st SBIRS Satellite Ready for May Launch, Officials Say," *Space News*, February 14, 2011, p. 7.

117. Stephen Clark, "SBIRS Missile Warning Craft Shipped to Cape Canaveral," *Spaceflight Now*, March 6, 2011, www.spaceflightnow.com; "SBIRS Satellite Arrives at Cape Canaveral for Launch," *Space News*, March 7, 2011, p. 3; "First SBIRS Satellite Still on Launch Track for Early May Launch," *Space News*, April 25, 2011, p. 9; Justin Ray, "Beginning a New Generation of Missile Early-Warning," *Spaceflight Now*, May 7, 2011, www.spaceflightnow.com; "Air Force's 1st Dedicated SBIRS Satellite Makes Orbit," *Space News*, May 16, 2011, p. 8; "First SBIRS Satellite Reaches Destination," *Space News*, May 23, 2011, p. 3; "Sbirs Checkout to Begin," *Aviation Week & Space Technology*, May 23, 2011, p. 13; "Missile Defense Satellite Transmits Initial Picture," *Global Security Newswire*, July 8, 2011, http://gsn.nti.org; Lockheed Martin, *SBIRS GEO*, n.d. The conclusion that SBIRS-GEO 1 is located at 116W is based on data appearing in the "Geostationary Satellite Log" of *Jonathan's Space Report*, www.planet4589.org/space/jsr/jsr.html, accessed January 28, 2012.

118. Stephen Clark, "Ariane 5 Rocket Lights Up Jungle with Perfect Launch," *Spaceflight Now*, September 22, 2011, www.spaceflightnow.com; Stephen Clark, "Money-Saving Missile Detection Sensor Powered On," *Spaceflight Now*, November 4, 2011, www.spaceflightnow.com. As of early 2012, satisfaction with the CHIRP results was sufficient to have the Air Force planning a follow-on effort. See Frank Morring Jr., "Slowly But Surely," *Aviation Week & Space Technology*, March 12, 2012, pp. 46-47; Titus Ledbetter III, "After Hosted Payload Success, U.S. Air Force Plans Follow-On," *Space News*,
April 16, 2012, pp. 1, 4.

119. "Missile Defense Agency Mulls Interim STSS Craft," *Space News*, April 23, 2007, p. 3; Government Accountability Office, *Assessments of Selected Weapon Programs*, (Washington, D.C.: GAO, 2007), p. 131; Jeremy Singer, "MDA Delays Launch of Missile Tracking Satellites to 2008," *Space News*, June 18, 2007, p. 11.

120. Justin Ray, "Delta 2 Rocket Launches Missile Defense Satellites," *Spaceflight Now*, September 25, 2009, www.spaceflightnow.com.

121. Government Accountability Office, *Actions Needed to Improve Transparency and Accountability*, GAO-11-372 (Washington, D.C.: GAO, 2011), pp. 96–97.

122. Ibid., p. 99; Stephen Clark, "STSS Demo Satellites Ready for Missile Defense Testing," *Spaceflight Now*, February 7, 2011, www.spaceflightnow.com; Turner Brinton, "STSS Satellites Demonstrate 'Holy Grail' of Missile Tracking," *Space News*, March 28, 2011, p. 10; Amy Butler, "Lights Out," *Aviation Week & Space Technology*, January 2, 2012, pp. 29–30.

123. Missile Defense Agency, Fact Sheet, "Precision Tracking Space System," April 12, 2011. For differing views on the cancellation of STSS, see Loren Thompson, "Why is MDA Moving So Slowly on Space Sensors?," *Space News*, March 14, 2011, p. 27; W. David Thompson, "MDA Was Right to Kill STSS," *Space News*, June 6, 2011, pp. 19, 21. An earlier discussion of problems with STSS was General Accounting Office, *Alternate Approaches to Space Tracking and Surveillance System Need to Be Considered*, GAO-03-597 (Washington, D.C.: GAO, 2003). In May 2012, the House Armed Services Committee recommended slowing down the PTSS program, recommending a substantial cut in funding. See Titus Ledbetter III, "U.S. House Panel Not Sold on Missile Tracking Constellation," *Space News*, May 14, 2012, p. 5.

124. "Agni-III's the Charm," *Aviation Week & Space Technology*, April 23, 2007, p. 12; "Military Tests Long-Range Missile," *Los Angeles Times*, April 13, 2007, p. A6; "Russia Says New ICBM Could Overcome U.S. Defenses," *Space News*, June 4, 2007, p. 8; "Russia: Ballistic Missile Test-Fired," *New York Times*, December 26, 2007, p. A11.

125. Chip Cummins and Jay Solomon, "Iranian Missile Test Escalates Tensions," *Wall Street Journal*, July 10, 2008, pp. A1, A11; Alan Cowell and William J. Broad, "Iran Launches 9 Missiles in War Games," *New York Times*, July 10, 2008, p. A10; Michael Scwirtz, "Rice Warns Iran as It Tests Missiles for a Second Day," *New York Times*, July 11, 2008, p. A8; William J. Broad, "Experts Point to Deceptions in Military Display by Iran," *New York Times*, July 12, 2008, p. A8.

126. David A. Fulghum, John M. Doyle, and Craig Covault, "Warhead Diplomacy," *Aviation Week & Space Technology*, July 14, 2008, pp. 60–61.

127. Nazila Fathi and Alan Cowell, "Iran Claims Success in Tests Firing Long-Range Missiles," *New York Times*, November 13, 2008, p. A12; "Russia Tests Topol ICBM," *Global Security Newswire*, September 3, 2008, www.nti.org; Russia Test-Fires Four Ballistic Missiles," *Global Security Newswire*, October 14, 2008, www.nti.org; "Russia Tests Nuclear-Capable Missiles," *Global Security Newswire*, December 1, 2008, www.nti.org.

128. Nazila Fathi and William J. Broad, "Iran Launches Satellite as U.S. Takes Wary Note," *New York Times*, February 4, 2009, pp. A1, A11; Turner Brinton, "Iran's Satellite Launch Viewed as Sign of Ballistic Missile Progress," *Space News*, February 9, 2009, p. 14; John M. Glionna, "North Korea Missile Trial May Test U.S.," *Los Angeles Times*, February 8, 2009, p. A11; Evan Ramstad, "North Korea Sets Date for Controversial Launch," *Wall Street Journal*, March 13, 2009, p. A5; Greg Miller, "U.S. Challenges N. Korea Satellite Claim," *Los Angeles Times*, March 27, 2007, p. A27; Choe Sang-Hun and David E. Sanger, "North Koreans Launch a Rocket over the Pacific," *New York Times*, April 5, 2004, p. 1, 10; William J. Broad, "Missile Flight Was a Failure, Trackers Say," *New York Times*, April 6, 2009, pp. A1, A8; Craig Covault, "North Korean Rocket Flew Further Than Earlier Thought," *Spaceflight Now*, April 10, 2009, www.spaceflightnow.com; David E. Sanger and Nazila Fathi, "Launching of Missile Shows Iran Is Advancing," *New York Times*, May 21, 2009, pp. A6, A10; Chip Cummins, "Iran Tests Missile, Launching Pre-Election Show of Defiance," *Wall Street Journal*, May 21, 2009, p. A8; Borzou Daragahi, "Iran Test Fires a Missile," *Los Angeles Times*, May 21, 2009, p. A24; "Iran Rocket Launch Called 'Success,'" *Aviation Week & Space Technology*, May 25, 2009, p. 18.

129. Craig Covault, "U.S. Military Spacecraft Aid Search for Missing Airbus," *Spaceflight Now*, June 1, 2009, www.spaceflightnow.com; Frances Fiorino, "The Riddle of Flight 447," *Aviation Week & Space Technology*, June 8, 2009, pp. 24–26.

130. "Russia Reports Successful Strategic Missile Test," *Global Security Newswire*, July14, 2009, http://gsn.nti.org; "Russia Tests Topol ICBM, *Global Security Newswire*, December 11, 2009, http://gsn.nti.org; "Russia Conducts Successful ICBM Test," *Global Security Newswire*, December 24, 2009, http://gsn.nti.org.

131. "Indian Ballistic Missile Test Deemed a Failure," *Global Security Newswire*, November 24, 2009, http://gsn.nti.org; Michael Slackman, "Iran Test-Fires Upgraded Version," *New York Times*, December 17, 2009, p. A22; Ramin Mostaghim, "Iran Test Launch Angers the West," *Los Angeles Times*, December 17, 2009, p. A37.

132. Secretary of State, Subject: Demarche Following China's January 2010 Intercept Flight-Test, January 12, 2010, www.telegraph.co.uk/news/wikileaks-files; Vladimir Isachenkov, "Russian Military Successfully Tests New Missile," October 7, 2010, www.washingtontimes.com; "India Test-Launches Nuke-Capable Missile," *Global Security Newswire*, November 29, 2010, http://gsn.nti.org.

133. "Iran Launches Missiles at Indian Ocean Targets," *Global Security Newswire*, July 11, 2011, http://gsn.nti.org; "Russian Sub Launches Nuke-Capable Missile in Test," *Global Security Newswire*, April 27, 2011, http://gsn.nti.org; "India Conducts Trial Launch of Agni 2 Missile," *Global Security Newswire*, September 30, 2011, http://gsn.nti.org; "Qadhafi Troops Launch Scud Missile, U.S. Says," *Global Security Newswire*, August 16, 2011, http://gsn.nti.org.

134. Missile Defense Agency, Fact Sheet, "The Ballistic Missile Defense System," July 19, 2011, www.mda.mil. On Aegis status in 2011, see Ronald O'Rourke, Congressional Research Service, *Navy Aegis Ballistic Missile Defense (BMD) Program: Background and Issues for Congress*, April 19, 2011.

135. Missile Defense Agency, Fact Sheet, "The Ballistic Missile Defense System"; Amy Butler, "Strategic Pause," *Aviation Week & Space Technology*, June 13, 2011, p. 49. For a critique of U.S. missile defense plans, see George N. Lewis and Theodore A. Postol, Massachusetts Institute of Technology, *A Technically Detailed Description of Flaws in the SM-3 and GMD Missile Defense Systems Revealed by the Defense Department's Ballistic Missile Test Data*, May 2010.

136. Missile Defense Agency, Fact Sheet, "The Ballistic Missile Defense System; James Kitfield, "The Long Road to Missile Defense," *Air Force Magazine*, March 2011, pp. 54–57; Government Accountability Office, *Assessments of Selected Weapon Programs*, GAO-11-233SP (Washington, D.C.: GAO, 2011), pp. 45–46.

137. U.S. Air Force, Fact Sheet, "SBIRS Mission Control Station," www.losangeles.af.mil/library/factsheets, accessed January 11, 2012; U.S. Air Force, Fact Sheet, "Space-Based Infrared Systems Wing," www.losangeles.af.mil/library/factsheets, accessed August 15, 2010; U.S. Force, Fact Sheet, Space Based Infrared Systems, August 2010, www.afspc.af.mil; Sanjay Allen, "Space Warning Squadron Reactivated," *Air Force Print News Today*, December 6, 2007, www.buckley.af.mil; Lockheed Martin, *SBIRS GEO*; information provided by Air Force Space and Missile Systems Center, January 27, 2012; Allison Day, "JTaGS Misawa sees 'launch'," *Flipside*, (Winter 2008), pp. 1F–3F.

138. Justin Ray, "Certification Work Continues on New Missile Warning Craft," *Spaceflight Now*, March 19, 2012, www.spaceflightnow.com; Titus Ledbetter III,

"Advanced SBIRS Sensor Capability Will Not Be Available Before 2016," *Space News*, March 26, 2012, pp. 1, 4.

139. Titus Ledbetter III, "Lockheed Denies Latest SBIRS Satellites Are Behind Schedule," *Space News*, April 9, 2012, p.14; Cristina T. Chaplain, General Accounting Office, *SPACE ACQUISITIONS: DOD Faces Challenges in Fully Realizing Benefits of Satellite Acquisition Improvements*, March 21, 2012, p. 4.

Appendix A. Space-Based Infrared Detection of Missiles

1. David Mosher and Raymond Hall, *The Future of Theater Missile Defense* (Washington, D.C.: Congressional Budget Office, 1994), p. 6.

2. Ibid.

3. Ibid.

4. Ibid., p. 7.

5. George P. Sutton, *Rocket Propulsion Elements: An Introduction to the Engineering of Rockets*, 6th ed. (New York: Wiley, 1992), p. 549.

6. Ibid., p. 555.

7. Maj. Michael J. Muolo, AU-18, *Space Handbook: An Analyst's Guide, Volume Two* (Maxwell AFB, Ala.: Air University Press, 1993), pp. 103–16.

8. Office of Technology Assessment, *SDI: Technology, Survivability, and Software* (Washington, D.C.: U.S. Government Printing Office, May 1988), p. 75.

9. Ibid., p. 86.

10. Ibid., p. 86 n. 21.

11. John Lester Miller, *Principles of Infrared Technology: A Practical Guide to the State of the Art* (New York: Van Nostrand Reinhold, 1994), p. 106.

12. Matthew Partan, *Soviet Assessments of U.S. Early Warning Technology Programs* (Cambridge: Center for International Studies, Massachusetts Institute of Technology, 1986), p. 52.

13. Jonathan Kidd and Holly Caldwell, *Defense Support Program: Support to a Changing World*, AIAA 92-1518 (AIAA Space Programs and Technologies Conference, March 24–27, 1992, Huntsville, Alabama).

14. Ibid.

15. Ibid.

Bibliographic Essay

The starting point for any book on the Defense Support Program is the open source literature that deals with DSP or events related to DSP, such as Soviet missile testing or Iraq's use of Scud missiles in the Persian Gulf War. There is only one other book devoted to DSP, Desmond Ball's *A Base for Debate: The US Satellite Station at Nurrungar* (Sydney: Allen & Unwin, Australia, 1987), and only a few articles in space journals on the topic (particularly Dwayne Day's three-part "Top Cover" series in *Space flight* and Nick Watkins's two-part treatment of MIDAS in the *Journal of the British Interplanetary Society* [1997–98]). However, the number of books and articles relevant to DSP is large, and the volume of articles in trade journals such as *Aviation Week & Space Technology, Aerospace Daily,* and *Space News* that have focused on various aspects of the MIDAS and DSP programs is considerable. In addition, unclassified official journals such as the Air Force Space Command's *Guardian* (formerly *Space Traces*) and the Naval Space Command's *Space Tracks* have carried informative articles on various aspects of the DSP program.

American press coverage of key events in the MIDAS and DSP programs was also helpful, and often provided the color lacking from official documents. Australian press coverage as well as *Hansard* accounts of parliamentary discussions were critical in developing an understanding of and describing the Australian debate over DSP and the Nurrungar station. In addition, congressional hearings are a source of valuable information, particularly on the early years of the MIDAS program, before the entire military space program was covered with the cloak of secrecy.

However, there was much more to be discovered about the MIDAS and DSP programs in already declassified documents or documents that could be obtained under the Freedom of Information Act. For as valuable as the published literature is, it also, as would be expected, leaves substantial gaps.

A key set of documents were the annual histories of the organizations involved in the research and development and operation of the MIDAS and DSP satellites—particularly what was known as the Air Force Space Systems Division in the early 1960s (and has undergone numerous name changes since), the Aerospace Defense Command, the Air Force Space Command, the U.S. Space Command, and the detachments operating the DSP ground stations. In addition, the yearly "Air Force in Space" histories that covered the 1960–74 period contained important sections on MIDAS, Program 461, and DSP. Of course, such documents often come to the FOIA requester with significant portions blacked out. Fortunately, multiple organizations are involved in the DSP program, and what one organization considers classified, another may not. The histories also come with extensive references to memos, briefing papers, and other original source material that may then be requested under the FOIA. Often, the ultimate result is the de facto restoration of some of the redacted material as well as information that goes beyond what appeared in the histories.

Of course, in some cases requests for documents resulted in complete denials. The undoubtedly fascinating story of the May 5, 1984, DSP detection of a space object reportedly can be found in a 300-page Air Force Space Command report. However, that organization will neither confirm nor deny the existence of the report.

In other cases the slowness of the FOIA process can mean that requested documents arrive too late to be incorporated in the book. An Air Force Office of Special Investigations study on "Australian Anti-Bases Groups" and a Naval Criminal Investigative Service study, "Australia: A Counterintelligence Assessment," have yet to arrive, as have several documents responsive to an FOIA request for August 1987 cables between the U.S. embassy in Australia and the State Department concerning the publication of Desmond Ball's *A Base for Debate* on the Nurrungar station.

In some cases the documents requested may be of the type that are periodically destroyed. In this category lie three messages, originating with the Strategic Defense Initiative Organization, the Space Systems Division, and the Air Force Space Command, on obtaining Australian government permission to activate an experimental package carried on DSP-14. Requests for the three messages produced a "no records" response.

There did, however, prove to be a large number of documents about DSP or relevant to DSP that could be obtained. Some, such as the CIA's National Intelligence Estimate's on Soviet strategic forces from the 1950s through 1984, were available because of the CIA's initiative in declassifying all or parts of those estimates. In addition, presidential libraries and the declassified papers of key Air Force officials such as Curtis LeMay and Thomas White produced significant memos. Others, such as a DIA study of the targeting of Iraqi short-range ballistic missiles during Desert Storm (*Mobile Short-Range Ballistic Missile Targeting in Operation DESERT STORM*, November 1991) were the products of FOIA requests.

Other Documents

Ad Hoc Group on Ballistic Missile Warning. *Evaluation of Ballistic Missile Early Warning Systems.* July 10, 1962.
Aerojet Electro Systems Inc. *SED Familiarization Manual—DSP SED Sensor—DSP Sensor Evolutionary Development (SED).* September 14, 1979.
AFSPACECOM Regulation 55–55. "Space-Based Sensor (SBS) Large Processing Station (LPS) and European Ground Station (EGS) Tactical Requirements Doctrine (TRD). September 30, 1992.
Air Force. *Space Based Infrared Systems (SBIRS) Pre-Solicitation Conference.* January 13, 1995.
Air Force Space and Missile Systems Center. *History of the Space and Missile Systems*

Center, 1 October 1998–30 September 2001. El Segundo, Calif.: Air Force Space and Missile Systems Center, n.d.

Air Force Space Command. Navy SLOW WALKER Support. May 9, 1983.

Air Force Space Command. Deputy Commander (Woomera) Controversy. April 20, 1990.

Air Force Space Command. *Air Force Space Command Intelligence Threat Assessment: 5th Defense Space Communications Squadron, 1st Space Wing, Woomera, Australia.* November 1, 1991.

Air Force Space Command. *SBIRS Overview Brief Combat Air Force Commander's Conference, 16–17 November 1998.* November 1998.

Air Force Space Systems Division. *DSP Desert Storm Summary Briefing.* Los Angeles AFB: SSD, June 1991.

Aru, Guido W., and Carl T. Lunde. *DSP-II: "Preserving the Air Force's Options."* El Segundo, Calif.: Aerospace Corporation, 1993.

Brown, Harold. Memorandum for the Assistant Secretary of the Air Force (Research and Development), Subject: MIDAS System. June 25, 1962.

Christie, Thomas P., William J. Barlow, Donald L. Ockerman, Elliot S. Parkin, Frank Sevcik, Grant Sharp, Charles M. Waespy, and Howard C. Whetzel. *Desert Storm Scud Campaign.* Alexandria, Va.: Institute for Defense Analysis, 1992.

DDR&E Ad Hoc Group. *Draft Report on MIDAS.* November 1, 1961.

Defense Science Board and Air Force Scientific Advisory Panel. *Acquisition of National Security Space Programs.* May 2003.

General Accounting Office. *Alternate Approaches to Space Tracking and Surveillance System Need to Be Considered,* GAO-03-597. Washington, D.C.: GAO, 2003.

Kellog, William W., and Sidney Passman. *Infrared Techniques Applied to the Detection and Interception of Intercontinental Ballistic Missiles.* Santa Monica, Calif.: RAND Corporation, 1955.

Kidd, Jonathan, and Holly Caldwell. *Defense Support Program: Support to a Changing World.* AIAA 92–1518. AIAA Space Programs and Technologies Conference, March 24–27, 1992, Huntsville, Alabama.

Office of the Secretary of Defense. *Status of the Space Based Infrared System Program.* March 2005.

Schnelzer, Major General Garry. *Air Force Space Sensor Study.* AFPEO/SP. April 12, 1993.

Space-Based IR Sensors/Technical Support Group. October 1993.

Sprague, Major Barkley G. *Evolution of the Missile Defense Alarm System (MIDAS) 1955–1982.* Maxwell AFB, Ala.: Air Command and Air Staff College, Air University, 1985.

Strategic Air Command. DSP-W Flight 9 False Reports. April 21, 1981.

Teague, Col. Roger. Air Force Space and Missile Systems Center. *SBIRS Transformational Capability.* November 30, 2006.

U.S. Space Command. Potential for Tactical Application of Ballistic Missile Launch Notification. June 20, 1990.

U.S. Space Command. *United States Space Command Operations Desert Shield and Desert Storm Assessment.* January 1992.

U.S. Strategic Command, *Theater Event System (TES) Architecture and Operations,* SD 523-2. May 3, 2004.

Ward, Robert, et al. *Desert Storm Reconstruction Report. Vol. 8, C3/Space and Electronic Warfare.* Alexandria, Va.: Center for Naval Analyses, June 1992.

Young, John Jr. "Infrared Augmentation Satellite (IRAS) Acquisition Decision Memorandum." December 1, 2008.

Index

Ababil-100 (Iraqi missile), 249–50
ABM. *See* Antiballistic missile system
ABM Treaty (1972), 85, 185, 202
Above-the-horizon (ATH) sensors, 73, 125, 126, 128
Abrahamson, James, 181, 182
ACDA (Arms Control and Disarmament Agency), 85
Activation Task Force (ATF), 53
ADC (Air Defense Command), 16, 21
Ad Hoc Group on Ballistic Missile Early Warning, 26
  report, 27–28, 31
Advanced Infrared Satellite System (AIRSS), 263–64, 266. *See also* Third Generation Infrared Surveillance Satellite (3GIRS) program
Advanced RADEC (AR), 124, 125, 130, 198, 199
Advanced RADEC Analysis Monitor (ARAM), 66
Advanced Research Projects Agency (ARPA), 11, 12, 14, 20, 44
Advanced Warning System (AWS), 88, 133, 180, 181, 186, 189
AEC (Atomic Energy Commission), 54
AEGIS combat systems, 231
Aegis Sea-Based Terminal Defense System, 273
Aegis SPY-1 radar, 260
Aegis warships, 260, 272, 273
Aeroject-General (company), 9, 33, 37, 43, 46, 71, 72, 89, 104, 105, 172, 190, 225
  DSP sensors, 125
  ground station construction, 50, 54
  Phase I payload, 65

and SBIRS, 223
  Series I payload, 14
  W-37 sensor, 36
Aerospace Corporation, 104, 107, 136, 197–99, 208, 219–20
Aerospace Daily, 56, 217
AFBMD (Air Force Ballistic Missile Division), 17, 18
Afghanistan, 108–9, 160, 229, 241–42
AFL-CIO (American Federation of Labor—Congress of Industrial Organizations), 96
A-4. *See* V-2 rockets
AFSATCOM. *See* Air Force Satellite Communications System
AFSC. *See* Air Force Systems Command
AFSCF. *See* Air Force Satellite Control Facility
AFSPC. *See* Air Force Space Command
AFTAC. *See* Air Force Technical Applications Center
Afterburner returns, 104–5
Agni (Indian missile), 160, 228, 272
Agni-II (Indian missile), 240, 241, 271–72, 272
Agni-III (Indian missile), 270
Ailes, Roger, 189
Airborne Instruments Laboratory, 26
Airborne Laser program (Air Force), 248, 260
Airborne Warning and Control System, 201
Air collision detection, 109
Air Defense Command (ADC), 16, 21
Air Defense Panel, 37
Air Force, 8, 88, 178, 186, 190
  Airborne Laser program, 248, 260

Air Force, *continued*
and ALARM, 213
Ballistic Missile and Space Committee,
    16, 17
and BMEWS, 29–30
Chief of Staff, 12, 21
Constant Source equipment, 167–68
and Desert Storm, 171
DSP and, 261–62
and FEWS, 194, 200–201, 202, 210
Foreign Technology Division (FTD), 71,
    72 (*See also* National Air Intelligence
    Center)
Logistic and Training Command, 53
and MIDAS, 14–15, 16, 20, 21, 23, 24, 26,
    28, 31, 33–34, 37–38, 39, 40
research and development, 11–12, 47, 60
and SBIRS, 218, 237–38, 242–46, 245,
    248, 252–58, 263–68, 269, 273–74
and SBIRS alternative, 257, 258, 263–64,
    266–67, 268
secrecy, 56–57
space programs, 11, 24–25
See also Strategic Air Command
Air Force Ballistic Missile Division
    (AFBMD), 17, 18
Air Force Magazine, 243
Air Force Maui Optical Station (AMOS),
    178
Air Force Satellite Communications System
    (AFSATCOM), 71, 151, 168, 171
Air Force Satellite Control Facility (AFSCF),
    59
Air Force Scientific Advisory Board, 253
Air Force Space and Missile Systems Center,
    200, 258, 267
Air Force Space Command (AFSPC), 105,
    124, 140, 143, 145–46, 149–50, 154,
    155, 190, 191, 242–43, 267, 268
Air Force Space Sensor Study (Schnelzer),
    194–96, 199
Air Force Systems Command (AFSC), 21,
    24, 26, 38, 43, 44, 48, 53, 75, 91, 186.
    See also Space Systems Division
Air Force Tactical Warning Study, 31
Air Force Technical Applications Center
    (AFTAC), 54, 66

Air France Flight 447, 271
Airglow, 128
Air National Guard Base (ANGB), 59
Air Research and Development Command
    (ARDC), 9
AIRSS. *See* Advanced Infrared Satellite
    System (AIRSS); Alternative Infrared
    Satellite System (AIRSS)
ALARM. *See* Alert, Locate, and Report
    Missiles
Alaska, 7, 13, 14, 43, 44, 82, 273
Aldridge, Edward C. Jr., 133, 145, 199–200,
    208, 216, 252
ALERT. *See* Attack and Launch Early
    Reporting to Theater
Alert, Locate, and Report Missiles
    (ALARM), 213–18
    cost, 215
    and DSP and FEWS comparison,
        214(table), 215
Al-Samoud missile (Iraq), 249–50
Alternative Infrared Satellite System
    (AIRSS), 257, 258, 263–65
Amery, Julian, 19
AMOS (Air Force Maui Optical Station),
    178
ANCHOR READY exercise (1990), 154, 155
Andrews, Duane, 184, 186–87, 192, 194, 197
Andrews Air Force Base (Md.), 151
ANGB (Air National Guard Base), 59
Antenna system, 151
Antiaircraft missiles, 71
Antiballistic missile (ABM) system, 47, 63,
    74, 230–33
Antibases campaign (Australia), 143–44, 224
Antisatellite (AS), 101, 108
Anti-ship missiles, 102, 105
"Application of Infrared Tactical
    Surveillance" (Aerojet), 72
AQUACADE (formerly RHYOLITE)
    satellite program, 51n, 68, 139
AR. *See* Advanced RADEC
ARAM. *See* Advanced RADEC Analysis
    Monitor
Arctic launches, 73–74
ARDC. *See* Air Research and Development
    Command

Arens, Moshe, 169
Argentina, 227
ARGUS program, 139, 264
Ariane 5 rocket, 268
Arleigh Burke–class guided missiles, 231
Arms Control and Disarmament Agency
    (ACDA), 85
Army Ordnance's Special Mission V-2, 1
    SECOR satellite, 44
    Special Forces Operational Detachment,
    157
    Strategic Defense Command, 189
    Success Radio receivers, 168
    THAAD system, 230–31
Arniston Missile Test Range (South Africa),
    160
ARPA. *See* Advanced Research Projects
    Agency
Arrow-2 ABM system (Israel), 232
Aru, Guido W., 197–99, 204, 206, 207,
    208–9, 210
Asa, Haim, 168, 169
ASAT (antisatellite), 101, 108
Ascension Island, 167
AS-4 (Soviet antiship missiles), 103–4, 105
ASIO (Australian Security Intelligence
    Organization), 138
ATF (Activation Task Force), 53
ATH. *See* Above-the-horizon sensors
Atkinson, J. H., 16
Atlantic Command, 124
Atlantic Missile Range, 12, 14, 15, 79(table)
Atlantic satellites, 69, 70(fig.), 79(table),
    100(table), 126, 128, 130, 162, 211,
    250(table)
Atlantis space shuttle, 114, 177–78
Atlas (ICBM), 4, 25
    E, 37
    I/Centaur, 206
    infrared detection, 29
    Soviet type, 26
    V, 268
Atlas-Agena rocket, 32, 42
    A, 14
    B, 18, 36
    D, 44
Atomic Energy Commission (AEC), 54

Attack and Launch Early Reporting to
    Theater (ALERT), 211–12, 229, 248
Atwood, Donald J., 185, 186
Auroral interference, 128
Australia, 25, 50–57, 68, 137–48, 155–56,
    218, 224
Australian National University Strategic and
    Defence Studies Centre, 139
Australian Security Intelligence Organization
    (ASIO), 138
Aviation Week, 55
Aviation Week & Space Technology, 76, 252,
    264, 266
AWS. *See* Advanced Warning system

Backfire Tu-22M (Soviet SNA), 102–3,
    117 (fig.), 273n38
"Background clutter," 29
Badger Tu-16/Tu-16C (SLRA), 103
Bahrain, 164–65(table)
Bailey, Joe, 200, 203
Baird-Atomic (company), 9, 18, 22, 25
Baker, James, 148
Balkans, 243
Ball, Desmond, 139, 140–42, 143, 147
Ballistic Missile Defense. *See* Antiballistic
    missile system
Ballistic Missile Defense Organization
    (BMDO), 233, 245, 258
Ballistic Missile Early Warning System
    (BMEWS), 7, 11, 13, 20, 23, 26, 28, 82,
    260, 272
    detection station proposals, 29–30
    effectiveness, 30–31, 58, 64
Ballistic missiles
    detection, 7, 8–9, 19, 214, 237–41,
    242(fig.)
    Iranian, 109
    North Korean, 212
    Soviet, 3, 4, 7
    See also Intercontinental ballistic missile;
        Intermediate-range ballistic missile;
        Medium-range ballistic missile; Short-
        range ballistic missile; Submarine-
        launched ballistic missile
*Base for Debate, A: The US Satellite Station at
    Nurrungar* (Ball), 140

Beale Air Force Base (Calif.), 272
Beazley, Kim, 32, 140, 142, 143, 145
Belgium, 233
Below-the-horizon sensors, 126
Bent-pipe arrangement, 167, 199, 224, 243, 248
Bersinger, Everitt V., 198
Bhangmeters, 79–80, 81
Blanking, 96, 99
BLUE EAGLE aircraft, 152
BMDO (Ballistic Missile Defense Organization), 233
BMEWS. *See* Ballistic Missile Early Warning System
Boatright, Marvin, 37, 45–46, 75
Boeing (company), 225, 245
Bombers, 5, 7, 20
Booen, Michael, 246
Boost-phase interception programs, 260
Boost Surveillance and Tracking System (BSTS), 181, 182–86
    Crew Support Vehicle (CSV), 183–84
    Survivable Ground Terminal, 183
"Borey"/Arctic Wind (Soviet SSBN), 226
Borkowski, Mark, 254
Bosnian Serbs, 229
Bow Mariner (ship) explosion, 251
Boyce, Christopher, 138
Brady, James, 96
Brashears, Mel, 182, 183
Braun, Wernher von, 1
Brezhnev, Leonid, 85
Brilliant Eyes system, 185, 186, 187, 192, 194, 200, 201, 202, 216, 217
Brilliant Pebbles system, 184, 186, 194
Brockway, Daniel, 57–58, 60
Brown, Harold, and MIDAS, 17, 19, 20, 23, 25, 26–27, 30, 31, 33–34, 37, 38–40, 41, 256n37
Brzezinski, Zbigniew, 98
BSTS. *See* Boost Surveillance and Tracking System
Buckley ANGB (Colo.), 59, 60, 66, 68, 106, 115, 126, 142, 154, 167, 199, 238, 248, 273
    BUCKLEY (Navy Space Command Detachment), 220
    MCS, 224

Bulava (Russian missile), 271, 272
Bulganin, Nikolai, 4
Bush, George, 147, 157, 162, 163, 259, 273n
Butts, John L., 101–2

California Institute of Technology, 9, 23
Canada, 7, 233
Canberra Times, 143
CANYON (satellite), 46n
Cape Canaveral (Fla.), 15, 22, 36, 219, 237, 238, 249, 261, 268
Carlson, James, 215
Carter, Jimmy, 85, 86, 98
Center for Strategic Stability and Y2K, 239
Central Intelligence Agency (CIA) satellites, 9, 51n, 138
Central Tactical Processing Element (CTPE), 212
CEP (circular error probable), 158
CGS. *See* United States, Ground Station
Challenger space shuttle, 129
Charyk, Joseph, 21, 31
Chechnya, 241
Cheney, Richard, 174
Cheyenne Mountain Complex (NORAD) (Colo.), 59, 69, 154, 167
Chilton, Kevin, 267
China, 49
    missile tests, 51n, 57, 100, 101, 130, 205, 226–27, 240, 241, 249, 250, 263, 272
    nuclear tests, 45, 70
Chu, David, 186
Churchill, Winston, 3
CIA. *See* Central Intelligence Agency
CINCEUR (European Command, Commander in Chief), 90
CINCPAC (Pacific Command, Commander in Chief), 90
CINCSPACE (Space Command, U.S., Commander in Chief), 232
Circular error probable (CEP), 158
Civil Applications Committee, 242
Civil defense, 35
Clinton, Bill, 231, 232, 234, 239n, 247
Cloud cover, 13, 65, 167, 204, 205, 207, 210, 240
COBRA BRASS space experiments, 224
COBRA DANE radar, 82, 260, 272

Coglitore, Sebastian, 213
Cohen, William, 239
Colmer, Margaret, 143
Colorado, 59, 60, 224
Columbia (space shuttle), 251
Combat Air Force Commander's Conference, 242–43
Combat Air Patrol, 201
COMINT. *See* Communications Intelligence (COMINT) satellites
Command, Control, Communications, and Intelligence (C3I), 81, 87, 89, 95, 142, 193, 231
Commercially Hosted Infrared Payload (CHIRP), 266, 268
Communications Intelligence (COMINT) satellites, 46n, 51n
Communications payload, 22
Communications Subsystem (CSS), 151
"Comprehensive Net Assessment and Military Force Posture Review" (PRM-10), 85–86
Computers, 166, 220
Condor II program (Argentina), 227
Condor II/Vector, 205
Congressional Budget Office, 245
Constant Source equipment, 167–68
Continental Ground Station (Buckley Air Force Base), 238
Conyers, John J., 206
Cooper, Henry, 168
Cooper, Thomas, 89
Cornhusker (Nebr.), 92, 149
CORONA (satellite reconnaissance program), 9, 35
Cosmos 398 (Russian lunar module test vehicle), 233
Cosmos 954 (Soviet ocean reconnaissance satellite), 107
"Countervailing strategy," 86–87
Coyle, Philip, 245, 246
Creswell, Jim, 209–10
Crew Support Vehicles (CSV), 151, 183–84
Cruise missiles, 204
CSS (Communications Subsystem), 151
CSS-5 missile (China), 249
CSS-7 (M-11) missile (China), 241
CSS-X-11 missile (China), 272
CSV. *See* Crew Support Vehicles

C3I. *See* Command, Control, Communications, and Intelligence
CTPE (Central Tactical Processing Element), 212
Cuba, 16
Curcio, J. A., 7, 9

DAB (Defense Acquisition Board), 193
DARPA. *See* Defense Advanced Research Projects Agency
Data Distribution Center (DDC), 69
Data-Processing Module, 89
  Shelter, 90
Davies, Barry, 146
DCI (Director of Central Intelligence), 235
DDC (Data Distribution Center), 69
DDR&E. *See* Director of defense research and engineering
Defense, Department of (DOD), 193
  Five-Year Force Structure report, 43
  and MIDAS, 31
  and SBIRS, 244, 245, 246, 248, 252, 264–65, 265, 267
  and SBIRS alternative, 257–58
  space-based reconnaissance and surveillance program, 9 (*See also* Defense Support Program)
  Under Secretary for Research and Engineering, 89 (*See also* Director of defense research and engineering)
Defense, Office of the Secretary of, 12, 202, 215, 252, 255, 263, 265
Defense Acquisition Board (DAB), 193
Defense Advanced Research Projects Agency (DARPA), 89, 154, 261–62
Defense Contract Management Agency, 265
Defense Intelligence Agency (DIA), 69, 107, 224
Defense News, 193
Defense Satellite Communications System (DSCS), 61, 69, 151, 168
Defense Science Board, 253
Defense Special Missile and Astronautics Center (DEFSMAC), 69
Defense Support Program (DSP), 55, 60, 63, 64–83, 89, 93, 95–96, 98–100, 105–9
  -1 series, 125, 130, 132, 136, 155, 162, 179

Defense Support Program (DSP), *continued*
  -II proposal, 198–99, 200, 201, 202–3,
    204, 205
  5-R, 125, 127
  6-R, 125–26
  Atlantic positioning, 123–24 (*See also*
    Atlantic satellites)
  -Augmentation (A), 74, 199, 201, 203,
    216, 250
  -E, 76, 82, 108, 124, 132, 149, 159
  effectiveness, 69–71, 95–96, 101, 108–9,
    123–24, 126, 129, 159, 173–75, 205,
    207, 214, 229, 249–50
  end of, and successor program, 237, 252,
    256, 258, 261–62, 266–67, 274
  events detected by, 240–41, 249–50,
    270–72
  future, advanced, 180, 208, 210, 212,
    213–14, 235–36 (*See also* Alert, Locate,
    and Report Missiles; Follow-on Early
    Warning Satellite; Space-Based Infrared
    System; Strategic Defense Initiative)
  ground processing, 238, 240, 242(fig.),
    248, 273
  launches, 248, 250(table), 256, 261–62
  launch from Atlantis, 114, 177–78
  launch problems, 237–38, 261–62
  launch time notifications, 129
  Midcourse Defense Segment and, 260
  numbering system for flights and satellites,
    237n
  onboard processing, 240, 241(fig.)
  operational span, 128–29, 203
  in orbit (1980s and 1990s), 126(table),
    128(table), 161–62, 212, 219
  payloads, 130–36
  Program Management Directive (PMD),
    92
  remodeled, 124–25, 130
  Scud warning, 119, 159, 166–67, 171–72,
    174–75, 201, 230
  sensors, 125, 196, 199, 216, 223, 235
  and shipboard system, 190
  status of, in 2012, 273–74
  stereo and mono data processing, 172, 189
  Tactical Warning/Attack Assessment
    Follow-On (TWAAFO), 187

thermal control failures, 211, 214
transition to SBIRS, 223
transition to SBIRS, Increments 1-3,
  224–25
uses of, 241–42, 243, 250–52, 271
-W, 82, 100
Y2K and, 238–39
Defense Systems Acquisition Review Council
  (DSARC), 88, 89
DEFSMAC (Defense Special Missile and
  Astronautics Center), 69
De Leon, Rudy, 245
Delta rockets
  II, 269
  IV, 249, 261, 262
Delta submarines (Soviet), 72, 100, 226
Denmark, 233
Denver Post, 58
Desert Shield, 157, 173, 243
Desert Storm/Gulf War (1991), 148, 163–75,
  190, 209–10, 229, 248
Detection Test Series (DTS), 38, 41, 42, 43
Deutch, John, 204, 217
DEW (Distant Early Warning Line project), 7
DF-4 missile (China), 249
DF-11 (Chinese short-range missile), 250
DF-15/M-9 (Chinese surface-to-surface
  missile), 227, 250
DF-21 (Chinese missile), 250
DF-31 and -41 (Chinese ICBMs), 227, 240,
  241, 249, 250
D'Hage, Adrian, 147
Dhahran (Saudi Arabia) barracks, Scud
  attack, 169–70
DIA. *See* Defense Intelligence Agency
DICBM (intercontinental ballistic missile,
  depressed trajectory), 47, 64
Dicks, Norman, 213
Dietz, Edward, 203, 206, 207, 210
Dineen, Gerald, 81
Director of Central Intelligence (DCI)
  Environmental Task Force, 235
Director of defense research and engineering
  (DDR&E), 12, 16, 17, 19, 25, 26, 39
  Ad Hoc Group on MIDAS, 20, 21, 22, 43
Director of Naval Intelligence (DNI), 101,
  103, 104

Discoverer (experimental satellite), 9, 11,
    16, 18
  radiometric mission (DRM), 38
Distant Early Warning (DEW) Line project, 7
DNI. *See* Director of Naval Intelligence
DOD. *See* Defense, Department of
Dominguez, Michael, 256
Domona nuclear facility (Israel), 169
Dong Feng missiles (China). *See under* DF
Donnelly Flats. *See* Alaska
Dornberger, Walter, 2
Dougherty, William A., 190, 191
Doyle, James H. Jr., 103
Drell, Sidney, 17, 20n
DRM (Discoverer, radiometric mission), 38
Druyun, Darleen, 213, 218
DSARC (Defense Systems Acquisition
    Review Council), 88, 89
DSCS. *See* Defense Satellite
    Communications System
DSP. *See* Defense Support Program
DSP-A. *See* Defense Support Program-
    Augmentation
DSP-II: "Preserving the Air Force's Options"
    (Aru and Lunde), 198
DSP Mobile Ground System, 273
DTS. *See* Detection Test Series
DUAL (stereo processing technique), 69
Dunn, Mike, 238
Duval, R. J., 50

Early warning programs, 7, 11, 12–13, 20,
    28, 50
  false returns, 97–98, 271n13
  U.S.-Russia cooperation in, 239n
Eastern Hemisphere satellites, 78, 79,
    80(fig.), 100, 126, 128(table), 129, 131,
    162, 211, 238, 250(table)
East Germany, 149, 150
ECHO (Navy Space Command
    Detachment), 220, 221
EC-135 LOOKING GLASS, 152
Edwards Air Force Base (Calif.), 15, 178
Eglin Air Force Base, 263
Egypt, 4, 71
Eisenhower, Dwight D., 6, 7
Electromagnetic pulse (EMP), 226

Electronic Industries Association, 168
Elliptical orbit, 72, 73, 217, 218, 223
EMP (electromagnetic pulse), 226
Energy, Department of, 80
Enthoven, Alain, 34
Environmental data, 235
Environmental Task Force, 242
European Command, Commander in Chief
    (CINCEUR), 90
European Ground Station. *See* Kapaun
European satellite, 130, 149
"Evaluation of Ballistic Missile Early Warning
    Systems" (Skifter et al.), 27–28, 31
Evaluation of the MIDAS R&D Program
    (DDR&E Ad Hoc Group), 22, 24
Evans, Gareth, 144, 147–48
Everett, Richard, 204, 205, 209, 213
Explosions detection (DSP), 108, 109, 173,
    210, 234–35

Faga, Martin, 184
Fairhill, Allen, 52
Falcon Air Force Base (Colo.), 191, 212, 220,
    229, 248
FAST WALKER (space object detector),
    107, 242, 271
FEWS. *See* Follow-On Early Warning
    Satellite
Fink, Daniel, 34, 40
Fire detection sensors, 235
Fires, DSP detection of, 242
Flax, Alexander, 34, 38, 40, 41, 44, 49
Flicek, Jerry F., 53
Flood, Daniel, 34
FOBS. *See* Fractional Orbital Bombardment
    System
Follow-On Early Warning Satellite (FEWS),
    189, 192–98, 199–206, 207–9, 210,
    216n
  costs, 201, 202
  criticism of, 197–99, 203, 204, 205, 208
  and DSP and ALARM comparison,
    214(table), 215
  and DSP comparison, 195–200, 201, 203
  effectiveness, 193, 200–201, 207, 208
  funding, 193
  and mobile launchers, 197

Follow-On Early Warning Satellite (FEWS), *continued*
   modifications, 201–2
   nuclear detection function removed, 202
   sensors, 196
   system, 194–95
F-117A stealth fighters, 151
Fort Greeley. *See* Alaska
Fractional Orbital Bombardment System
   (FOBS) (Soviet), 46–48
France, 232, 233
   nuclear tests, 45, 70
Fraser, Malcolm, 138
"Freedom of Entry" custom (Australia), 144
Freeth, Gordon, 53
FROG (Iraqi missile), 249
FROG-7s (surface-to-air missiles), 159
FSS-7 radars, 57, 58
FTD. *See* Air Force, Foreign Technology
   Division
Fubini, Eugene, 34
Fuel. *See* Liquid fuel; Solid fuel
Fylingdales Moor. *See* Yorkshire

Gamma-ray detectors, 66, 226
Gandhi, Rajiv, 160
GAO. *See* General Accounting [Government
   Accountability] Office
Gascoigne, Martin, 156
Gas flare and fire, 76–77
Gates, Robert, 235
GCN (Ground Communications Network),
   89, 155
General Accounting [Government
   Accountability] Office (GAO), 168, 193,
   244, 246, 247, 253, 257–58, 263, 265,
   266, 269, 273, 274
Geodetic satellite (SECOR), 44
GEOGES (Geosynchronous Earth Orbit
   Infrared Gap Filler System), 265
Geostationary orbit, 49, 68, 72, 124, 223
Geosynchronous Earth Orbit Infrared Gap
   Filler System (GEOGES), 265
Geosynchronous orbit, 57, 68, 199, 217
German scientists, 1, 2, 5
Germany, 23. *See also* East Germany;
   Kapaun; Nazi Germany; West Germany

Ghauri-II missile (Pakistan), 240
Gilpatric, Roswell, 30, 31, 33
Global Positioning System (GPS), 199
Goebbels, Joseph, 1
Goldwater, Barry, 98
Gorbachev, Mikhail, 147
Gorodomyla Island (Lake Seliger), 2
Gorshkov, Sergei G., 102
Gorton, John, 50, 51
Government Accountability Office. *See*
   General Accounting [Government
   Accountability] Office
GPS (Global Positioning System), 199
Grand Forks (N.D.), 231
Granger, Todd, 238
Greece, 233
Greeley ANGS (Colo.), 224
Greenhouse gases, 235
Greenland, 7, 14, 272
Gregory, Fred, 177
Groettrup, Helmut, 2
Ground Communications Network (GCN),
   89, 155
Group for the Investigation of Jet Propulsion
   (GRID) (Soviet Union), 2
GRU (Soviet Union, Chief Intelligence
   Directorate), 153
Grumman (company), 89, 183, 223
Guam, 67
Gulf War/Desert Storm (1991), 148, 163–75,
   190, 209–10, 229, 248

Haig, Alexander, Jr., 100
HAIRPIN-3 test (Australia), 156
Hale, John, 191
Hall, Keith, 215, 217
Hall, Roger L., 203–4
Hamel, Michael, 258
HAMMER RICK (communications link),
   169
Hammond, Robert, 189
Hard, Donald, 196–97
Hart, Gary, 98
Hartinger, James V., 81, 105, 124
Hassluck, Paul, 53
Hawaii, 67
Hawk Air Defense System, 233

Hawke, Bob, 139, 140, 141, 144
Hayden, Bill, 52, 139–40
Hayman Forest Fire (Colorado), 251
Heat detection, 7, 8. *See also* Infrared
    emissions detection; Space-Based
    Infrared System (SBIRS)
Hennen, Tom, 178
HERITAGE sensor, 74, 108, 216, 217, 223
Herriges, Darrell, 220
Herschel, William, 252n35
Highly elliptical orbits, 217, 218
High Resolution Multi-Spectral Instrument
    (HRMSI), 217
Hinckley, John, Jr., 96
Hirsch, Fred H., 150
Holloman Air Force Base (N.Mex.), 151,
    153, 154, 155, 273
Holloway, James L., III, 102
Horn, Kenneth, 104
Horner, Charles, 166, 167, 199, 200, 202,
    203, 206, 207, 208, 210, 211, 232, 233
Hostetter, H. C., 221
House Appropriations Committee, 34, 185,
    194, 196, 214, 244
  Defense Subcommittee, 214, 246
House Armed Services Committee, 77, 88,
    216
House Committee on Government
    Operations, 206
HRMSI (High-Resolution Multi-Spectral
    Instrument), 217
Hughes Aircraft Corporation, 43, 223
Hussein, Saddam, 157, 161
Hydrazine, 76
Hydrogen-fluoride chemical laser, 76
Hydrogen warhead, 4

IAC (Intelligence Advisory Committee), 105
IBM company, 89, 126, 155
ICBM. *See* Intercontinental ballistic missile
ICBM Detection System (Lockheed), 13
ICBM Force Modernization Study, 86
IDA. *See* Institute for Defense Analyses
Impact Sensor, 78
Inclined circular orbit, 202
India, 130, 160, 228, 240, 241, 251, 270,
    271–72, 272

Inertial upper stage (IUS), 125, 162, 177
Infrared Augmentation Satellite (IRAS), 265
Infrared emissions detection, 7–8, 9, 12, 13,
    15, 16, 22–23, 25, 27, 29, 36, 57, 65–66,
    67, 71, 99–100, 107, 118, 223, 237–41,
    242(fig.), 252n35
  for ABMs, 74
  electromagnetic spectrum, 239
  multiple sensors, 240
  and short-range missiles, 209
  system, 18, 27, 34, 204, 295n31
  telescope, 65, 118
  thermal control, 125
  See also Defense Support Program; Missile
    Defense Alarm System; Space-Based
    Infrared System (SBIRS)
Institute for Defense Analyses (IDA), 206,
    208, 209
Integrated Test Range (Orissa), 272
Intelligence Advisory Committee (IAC), 5
Intercontinental ballistic missile (ICBM), 15,
    25, 85, 97
  Chinese, 227, 240, 241, 249, 250
  depressed trajectory (DICBM), 47, 64
  detection, 8–9, 19–20, 40, 45, 49, 57, 58,
    63, 64–71, 74, 101, 161, 236, 237
  impact area prediction, 74
  launch routes, 26, 31
  launch warning time, 69, 129
  mobile launchers, 5
  Russian, 240, 241, 249, 250, 270–71
  Soviet, 4, 5, 6–7, 16, 21, 25, 26, 27, 45, 47,
    63–64, 74, 101, 161, 226
  throw weight, 63
  tracking through ultraviolet emissions, 131
  trajectory, 74
  See also Atlas; Titan
Intermediate-range ballistic missile (IRBM),
    19, 25, 228–29
  ABM system, 232
  arms control talks, 123
  detection, 72, 123, 227
  Soviet, 100, 123
  See also Scud missiles
Iran, 109, 123, 228, 240, 241, 249, 250, 270,
    271, 271–72, 273
Iran-Iraq War (1980–1988), 159, 229

Iraq, 109, 119, 120, 130, 157, 158–59, 162, 163, 228. *See also* Desert Storm/Gulf War

Iraq War, second
  DSP in, 249–50
  Patriot missiles in, 261

IRAS (Infrared Augmentation Satellite), 265

IRBM. *See* Intermediate-range ballistic missile

Israel, 80, 130, 273
  and Scud missiles, 158, 163, 164–65(table), 169

Italy, 233

IUS. *See* Inertial upper stage

Jackson, Henry M., 6

James, Daniel, Jr., 78

Jamming, 130

Japan, 12, 13, 232–33, 260

JCS. *See* Joint Chiefs of Staff

JDSCS. *See* Joint Defence Space Communications Station

Jericho-II missile (Israel), 160

JL-2 (Chinese SLBM), 226–27, 241

JOGGER reporting system, 105–6

Johns Hopkins University Applied Physics Laboratory, 204, 223

Johnson, Roy, 11

Joint Chiefs of Staff (JCS), 124, 161
  and arms limitation, 85
  and DSP satellites, 99
  and FOBS, 47
  and MIDAS, 16, 30, 31, 37, 38, 39

Joint Data Exchange Center (Moscow), 239n

Joint Defence Facility (Nurrungar), 144, 147, 148, 238

Joint Defence Space Communications Station (JDSCS) (Nurrungar), 54

Joint Defence Space Research Facility (Pine Gap), 51

Joint Requirements Oversight Council (JROC) (JCS), 193, 199

Joint Special Operations Command, 157

Joint Surveillance Target Attack Radar System (JSTARS), 171

Joint Tactical Ground Station (JTaGS), 189–90, 191–92, 219, 220, 221, 230, 273

JOTS II screen, 169

JROC. *See* Joint Requirements Oversight Council

JSTARS (Joint Surveillance Target Attack Radar System), 171

JTaGS. *See* Joint Tactical Ground Station

JUMPSEAT satellites, 59, 74

Kamchatka (Soviet Union), 45, 100, 101, 241, 250, 270

Kaminski, Paul, 219

Kapaun (Germany), 128, 130, 149–50, 167, 224, 238, 248

Kapustin Yar (Soviet test range), 3, 5, 45, 250, 271

Karman, Theodore von, 9

Kazakhstan test site. *See* Tyuratam

Kehler, Robert C., 262, 267, 268

Kelley, David, 207, 209

Kellogg, William, 7, 8, 9

Kennedy, John F., 97

Kerr, John, 138

KH-9 and -11 (imagery satellites), 108, 127

Khrushchev, Nikita, 4, 16

Kh-22/Kitchen (Soviet AS missile), 104

Kidd, John R., 134, 198, 200

King, William G., Jr., 9, 109

Kirkbride (England), 19

Kissinger, Henry, 137

Knapp, James, 155

"Kneecap." *See* National Emergency Airborne Command Post (NEACP)

Knopow, Joseph, 8

Kodiak. *See* Alaska

Korla Missile Test Complex (China), 272

Korolev, Sergei, 1–2, 3, 4

Kourou (French Guiana), 268

Krieg, Kenneth, 257

Kuter, Laurence S., 16

Kutyna, Donald, 194n

Kuwait, 157, 161. *See also* Desert Storm/Gulf War

LADS (Low-Altitude Demonstration Satellite), 225

Lagos (Nigeria), munitions explosion in, 250

Laird, Melvin, 57n, 63, 73

Lake Seliger (Soviet Union), 2
Lane, Laurence W., Jr., 144
Langley, Samuel Pierpont, 252n35
Lapin, Ellis, 71
Laser beacons, 220
Laser blinding, 76, 77, 118
Laser Crosslink System (LCS), 131–36, 198
Lasers, 76–77, 131–36, 190n
Latham, Donald C., 93, 133
Launch on warning, 256n37
Lawrence Livermore National Laboratory, 17
Lead sulfide detectors, 7
Lee, Andrew Daulton, 138
Lee, Robert M., 21
Lehesten (Poland), 1
LeMay, Curtis, 21, 23, 30
Lewis, Jerry, 246
Libya, 228, 229
Limited Reserve Satellites (LRSs), 78, 82,
    100, 126
Lincoln Laboratory (MIT), 204
Linked Operations–Intelligence Center,
    Europe (LOCE), 233
Liquid fuel, 25, 29
    Soviet, 45
Liquid oxygen/kerosene (LOX-RP) fuel, 29,
    45
Lockheed (company), 46
    infrared detection system, 13
    -Martin, 225, 244, 254, 256, 264, 265,
        267, 268, 274
    Missile Systems Division, 8, 9
Lockheed Missile and Space Company, 43,
    182, 184, 185, 223
Long-range aircraft, 5, 102, 103
LOOKING GLASS (EC-135), 152
Lord, Lance, 256
Los Alamos National Laboratory, 179
Low-Altitude Demonstration Satellite
    (LADS), 225
Low-earth orbit, 217, 218
Lower-Tier. See Theater Ballistic Missile
    Defense System
Lowery, Craig, 237
Lowry Air Force Base (Colo.), 60–61
LOX-RP. See Liquid oxygen/kerosene fuel
LRSs. See Limited Reserve Satellites

Lunar module test vehicle (Russian), 233
Lunde, Carl T., 197–99, 204
Lynn, William J., 202

Magnetospheric Plasma Analyzer (MPA),
    179
Mahon, George, 35
Malenkov, Georgi, 2
Malyshev, Vyacheslav A., 3
Mangold, Sanford, 206, 207, 210
Marshall Islands, 247, 261
Maryland, 151
Mas, Newton, 174
MASINT (Measurement and Signature
    Intelligence), 224
Massachusetts Institute of Technology
    (MIT), 17, 58, 204
McComas, David, 179
McDonnell-Douglas Corporation, 131, 133,
    135
McElroy, Neil H., 7
McLucas, John, 54, 55, 68
McMillan, Brockway, 24, 25, 26, 27, 30
McNamara, Robert S.
    and MIDAS, 18, 19, 20, 21, 26, 27, 28, 29,
        30, 31, 37, 38, 39, 42, 47, 256n37
MCS (Mission Control Station), 224, 225,
    238, 248, 273
MCT (Mobile Communications Terminal),
    151, 152
MDA. See Missile Defense Agency
MDM (Mission Data Message), 199
Measurement and Signature Intelligence
    (MASINT), 224
Medium Extended Area Defense System, 233
Medium-range ballistic missile (MRBM), 19
    detection, 39, 41, 46, 70, 161
    Soviet, 70
Medvedev, Dmitry, 239n, 271
M88-1 Battleview and Maritime experiment,
    178
MEL (mobile erector launcher), 158
MENS ("Mission Element Need Statement
    for Improved Missile Warning and
    Attack Assessment"), 88
Meteoroids, 234–35
meteors, DSP detection of, 242, 250–251

MGSU (Mobile Ground Support Units),
    154
MGT. *See* Mobile Ground Terminal
MIDAS. *See* Missile Defense Alarm System
Midcourse Defense Segment, 260
Midcourse Space Experiment (MSX),
    223–24, 234
Midget. *See* Mobile Ground Terminal
Mikoyan, Anastas, 4
*Military Strategy and Force Posture Review:
    Final Report* (PRM-10), 86
*Military Uses of Tactical Warning of an Attack
    on CONUS* (Rand), 30–31
Millikan, Clark, 23
MILSTAR communications system, 199,
    210, 217
Miniature Sensor Technology Integration
    (MSTI) sensors, 217, 224
Minuteman (ICBM), 22, 36, 247, 269
Misawa Air Base (Japan), 273
Missile Attack Conference, 98
Missile Defense Agency (MDA), 258, 269
Missile Defense Alarm System (Program
    239A, Program 461) (MIDAS) satellites,
    11, 12, 13, 14, 33, 111, 112, 236
  2 (Series I, Flight 2), 14–15
  3 (Series II), 18
  4 (Series II, Flight 2), 21–22
  5, 25
  6, 32
  7 (Series III, Flight 2) (1963), 36–37
  8, 37
  9 (Series III, Flight 4), 37
  costs and funding, 19–20, 21, 24, 28, 34,
    38, 39, 41, 42–43
  development, 16–18, 20, 21, 24, 25–27,
    33, 38, 83
  effectiveness, 14–15, 17, 18, 20–21, 22–24,
    27, 28, 31–32, 34–35, 36, 38–48, 57,
    256n37
  launch listings, 250(table)
  model, 111
  success, 36–37
  and vulnerability to Soviet
    countermeasures, 122–23, 254n26
Missile silos, 20, 28
  identification, 125

Mission Control Station (MCS), 224, 225,
    238, 248, 273
Mission Data Message (MDM), 199
"Mission Element Need Statement (MENS)
    for Improved Missile Warning and
    Attack Assessment" (Air Force), 88
Mitex, 262
MITRE Corporation, 204
M-9 and -11 (Chinese missiles), 205, 227,
    241, 250
Mobile Communications Terminal (MCT),
    151, 152
Mobile erector launcher (MEL), 158
Mobile Ground Support Units (MGSUs),
    154
Mobile Ground System, 273
Mobile Ground Terminal/Midget (MGT),
    91, 92–93, 151, 152, 154, 155–56, 273
Mobile launchers, 5
Mobile Multi-Mission Processors, 273
Mobile terminals, 133
Molniya orbit, 49, 57, 74
Monahan, George L., Jr., 184
Money, Art, 246
Moorman, Thomas S., Jr., 174, 194, 211,
    213, 214, 215
Mosaic Sensor Program (MSP), 88, 89
MOS/PIM (Multi-Orbit, Performance
    Improvement Spacecraft), 76
MPA (Magnetospheric Plasma Analyzer),
    179
MRBM. *See* Medium-range ballistic missile
MSP (Mosaic Sensor Program), 88, 89
MSTI (Miniature Sensor Technology
    Integration) sensors, 217, 224
MSX. *See* Midcourse Space Experiment
Multi-Mission Mobile Processor, 219
Multi-Orbit, Performance Improvement
    Spacecraft (MOS/PIM), 76
Multi-Purpose Tanker, 184
Multispectral data, 224
Mururoa (French Pacific test site), 70
MX program, 147

National Air and Space Intelligence Center,
    274
National Air Intelligence Center, 107

National Command Authorities (NCA), 87
National Emergency Airborne Command
    Post (NEACP), 90, 151
National Intelligence Estimate (NIE), 5, 6,
    16, 47, 102
National Military Command Center, 69
National Missile Defense Independent
    Review Team, 247
National missile defense system, 231–32,
    300n33
  Bush administration changes in, 259–60
  criticisms of, 247
  issues in, 247–48
  restructuring of, 260
  status of, in 2011, 272–73
  testing of, 247
National Reconnaissance Office (NRO), 51n,
    59, 73, 107, 127, 216, 217, 220, 224,
    252, 256, 262, 265
National Security Agency (NSA), 69, 153
National Security Council (NSC), 137
National Security Decision Directive 12
    (NSDD-12) and 13(NSDD-13), 95
National Security Decision Memorandum
    242 (NSDM-42), 86, 87
National Security Directive 54, 162
National Test Facility (Colorado Springs), 199
National Times (Australia), 139
NATO. See North Atlantic Treaty
    Organization
Naval Collection (magazine), 102
Naval Research Laboratory, 7, 80
Navy, 178
  and Backfire detection, 105–6
  deployment, 102
  and MIDAS, 36
  missile defense systems, 231, 247–48
  satellites, 217
  SEAL Team 6, 157
  Space and Naval Warfare Systems
    Command, 189
  Tactical Ground Systems (TaGS), 189
  Tactical Receive Equipment (TRE), 167,
    169
  Theater Wide program, 247–48, 260
  Wide Area missile defense program, 259
Navy Area Defense, 248

Nazi Germany, 1–2
NCA (National Command Authorities), 87
NEACP (National Emergency Airborne
    Command Post), 90, 151
Nebraska, 92, 149
Netanyahu, Benjamin, 232
Neutron detector, 66
New Hampshire, 67
New Mexico, 151
New York Times, 15, 22, 64, 65, 66, 240
NIE. See National Intelligence Estimate
NII-88 (Scientific Research Institute 88), 2, 3
Nike-Javelin rockets, 24–25
Nixon, Richard M., 85
Nockels, Jim, 145, 146
Nodong-1 missiles (North Korea), 212, 228
North American Aerospace Defense
    Command (NORAD), 13, 16, 20, 69,
    97, 100, 154, 231
  commander in chief (CINCNORAD), 90,
    124
  See also Cheyenne Mountain Complex
North Atlantic Treaty Organization (NATO),
    103, 150, 233
North Dakota, 231
North Korea, 212, 214, 228, 232, 240, 250,
    251, 260, 271, 273
Northrop-Grumman (company), 245, 256,
    258, 261, 264, 267, 269
Norway, 13, 233
NRO. See National Reconnaissance Office
NSA (National Security Agency), 69, 153
NSC (National Security Council), 137
NSDD (National Security Decision
    Directive)-12 and -13, 95
NSDM (National Security Decision
    Memorandum)-242, 86, 87
NTPR (Nuclear Targeting Policy Review), 86
Nuclear attack defense, 86–88, 89–93
Nuclear ballistic missile submarine (SSBN)
    (Soviet), 72, 226
Nuclear detonation (NUDET)
  detection, 40, 44, 45, 79–81, 124, 125,
    130, 142n
  by DSP satellites, 66, 70–71, 161
  and FEWS, 202
  GPS endo-atmospheric, 199

Nuclear strike orders card, 97
Nuclear Targeting Policy Review (NTPR), 86
Nuclear warfighting, 205
Nuclear warheads, 3, 5, 86
Nuclear weapons, 147
"Nuclear Weapons Employment Policy"
    (NSDD-13), 95
"Nuclear Weapons Employment Policy" (PD/
    NSC-59), 87
NUDET. *See* Nuclear detonation detection
Nunn-McCurdy reviews, 252, 256–57
Nurrungar (Australia), 54–55, 66, 67, 68,
    105, 116, 137, 138, 139, 140, 142–43,
    144, 147, 148, 154, 156, 159, 248,
    280n31
    outages, 81
Nurrungar News, 143

Obama, Barack, 239n, 273n
Office of Defense Mobilization, 6
Office of Special Investigations (OSI), Air
    Force, 146
Office of Systems Analysis, 34
Offutt Air Force Base (Nebr.), 92
O'Grady, Sean, 243
OGS (Overseas Ground Station), 50
Omid spacecraft (Iran), 271
Omni (magazine), 106–7
Onyx mainframe computers, 220
Operationally Responsive Space Office, 267
Operations Security Workshop (1985),
    152–53
Optical sensors, 226
OSI. *See* Office of Special Investigations
Otis Air Force Base (Mass.), 82
Outages, 81, 197
Overseas Ground Station (OGS), 50, 238
Over-the-horizon radar, 28, 31, 58, 126
    Commander in Chief (CINCPAC), 90
    Pacific Command, 152

Pacific Missile Range, 12, 18, 21, 25, 37,
    79(table)
Pacific satellites, 69, 70(fig.), 78, 79, 80(fig.),
    100(table), 126, 128, 129, 131, 161,
    162(table), 211, 238, 250(table)
Packard, David, 73

PAC-2 and -3 missiles, 230, 248, 260, 261,
    273
Paige, Emmett, Jr., 214
Pakistan, 5, 27, 228, 240, 250
Palmikhim (Israel), 160
Panel on Reducing Risk in BMD Flight Test
    Programs, 247
Panofsky, Wolfgang K. H., 17, 25, 26
Partial Test Ban Treaty (1963), 70
PASS (Phased Array Subsystem), 151
Passman, Sidney, 7, 8, 9
Patriot air defense missiles, 120, 165–69,
    212, 221, 250
    effectiveness, 168–69, 260–61, 286n47
    notice time, 168
    PAC-2, 230, 260
    PAC-3, 230, 248, 260, 261, 273
    upgrade, 230
Paulikas, George, 208
PAVE PAWS SLBM radar, 81–82
Payton, Gary, 257, 258
PD-18 (Presidential Directive 18), 86
PD/NSC-59 (Presidential Directive/National
    Security Council), 87
Peenemunde (Germany), 1, 2
Penkovskiy, Oleg, 21
People's Republic of China. *See* China
Perry, William J., 92
Peters, Whitten, 245
Peterson Air Force Base (Colo.), 155
Phased Array Subsystem (PASS), 151
Philco-Ford (company), 54
Philippines, 27
Pine Gap (Australia), 51, 53, 68, 116, 138,
    139, 224
Pinetree Line of Radars (Canada), 7
PIONEER SS-20 (Soviet IRBM), 123
"Planning the Employment of Nuclear
    Weapons" (NSDM-242), 86, 87
Plasmas, 179
Plesetsk (Soviet Union), 45, 241, 270
PMD (Defense Support Program, Program
    Management Directive), 92
Point Arguello (Calif.), 18
Poland, 1
Polaris (SLBM), 22, 36
Polaris submarine, 20

Polar orbit, 12, 21, 44
POLO HAT exercise (1986), 154
Porter, Frederick S., Jr., 63
Powell, Colin, 211
Power, Thomas S., 12, 13
PRC (People's Republic of China). *See* China
Precision Tracking Space System (PTSS), 269–70
"Preliminary Analysis of Project Hot Spot IR Signals" (Air Force), 72
Presidential Directive 18 (PD-18), 86
President's Scientific Advisory Committee (PSAC)
    Advanced ICBM Panel, 25
    Early Warning Panel, 12–13
    and MIDAS, 17
Prince Edward Island, 81
PRM-10 (Presidential Review Memorandum), 85–86
PROGNOZ (Russian satellites), 233
Program 239A/Program 461. *See* Missile Defense Alarm System
Project HOT SPOT, 71–72
Project WEST FORD, 22
PSAC. *See* President's Scientific Advisory Committee
PTSS. *See* Precision Tracking Space System
Puerto Rico, 154
PYRAMIDER (CIA satellite), 138

Qadhafi, Moammar, 272
Quinn, Thomas, 206, 214
Quirk, Jeffrey, 208–9

Radars, 7, 11, 13, 27–28, 31–32, 57, 58, 81–82, 166, 171, 231
RADEC. *See* Radiation detection
Radiance wavelengths, 16–17, 210
RADIANT AGATE (navy satellite), 217
RADIANT IVORY, 190
Radiation detection (RADEC) I and II, 66, 118, 124, 125, 198, 199
Radio cross-link, 133, 134
Radiometric flights (RM-1 and RM-2), 16, 18, 41
RAMOS (Russian-American Observation Satellite), 233

Rand, Ron, 129
RAND Corporation, 7–8, 30
Randolph, Bernard F., 134
Ray, Robert, 155, 174
Raytheon (company), 225, 245, 258, 266, . 268
Reagan, Ronald, 95, 96–97, 141, 180
Rearden, Timothy J., Jr., 36
Redundant warning capacity, 256n37
Reese, Robert S., 248–49
Relay Ground Station–Europe (Menwith Hill, United Kingdom), 238, 248
Relay Ground Station–Pacific (Pine Gap), 238, 248
Report to the Defense and Intelligence Committees of the Congress of the United States on the Status of the Space Based Infrared System Program, 255
Required Operational Capability (ROC), 91
Requirements Review Group (RRG), 91
Research and development, 11–12
Research Test Series (RTS), 38, 40, 41, 42, 43–44, 45, 46
"Responding to Iraqi Aggression in the Gulf" (National Security Directive 54), 162–63
RHYOLITE. *See* AQUACADE
Rice, Donald, 129, 209
Rice, W. R., 53
Richards, Bob, 71
Riepe, Quentin, 9, 11
Rinehart, John, 6
RM-1/RM-2. *See* Radiometric flights
ROC (Simplified Processing Station, Required Operational Capability), 91
Rocketry, 1–2, 3, 24–25. *See also* Ballistic missiles
Rodong missile (North Korea), 240
R-1–3 (Soviet V-2 variant), 2, 3
Ross, Frank, 60
RRG (Requirements Review Group), 91
RS-24 Yars (Russian ICBM), 270, 271
R-7 (Soviet ICBM), 4, 5, 6–7, 25
RSM-54 Sineva (Russian hybrid missile), 270
RTS. *See* Research Test Series
Ruina, Jack P., 20, 21, 22, 23, 24, 256n37
Rumsfeld, Donald H., 77
Runco, Mario, Jr., 178

Rush, Kenneth, 73
Russell, Richard, 6
Russia, 226, 233
    and Joint Data Exchange Center, 239n
    missile tests, 240, 241, 249, 250, 270–71,
        272
    TU-154 airliner explosion, 242
    Y2K issues and, 238–39
Russian-American Observation Satellite
    (RAMOS), 233

SABRS (Space and Atmospheric Burst
    Responding System), 226
SAC. See Strategic Air Command
Sadler, Steven, 207, 209
SA-5 GAMMON (S-200) missile, 242
SAIC. See Science Applications International
    Corporation
SALT II. See Strategic Arms Limitation
    Talks, II
SAMSO. See Space and Missile System
    Organization
SA-N-1 and -3 (Soviet naval missiles), 108
Sanderson, J. A., 7, 9
Sandia National Laboratory, 54, 66, 125,
    224, 264
Sandia RADEC Analysis Monitor (SRAM),
    66
Sanum Dong Research and Development
    Facility (North Korea), 228
Saratoga (U.S. warship), 190
SATAN (Soviet ICBM), 74, 271
Satellite Communications Module, 89, 90
Satellite Control Facility Tracking Station
    (Alaska), 43
Satellite Early Warning System (SEWS), 57n
Satellite Mission Survivability Study
    (Pentagon), 141–42
Satellite Operations Center (SOC), 66
Satellites, 6, 8–9, 38, 44, 58
    altitude sensor, 99, 113
    blanking, 96, 99
    command and control, 68
    communications, 51n, 138
    communications (Soviet), 49
    coverage, overlapping, 69, 70(fig.)
    and espionage, 138–39

First (Sputnik), 6
ground stations, 50–57, 58–61, 79, 89,
    123, 128, 132, 137–56, 189–90, 218,
    224
high-altitude, 201, 223
laser blinding, 76
launch listing, 250(table)
low-altitude, 201, 223, 225
LRSs, 78, 79(table), 82, 126
next-generation programs, 242–43
operational time, 42, 74–75, 79, 130
orbits, 49, 57, 72, 217
photographic reconnaissance, 8, 35, 59
positions, 100
reentry, 107–8
size, 130
and solar outages, 197
Soviet, 107, 153
thermal control failure, 211
VELA, 79–81
view of earth, 79, 80(illustration)
See also Defense Support Program; Follow-
    On Early Warning Satellite; Missile
    Defense Alarm System satellites
Satish Dhawan Space Center (India), 251
SA-2 and -3 (antiaircraft missiles), 71, 229
Saudia Arabia, 120, 157, 163, 164–65(table),
    169–70
SAVAGE (Soviet ICBM), 45
SBEWS (Space-Based Early Warning
    System), 206
SBIRS. See Space-Based Infrared System
SC-19 (Chinese missile), 272
Schmidt infrared telescope, 65
Schnelzer, Garry, 194–96, 197, 218
Schriever, Bernard A., 11, 21, 24, 28–29, 30,
    220, 256n37
Schriever Air Force Base (formerly Falcon)
    (Colo.), 220, 269, 273
Schwartzkopf, Norman, 171
Science Applications International
    Corporation (SAIC), 266, 268
Scientific Advisory Committee (Office of
    Defense Mobilization), 6
Scientific Research Institute-88 (NII-88)
    (Soviet Union), 2, 3
Scud missiles, 72, 119, 158–61, 197, 229

in Afghanistan war, 160
attacks on, 171
detection, 170, 171–72, 174–75, 179, 197, 205, 223
interception of, 273
Iranian, 228–29, 270
Iraqi use, 158–60, 163
launchers, 158, 171, 173
launchers, mobile, 158, 163–67, 169–71, 173, 228
launches, 164–65(table), 172–73, 230
Libyan, 272
North Korean, 212, 214
Russian, 241
Soviet, 158
Syrian, 250
See also Patriot air defense system
SDI. See Strategic Defense Initiative
SDIO. See Strategic Defense Initiative Organization
Sea-Based Midcourse missile defense component, 260
Sea Power of the State (Gorshkov), 102
Search and rescue, DSP satellites in, 243
SECOR (Army geodetic satellite), 44
Secure Reserve Force Study, 86
Security Resources Panel subcommittee (Scientific Advisory Committee), 6
SED. See Sensor Evolutionary Development
Sejil and Sejil-2 missiles (Iran), 270, 271, 272
Semyorka/SS-6 (Soviet ICBM), 4–5, 6–7, 25
Senate Appropriations Committee, 133
Senate Armed Services Committee, 6, 18
Senate Committee on Aeronautical and Space Sciences, 11
Sensor Evolutionary Development (SED), 113, 125, 130, 198
SENTRY (radio-return reconnaissance satellite project), 9, 13, 14
Sergeyev, Igor, 239
SES Americom, 266, 268
Severomorsk (Soviet Union), 108
SEWS (Satellite Early Warning System), 57n
Shahab-3 missile (Iran), 240, 241, 249, 250, 270
Shaheen-2 missile (Pakistan), 250
Shapiro, Sumner, 103

Sharaga, 2
Shemya Island (Alaska), 82, 260
Ship-to-ship missiles, 108
Shoemaker, Eugene M., 235
Shomron, Dan, 168, 169
Short-range attack missile (SRAM), 147
Short-range ballistic missile (SRBM), 209
South Africa, 160, 161, 209
Shuangchengzi Space and Missile Center (China), 272
Simplified Processing Station (SPS), 89–90, 91–92, 149
Sineva (Russian SLBM), 271, 272
Single Integrated Operational Plan (SIOP), 87
SIOP (Single Integrated Operational Plan), 87
Sixty Minutes (TV program), 210
Skaggs Island (Calif.), 167
Skifter, H. R., 26, 27, 31
Skylab (space station), 108
SLBM. See Submarine-launched ballistic missile
SLOW WALKER Reporting System (SWRS), 105, 106, 117 (caption), 156, 212, 242
SLRA. See Soviet Long-Range Aviation
SM-3 (interceptor missiles), 273
Smith, F. H., 23
Smith, Jeff, 268
Smithsonian Astrophysical Observatory, 6
"Smoky Joe" (U-2D), 18
SMTS (Space and Missile Tracking System), 219
SNA (Soviet Naval Aviation), 102, 103
SNIE (Special National Intelligence Estimate), 6–7
SOC (Satellite Operations Center), 66
Sodium flares, 15
Solid fuel, 22, 36, 38, 45
SOPA (Synchronous Orbit Particle Analyzer), 179
Soper, Gordon, 192
SOR (Specific Operational Requirement), 40
Sound Surveillance System (SOSUS) (navy), 105n
South Africa, 80, 130, 160–61, 209

South Korea, 212, 221, 260, 273
Soviet Capabilities and Probable Programs in the Guided Missile Field (NIE), 5
Soviet Capabilities for Strategic Attack (NIE), 47
Soviet Long-Range Aviation (SLRA), 103
Soviet Naval Aviation (SNA), 102, 103
Soviet Union, 97, 147
  and Afghanistan, 160
  ballistic missiles, 3, 4–5, 6–7, 46–48, 51n, 57, 72, 100–101, 108, 130
  ballistic missile time to reach United States, 7
  bombers, 7
  Chief Intelligence Directorate (GRU), 153
  and Cuba, 16
  and Egypt, 4
  electronic intelligence ships, 54
  and German rocket program and scientists, 1, 2, 3
  intelligence agents, 153
  naval weapons depot explosion, 108
  Navy, 72, 102, 103
  propellant plant explosion, 109
  satellites, 6, 49
  space program, 6, 108
  special operations forces (SPETsNAZ), 153
  U.S. targets in, 86
  See also Russia
Space and Atmospheric Burst Reporting System (SABRS), 226
Space and Atmospheric Burst Reporting System (SABRS) Validation Experiment (SAVE), 261
Space and Missile System Organization (SAMSO) (Air Force), 51, 55, 58, 68, 71, 72, 89
Space and Missile Systems Center Infrared Space Systems Directorate, 268
Space and Missile Tracking System (SMTS), 219
Space-Based Early Warning System (SBEWS), 206
Space-Based Infrared System (SBIRS), 204, 218–19, 223, 224–26, 237
  alternative system, development of, 257, 258, 263–64, 266–67, 268

certification of, 252–53, 265–66, 266n, 267
component deliveries, 255, 256, 263, 267
contractors for, 245, 258–59
data sharing, 233
effectiveness, 229, 273–74
events detected by, 270–71
GEO-1, 263, 265, 268, 273, 273–74
GEO-2, 274
GEO-3, 267, 274
GEO-4, 274
GEO-5, 274
GEO-6, 274
ground processing, 273, 274
HEO-1, 262–63, 265–66
HEO-2, 265
HEO-3, 267
High, 121, 218, 225, 237, 243–44, 252–58, 263–68
improvements over DSP, 258
launches and launch plans, 257, 259, 262–63, 266, 268, 274
Low, 121, 122, 218, 219, 225–26, 231–32, 243, 244–46, 253, 258–59 (See also Space Tracking and Surveillance System)
Midcourse Defense Segment and, 260
Mission Control Station (MCS), 224, 225, 238, 248
missions of, 258
planning for, 242–43, 245
problems and cost overruns in, 243–46, 252–58, 263–68
sensors, 236
status of, in 2012, 273–74
testing of, 254
Space Command, U.S., 161, 162, 191
  Center (SPACE), 167
  Commander in Chief (CINCSPACE), 232
Space News, 244, 245, 252, 254, 256, 264, 266
Space program. See under Soviet Union; United States
Space shuttle, 114, 125, 129, 237, 251
Space Systems Division (SSD) (AFSC), 20, 23, 24, 33, 43, 134, 135, 170, 185
Space Technology Laboratories, 43
Space Tracking and Surveillance System (STSS), 259, 269–70
  Advanced Technology Risk Reduction spacecraft, 269

Space Tracks, 243

Space Warning Squadrons (SWSs), 220, 248, 273

Spain, 233

SPANKER (Soviet ICBM), 74

Special National Intelligence Estimate (SNIE), 6–7

Specific Operational Requirement (SOR) (Air Force) Number 209, 40

Spectrum, 246

Spectrum-Astro (company), 225–26, 245, 258, 259

SPETsNAZ (Soviet Union, special operations forces), 153

Sphera (Russian rocket), 205n

SPIRIT III (infrared payload), 223

SPRINT ABM, 74

SPS. *See* Simplified Processing Station

Sputnik (Soviet satellite), 6

SPY-1 radar, 272

SRAM.

See Sandia RADEC Analysis Monitor; Short-range attack missile

SRBM. *See* Short-range ballistic missile

SS-4 (Soviet MRBM), 108

SS-6 SAPWOOD/Semyorka, 4–5, 6–7, 25

SS-7 (Soviet missile), 70

SS-9 SCARP (Soviet ICBM), 47, 63
   Mod 3, 64

SS-11 SEGO (Soviet ICBM), 45, 63, 64, 70, 100

SS-13 SAVAGE (Soviet ICBM), 45, 63, 64

SS-17 SPANKER (Soviet ICBM), 74

SS-18 SATAN (Soviet/Russian ICBM), 74, 271

SS-19 STILLETO (Soviet/Russian ICBM), 74, 249, 250

SS-20 PIONEER (Soviet IRBM), 123

SS-21 (Soviet/Russian short-range missile), 203–4, 209, 241

SS-22 (Soviet IRBM), 100

SS-24 (Soviet ICBM), 109

SS-25 Topol (Soviet/Russian ICBM), 226, 241, 270, 271

SS-27 Topol-M (Soviet/Russian ICBM), 70, 226, 240, 241, 250, 270–71

SSBN (nuclear ballistic missile submarine), 72, 226

SSD. *See* Space Systems Division

SS-N-3 (Soviet missile), 108

SS-N-5 SERB (Soviet SLBM), 58

SS-N-6 (Soviet SLBM), 45, 58, 67, 69, 167

SS-N-8 (Soviet SLBM), 70, 72, 78, 125, 226

SS-N-12 (Soviet SLBM), 108

SS-N-20 (Soviet SLBM), 101, 226

SS-N-22 (Soviet SLBM), 108

SS-X-6. *See* SS-9

St. Louis Post-Dispatch, 134

Stalin, Josef, 3

Stanford University, 17

Star sensors, 99, 118, 124, 130

Star tracking telescope (DSP), 106

Star Wars. *See* Strategic Defense Initiative

Stefula, Joseph, 106, 107

Stereo processing, 172, 189

Stevens, Ted, 149

Stewart, Robert, 190

STILLETO (Soviet ICBM), 74, 249, 250

Strategic Air Command (SAC), 6, 7, 12, 13–14, 20, 67, 69, 98, 124, 152
   headquarters, 92

Strategic Arms Limitation Talks (SALT) (1969), 85
   II, 48

Strategic Defense Initiative (SDI), 131, 133, 180, 186

Strategic Defense Initiative Organization (SDIO), 141, 168, 181, 182, 184, 194

"Strategic Forces Modernization Program" (NSDD-12), 95

Strategic Rocket Forces (Russia), 226

STRELA satellites (Soviet), 153

Struck, Ed, 220

STSS. *See* Space Tracking and Surveillance System

S-200 (SA-5 GAMMON) missile, 242

Submarine-launched ballistic missile (SLBM), 19, 85
   Chinese, 101, 226–27, 241
   detection, 26, 32, 39, 41, 45, 46, 57, 58, 60, 68, 69, 72–73, 74, 81, 97, 161, 199
   launch warning time, 129
   Russian, 241, 270, 271, 272
   Soviet, 45, 57, 58, 60, 68, 70, 72, 100, 101, 161, 226
   See also Polaris

Submarines, 20
  detection, 105, 125
  Soviet, 45, 58, 72, 100, 105, 226
Subsystem G (WS-117L), 9, 11
Success Radio receivers, 168
Suitable Piece of Real Estate, A: American Installations in Australia (Ball), 139
Sunnydale (Calif.) (AFSCF), 59, 60
Sun's reflection off ice and snow, 96
Surface-to-air missiles, 108
Surface-to-surface missiles, 159, 227
Survivable/Endurable Mobile Ground Terminal program, 273
SWRS. *See* SLOW WALKER Reporting System
SWSs (Space Warning Squadrons), 220, 248, 273
Symington, Stuart, 6
Synchronous Orbit Particle Analyzer (SOPA), 179
Syria, 71, 250

TACDAR (Tactical Detection and Reporting), 220
Tactical Detection and Reporting (TACDAR), 220
Tactical Event Reporting System (TERS), 106, 167
"Tactical Exploitation of DSP" (Space Command), 191
Tactical Information Broadcast Service, 221
Tactical Related Applications (TRAP), 167, 168, 212
Tactical Surveillance Demonstration (TSD), 189, 190, 191, 208, 214, 221
  Enhanced (TSDE), 221
Taepo-Dong 1 and 2 missiles (North Korea), 212, 228, 240
TaGS (Navy, Tactical Ground System), 189
Taiwan, 227
TALON SHIELD (Air Force), 191–92, 196, 199, 201, 206, 207, 211–12
TALON SWORD (Air Force), 192
Target Radiation Measurement Program (TRUMP), 24
Target-via-missile (TVM) guidance, 168
Tashkent-50. *See* Tyuratam

Taylor, Ed, 71
Taylor, Maxwell, 29, 39
TBMDS (Theater Ballistic Missile Defense System), 231
TCE (Three-Color Experiment), 131
Teague, Roger, 258, 268
Teets, Peter, 252, 254, 255–56
Telemetry, 16
Teller, Edward, 6
TELS. *See* Transporter-erector-launchers
Terminal Defense Segment of national missile defense system, 260
Terminal High Altitude Area Defense, 260, 273
Terra Scout (navy), 178
Terrorism, 157
TERS (Tactical Event Reporting System), 106, 167
THAAD (Theater High Altitude Area Defense System), 230–31
Theater Ballistic Missile Defense System (TBMDS) (navy), 231
Theater High Altitude Area Defense (THAAD) system (army), 230–31, 248
Theater Intelligence Broadcast System (TIBS), 212
Theater Wide program (Navy), 247–48, 260
Third Generation Infrared Surveillance Satellite (3GIRS) program, 264–65, 266
Three-Color Experiment (TCE), 131
3GIRS. *See* Third Generation Infrared Surveillance Satellite (3GIRS) program
Thule. *See* Greenland
Thurston, Dave, 213
TIBS (Theater Intelligence Broadcast System), 212
Ticonderoga-class guided missile cruisers, 231
Titan (ICBM), 15, 22, 25, 75
  I, 29
  II, 25, 29, 36
  III, 56, 64
  IIIB, 49
  IIIC, 48, 49, 66, 68, 69, 78
  IV, 127, 162, 182, 204, 237
  IVB, 238, 249
  IV/IUS, 162

34D/IUS (inertial upper stage) system, 125
34D/transtage, 127
Tokaty-Tokaev, Georgi, 5
Topol-M missile (SS-27; Soviet/Russian
    ICBM), 70, 226, 240, 241, 250, 270–71
Topol SS-25 (Russian ICBM), 226, 241,
    270, 271
Transportation Subsystem (TSS), 151
Transporter-erector-launchers (TELs), 158,
    170, 171
TRAP. See Tactical Related Applications
TRE (Navy, Tactical Receive Equipment),
    167, 169
Truman, Harry, 3
TRUMP (Target Radiation Measurement
    Program), 24
TRW Systems (company), 43, 46, 54, 67,
    127, 134, 135, 138, 217, 223, 225, 245,
    246, 258, 259
TSD (Tactical Surveillance Demonstration),
    189, 190, 191, 208, 214, 221
TSDE (Tactical Surveillance Demonstration,
    Enhanced), 221
Tsiolkovsky, Konstantin E., 2
TSS (Transportation Subsystem), 151
TT-6 and -7 (Soviet ICBM), 70
Turkey, 5, 12, 81, 163, 170
Tu-16/Tu-16C Badger (SLRA), 103
Tu-22M Backfire (SNA), 102–3, 117 (fig.),
    273n38
TVM (target-via-missile guidance), 168
TWAAFO (Defense Support Program,
    Tactical Warning/Attack Assessment
    Follow-On), 187
TWA Flight 800 explosion detection, 109
Tyler, Patrick, 97
Typhoon submarine (Soviet), 101
Tyuratam (Kazakhstan), 4, 5, 45, 47, 271

UFOs, 106–7
United Kingdom, 14, 233
    and MIDAS, 18–19
United Nations Resolution 678 (1990), 157
United States, 147
    and German scientists, 1, 5
    Ground Station (CGS), 57, 58–61, 96,
        123, 136
    missile program, 4, 15
    nuclear war program, 86–88
    and Russia, 233–34
    and Soviet bombers, detection of, 7
    and Soviet missile program, intelligence
        on, 5–7, 12, 14
    space program, 11, 12
Upper tier ballistic missile defense system
    (navy), 231, 260
"U.S. National Strategy" (PD-18), 86
Ustinov, Dmitriy, 3
U-2 aircraft, 5, 13, 29
    D (modified/"Smoky Joe"), 18

Vandenberg Air Force Base (Calif.), 15, 36,
    44, 67, 92, 247, 260, 262, 265, 269, 273
Vanguard (satellite-to-be), 6
VELA HOTEL nuclear detonation detection
    satellite program, 40, 44, 79–81
Visible Ultraviolet Experiment (VUE), 131
volcanic activity, DSP detection of, 251–252
Volcano Hazards Program (U.S. Geological
    Survey), 251
VORTEX satellite, 107
V-2 (Vengeance Weapon Number Two) (Nazi
    Germany) rockets, 1, 2
VUE (Visible Ultraviolet Experiment), 131

Wade, James, 76
Warner, John, 255–56
Warren, F. E., Air Force Base (Wyo.), 224,
    225
Warsaw Pact countries, 149, 150
Washington Post, 97
Waverly Air Force Station (Iowa), 58–59
Weapons control computer, 166
Weapons System 117L (WS-117L) programs,
    9. See also CORONA; Discoverer;
    SENTRY
Weisner, Jerome, 12, 17, 25
Welch, John J., Jr., 134, 186
Welch, Larry, 247
Western Europe, as target, 123–24
Western Hemisphere satellites, 69, 79(table),
    100, 126, 128, 131
West Germany, 149–50
Wheeler Island test range (India), 270, 271

White, Thomas D., 12, 13, 16
White Sands Missile Range (N.Mex.), 189, 190, 230
Whitlam, Gough, 51–52, 56, 137, 138
Wide Area missile defense program, 259
Woodward, Patrick, 97
Woomera (Australia), 51, 53, 54, 106, 144
Worden (U.S. warship), 169
Worden, Simon P., 235
World War II. *See* Nazi Germany
Wright Field (Ohio), 9
Wright-Patterson Air Force Base, 274
WS-117L. *See* Weapons System 117L programs
Wuzhai Missile and Space Center (China), 241
Wynne, Michael, 254, 264
Wyoming, 224

X-band radar, 260, 272, 273
X-ray sensors, 226
X-ray spectrometer, 66

Yankee-class submarine (Soviet), 45, 58
Yarymovych, Mike, 50
Yekaterinburg (Russian submarine), 272
Yeltsin, Boris, 239n
Yemen, 229
Yockey, Donald, 193
Yom Kippur War (1973), 71, 229
York, Herbert, 17
Yorkshire (England), 7
Young, Bill, 214
Young, Thomas, 254
Yousef, Mohammad, 108
Y2K issues, Defense Support Program and, 238–39

Zeiberg, Seymour L., 87, 88, 93
Zhukov, Georgi K., 4
Zubiel, Nina, 172
Zuckerman, Solly, 19
Zuckert, Eugene M., 23, 24, 26, 28, 30, 31, 37, 38